Alchemy

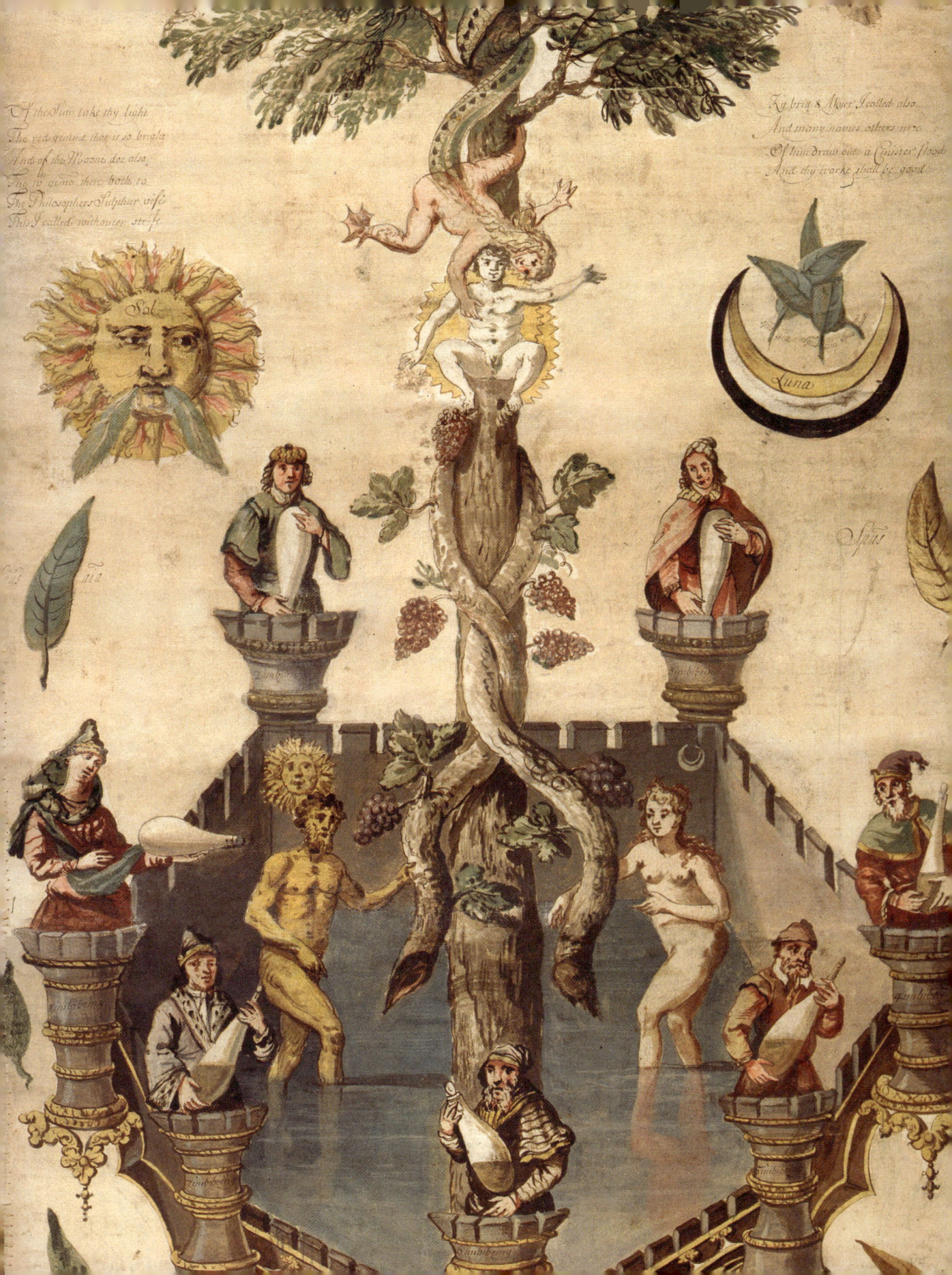

Alchemy

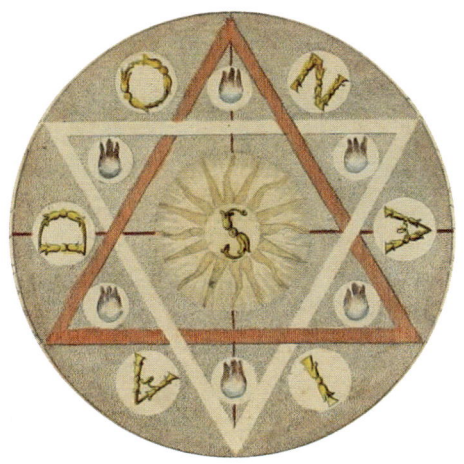

AN ILLUSTRATED HISTORY OF
ELIXIRS, EXPERIMENTS, AND THE
BIRTH OF MODERN SCIENCE

PHILIP BALL

YALE UNIVERSITY PRESS
New Haven and London

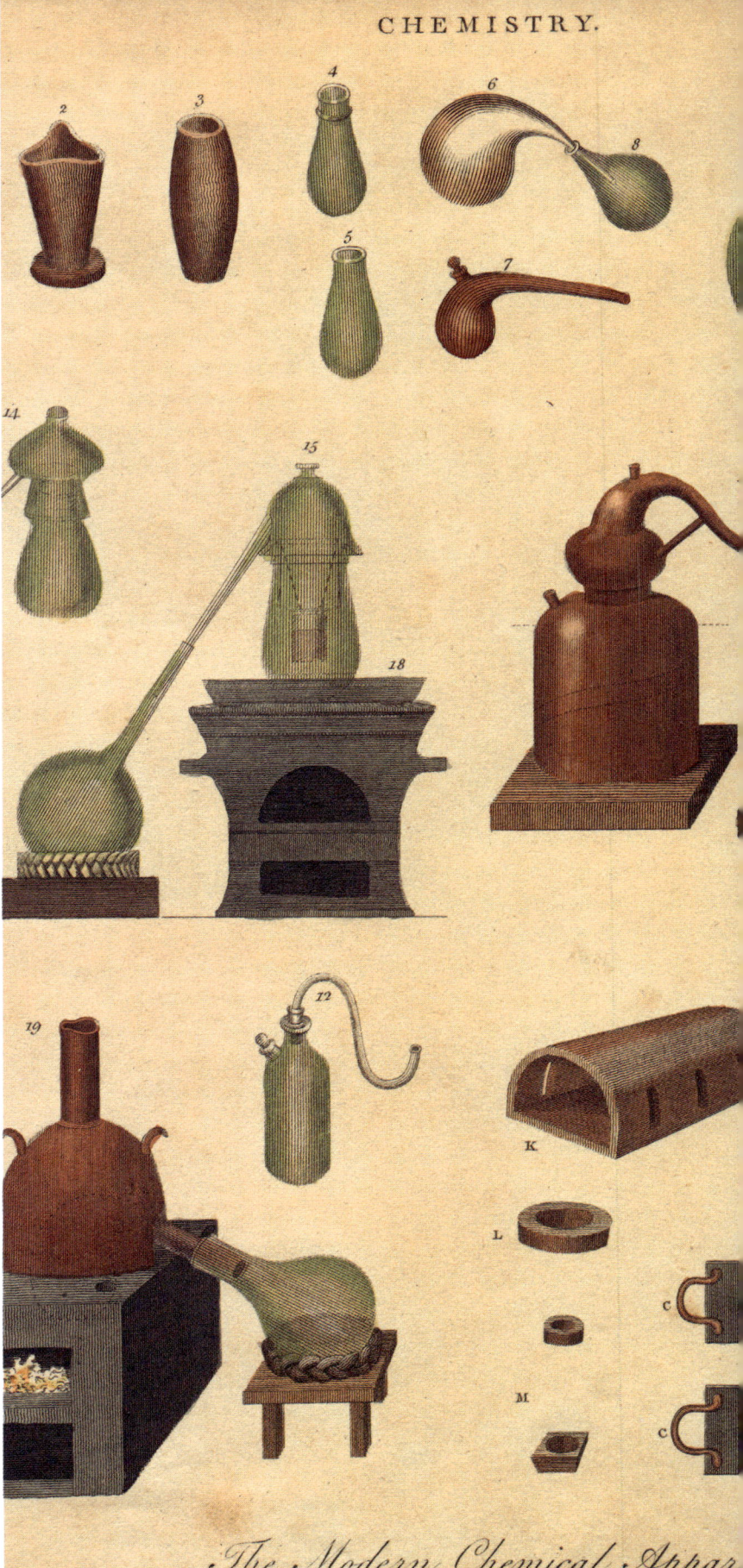

PREVIOUS PAGES Page 1: A carved ivory netsuke of a rabbit stirring an elixir, signed Ho kaku, from Japan, Edo or Meiji period, 19th century; Page 2: *The Book of Secrets*, detail from the Ripley Scroll, watercolor on paper, c. 1600; Page 3: An alchemical cosmology, by the Reverend William Law (1686–1761), taken from *The Works of Jacob Behmen, the Teutonic Theosopher* (1764).

RIGHT A detail from *The Modern Chemical Apparatus. No. 1*. Colored engraving by J. Pass after Henry Lascelles for the entry on *Chemistry*, from *Encyclopaedia Londinensis*; or, *Universal Dictionary of Arts, Sciences, and Literature* (1800).

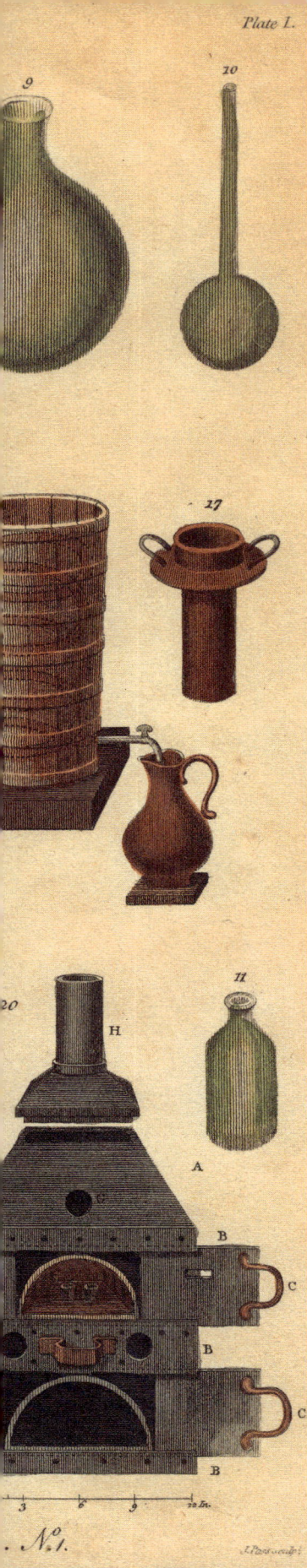

Contents

Introduction — 6

1 **BLACK EARTH** — 14
The Origins of Alchemy

2 **ELIXIRS** — 38
Alchemy in the East

3 **CHRYSOPOEIA** — 62
The Quest for Gold

4 **BOOKS OF SECRETS** — 86
The Uses of Alchemy

5 **PUFFERS** — 112
The Alchemical Laboratory

6 **THEATER OF THE WORLD** — 140
The Chemical Philosophy

7 **ALCHEMICAL WARS** — 166
Controversies of Alchemy

8 **CRUCIBLES** — 196
From Alchemy to Chemistry

9 **TRANSMUTATIONS** — 222
Alchemy in Culture

Index — 250
Further Reading — 254
Credits — 255

Introduction

Alchemy is one of the most versatile, allusive, and fertile products of the human imagination. With one foot in practical crafts, drawing on the skills and knowledge of artisans, and the other in speculative philosophies about the nature of the world, it is a phase in the history of ideas that laid some of the foundations of modern science while also providing inspiration to artists of all persuasions.

Within the walls of Prague Castle is a narrow street known as Golden Lane. It looks like a set from a fantasy film, with tiny, low-roofed sixteenth-century houses painted in bright fairy-tale colors. The street was constructed by the Holy Roman Emperor Rudolf II, originally to accommodate his guardsmen. The name derives from the fact that several of the houses were occupied by goldsmiths, some of whom were also considered to be alchemists.

RIGHT The atmospheric Golden Lane (Zlatá ulička) inside the bailey of Prague Castle, in a photograph dated c. 1920–29. The lane is lined with houses once occupied by goldsmiths, and has long been associated in tradition with the alchemists of the court of Rudolf II.

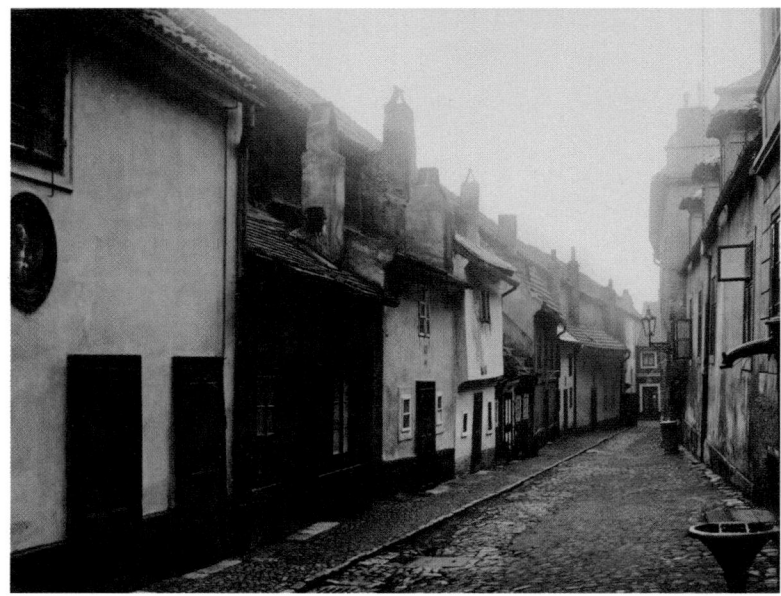

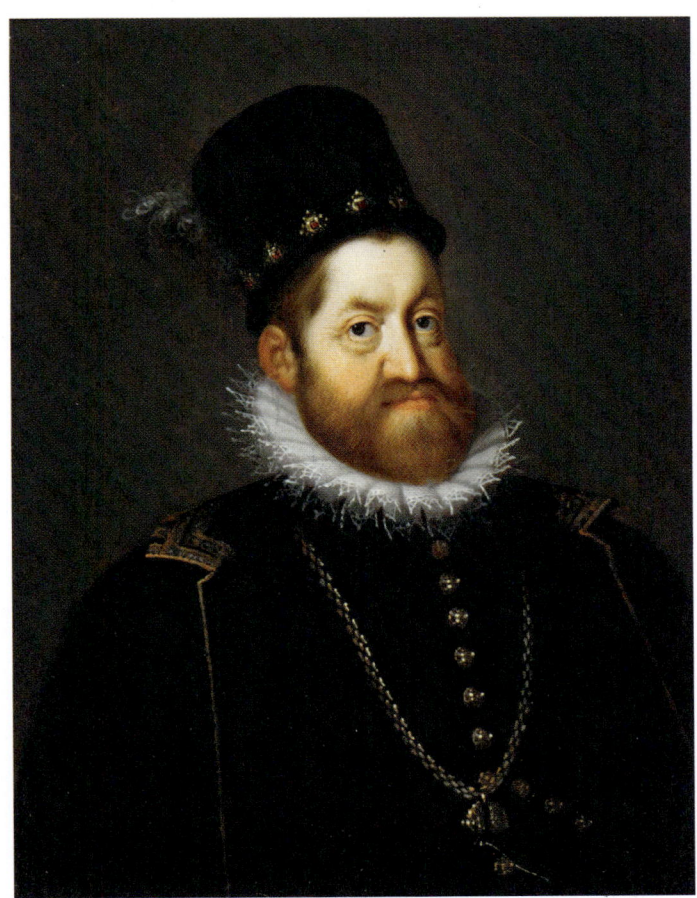

RIGHT A portrait of the Holy Roman Emperor Rudolf II (1552–1612), a patron of the occult arts, by Joseph Heintz the Elder, 1592. Rudolf's court attracted astronomers such as Tycho Brahe and Johannes Kepler, artists including Giuseppe Arcimboldo, and eminent alchemists such as Michael Maier and Michael Sendivogius.

Such artisans, seeking the secret of transmuting base metals into gold, found a warm welcome in Rudolf's court: alchemists, astrologers, seers, and mystics of all sorts congregated in early seventeenth-century Prague. In truth, Rudolf's personal alchemists were housed inside the castle itself, not in Golden Lane—but still there is probably no location in the world that better embodies the atmosphere of romance that even today surrounds alchemy.

Some of the most famous names associated (rightly or wrongly) with alchemy were even in their own time reputed to be wizards or necromancers: Albertus Magnus, Paracelsus, Cornelius Agrippa. The legendary Faustus practiced the gold-making art, and that discipline pervades Johann Wolfgang von Goethe's retelling of the old tale in the early nineteenth century. Alchemy holds the allure of forbidden knowledge, and our certainty today that its central quest—to transform other metals into gold—was futile only adds a

INTRODUCTION 7

quixotic piquancy to its attraction. Meanwhile, the reimagining of alchemy during the nineteenth and early twentieth centuries as a search for inner spiritual knowledge, however mistaken as a historical account of what alchemists actually did or thought, has only added to its contemporary appeal. It seems to speak of a time when efforts to understand the world did not have to capitulate to what is often regarded as the cold, impersonal rationalism of modern science, but instead involved thrilling flights of the imagination.

While this exotic image may sustain interest in alchemy today, making it an enduring cultural trope long after any serious scholars abandoned attempts at transmutation, it is also something of a burden to historians of science who strive to understand what

RIGHT Under Rudolf II, the imperial court of Prague became a hub of alchemical activity. In this 17th-century painting, *The Alchemist* by Pieter Gerritsz van Roestraten, an alchemist seeks instruction in a text while the young boy at his feet tends the furnace. Such scenes became popular in Dutch painting at that time; this work is now housed in the collection at Prague Castle.

OPPOSITE An 18th-century aquatint print of the Elizabethan mathematician and philosopher John Dee and his assistant Edward Kelley, who spent time in Bohemia, where they reportedly used a red powder to turn mercury into gold in front of Rudolf II. Here Kelley is shown raising a specter of the dead in an act of necromancy. Alchemy was commonly associated with such dark and forbidden arts.

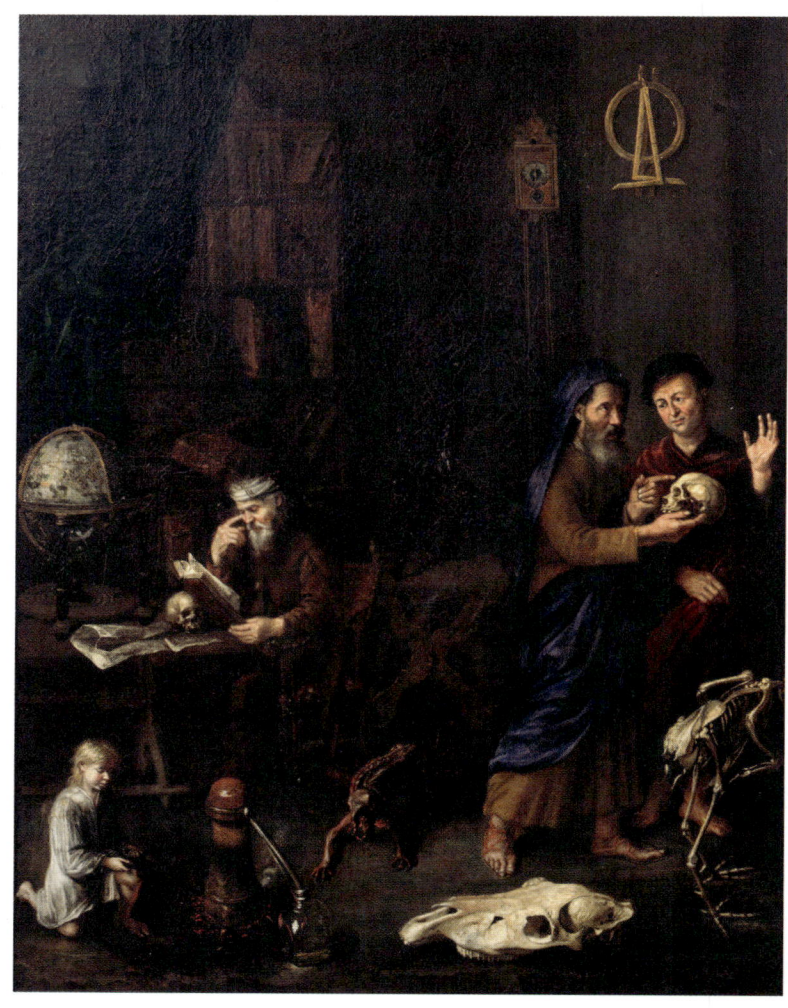

Necromancy.

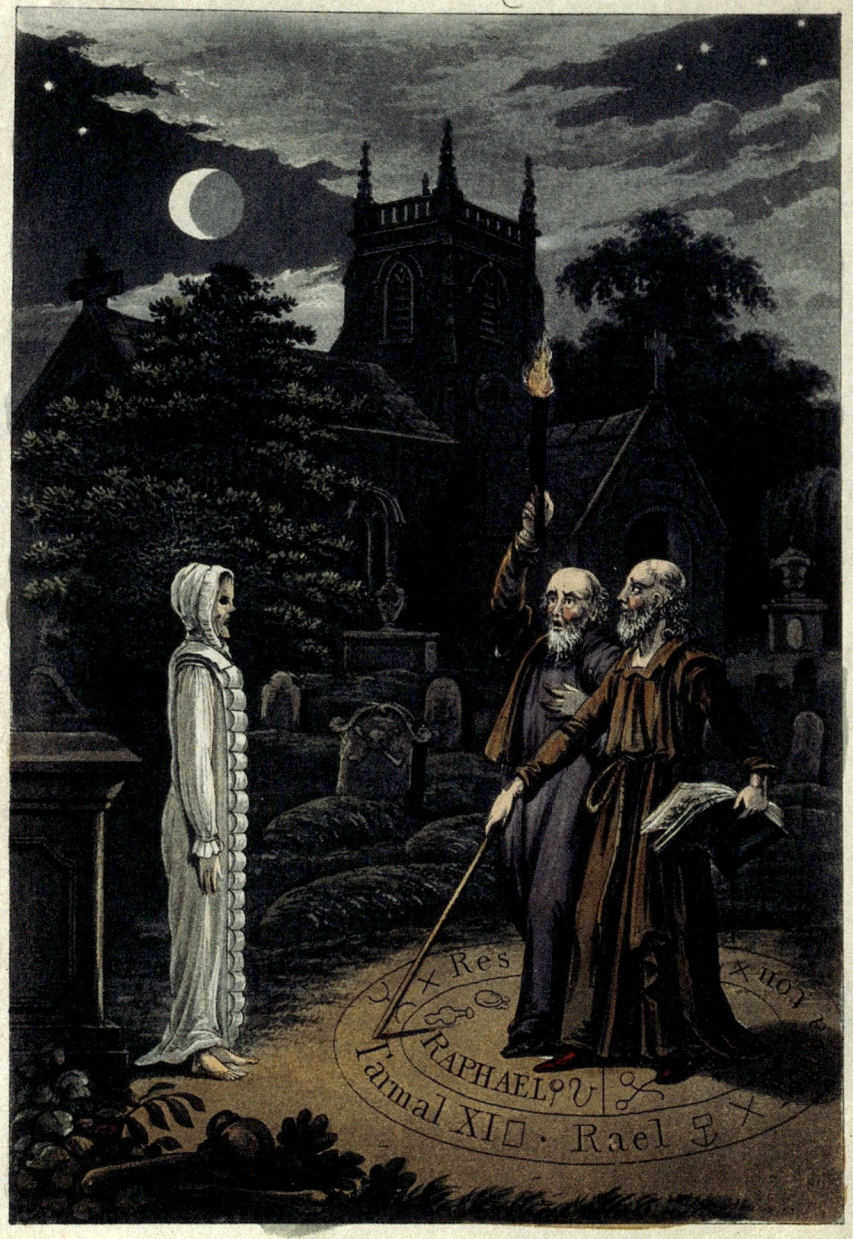

Edward Kelly, a Magician, raising the Ghost of a Person lately deceased, in the Church Yard of Walton-le-dale, Lancaster.

Vide page 229.

LONDON.
Published by William Charlton Wright, 65, Paternoster Row.

alchemy was really all about. Far from being a romantic prelude to science, alchemy was for many centuries thoroughly entrenched in the natural philosophy of its times—a component of what would become science. Yet there has never been a time when alchemy was not tainted with disrepute: it was often suspected (not without reason) of harboring cheats and swindlers whose claims to make gold were based in trickery. If they were not mere frauds, then alchemists were sometimes accused (again with some justification) of hiding their secrets behind a veil of arcane terminology, designed not (as we now expect of scientists) to communicate truths about nature but to obscure them from anyone who did not belong to a select elite.

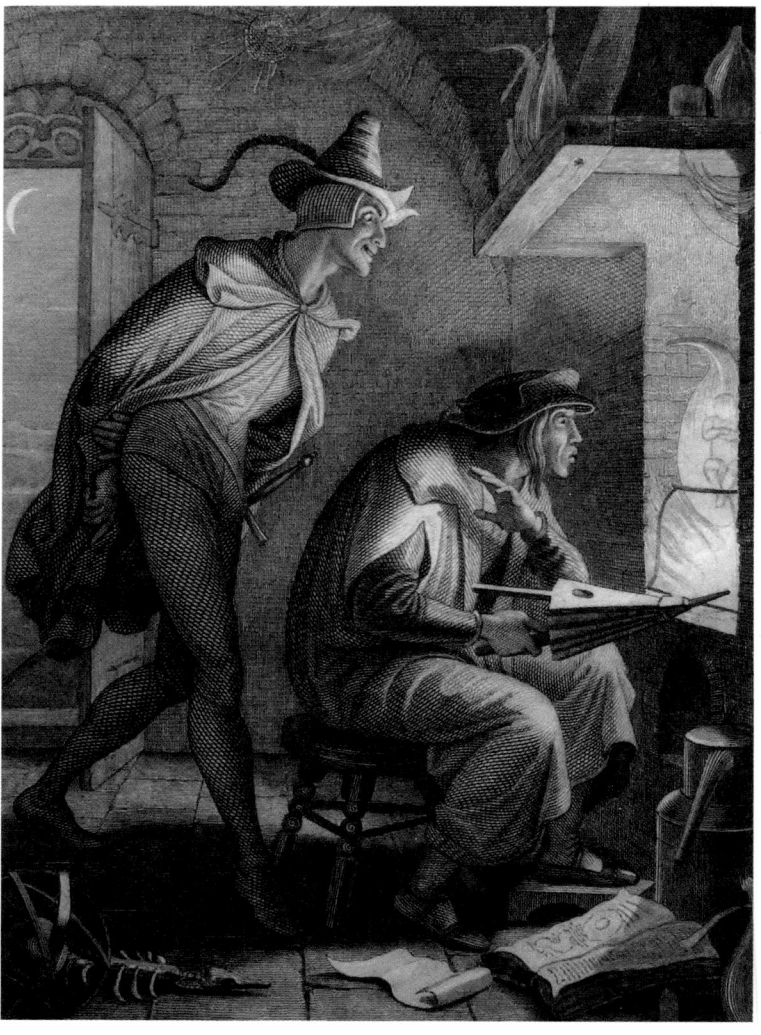

RIGHT Faust's assistant Wagner creates an homunculus, or artificial being, by alchemy in this engraving after a drawing by Jacques Alfred van Muyden, from the second part of Goethe's *Faust* (Stuttgart and Tübingen, 1840). As seekers of hidden—some would say forbidden—knowledge, alchemists were often associated with the Faust legend, which goes back at least to the Middle Ages. The alchemical literature contained various recipes for making an homunculus.

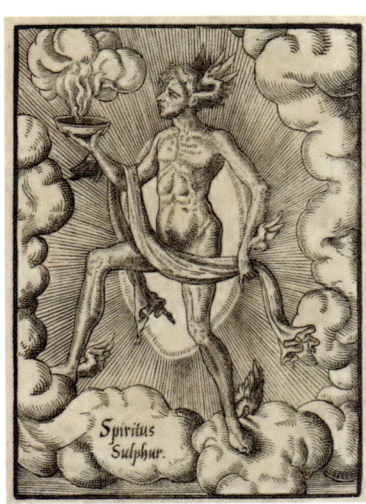

ABOVE The alchemist's laboratory could be dangerous: this woodcut by Jost Ammann in Leonhard Thurneysser zum Thurn's *Quinta Essentia* (Leipzig, 1574) depicts the corrosive "spirit of sulfur" (the fumes of sulfuric acid).

There is another reason, however, why alchemy was, until the late twentieth century, denied its proper place in the history of science: it deals primarily with what the ancients called art—*techne*, in ancient Greek, which was concerned with *making*. Alchemy was at least as much a practical chemical technology as a theoretical system for understanding the composition of nature. Yes, many alchemists were determined to make precious gold, but they also made medicines, pigments and dyes, household substances like lime, alkalis, and acids, and much else besides. The history and philosophy of science have long struggled to incorporate this applied aspect, being much more at home with efforts to theorize about the physical world (as in physics) or to catalogue and rationalize what we find in it (as in biology). Alchemy seemed to lack the grandeur of attempts to explain the heavens; it was apparently conducted in grimy, smelly laboratories, exposing it to snobbish intellectual ridicule as a mere manual art. Yet we can understand the role alchemy played in the development of science only if we admit applied and practical experimentation into the picture.

Although the allegorical and symbolic codes employed by alchemists to ensure that only fellow adepts could interpret their works create difficulties for historians today trying to figure out what they were really doing and saying, that secretive tradition had the virtue of spawning some of the most glorious imagery in the history of science. Later alchemical texts were sometimes profusely illustrated, and color itself played a central part in both the practice and the iconography of the discipline. Some of these images may look fanciful, obscure, and superfluous, but in fact alchemy acquired a kind of visual language that foreshadows the strongly visual aspect of modern chemistry, in which conventions for depicting molecules and their changes speak instantly and eloquently to those who know how to read them.

Alchemical imagery speaks more widely, however. While it typically encodes—however fancifully—actual chemical processes conducted in the workshop, its meaning goes deeper.

Alchemy was apparently conducted in grimy, smelly laboratories, exposing it to snobbish intellectual ridicule as a mere manual art.

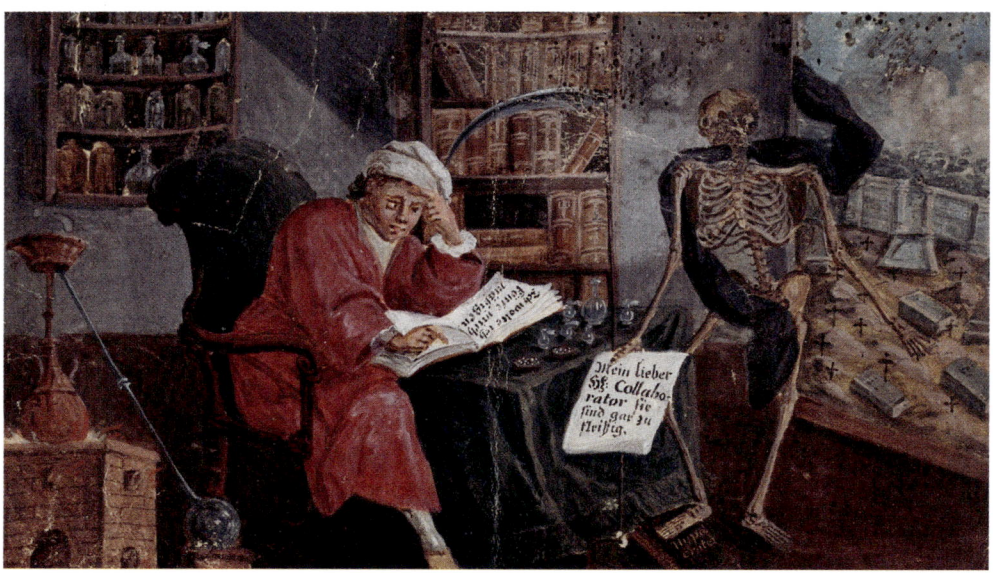

ABOVE Alchemical practice could demand levels of concentration and labor that were potentially injurious to health. In this undated early German gouache, an alchemist, seated by a distilling apparatus known as an alembic, concentrates on his book, on which is written "I wish I could moderate myself." Beside him, Death, standing on the work of the legendary Greek physician Hippocrates, responds: "My dear Herr Collaborator, you are too hard-working."

The widespread use of allegory and symbolism wasn't just a ploy to make this knowledge hard for outsiders to decrypt; it reflects a view of the cosmos that, while seemingly alien to modern thought, was part of the intellectual mainstream of its time.

These allegories were not a mere matter of one thing representing another; rather, they encode a complex web of associations thought to hold the key to the innermost workings of nature. Metals were not simply represented by planets, for example, but were thought to be genuinely influenced by them. This vision of deep correspondences no longer has a place in scientific thought, but that need not prevent us from acknowledging its beauty—and the soaring ambition of its attempt to discern a unified view of nature and thus to reveal God's plan. In asserting and exemplifying this framework, alchemy and its associated imagery has long offered a rich source of inspiration for the wider culture of the arts and humanities.

At the same time, alchemy provided genuine knowledge and fruitful speculation. It supplied recipes for making useful substances, and methods and apparatus for doing so. It commended experimentation guided by careful quantification—a hallmark of modern science. It furnished medicines, and though many were of little real value for healing, the important idea that specific diseases demand specific chemical remedies, a central aspect of modern pharmacology, has roots in alchemy. Finally, it encouraged speculative theories of matter and its transformations, promoting

RIGHT Some alchemical texts were lavishly illustrated, with color playing a central part in the iconography and practice of the art, as evidenced in this depiction of "Fire" from a North Italian manuscript, *Musaeum hermeticum* (Hermetic museum; 1692), reproducing the work of the pseudonymous alchemist Basil Valentine among others.

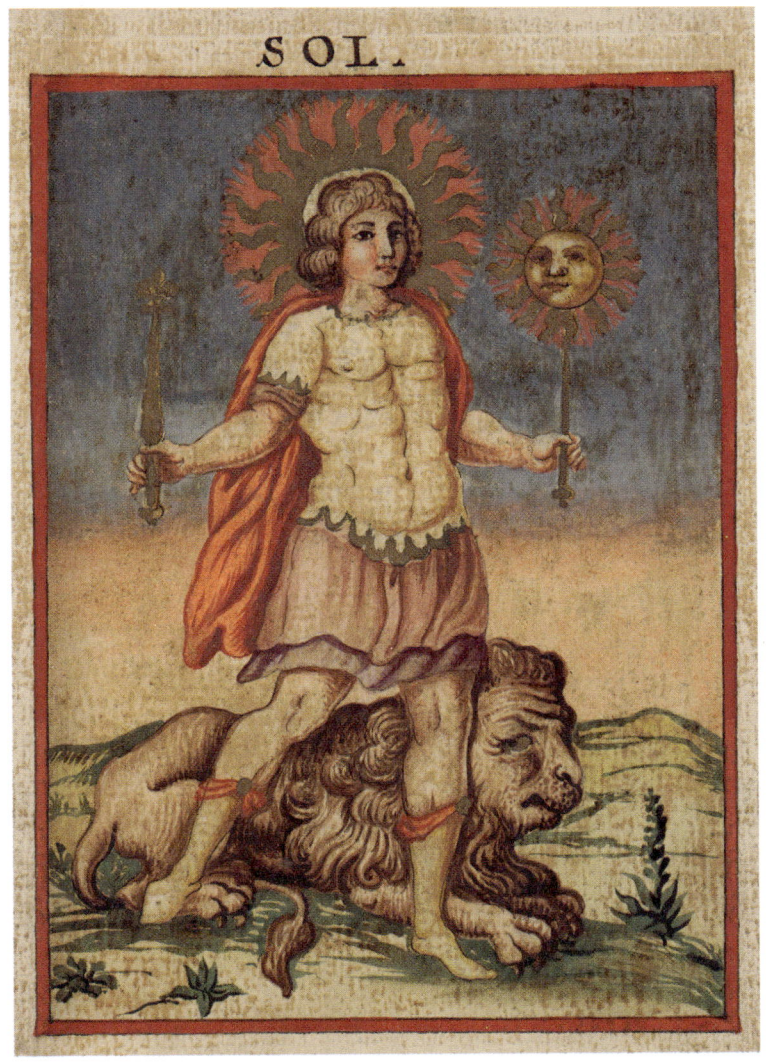

the idea that matter is composed of microscopic particles that may be considered the conceptual forerunners of the modern atom.

Far from being unmitigated, mystical, and baseless folly, then, alchemy was an important stage in the development of thought about the world. As we will see, its historical image problem is not the result of any inherent absurdity or dishonesty, but rather a narrative constructed for a rhetorical purpose that is no longer necessary or useful. We should feel free to enjoy alchemy for what it was, and for what it provided to science and to culture. I hope this book will help you do that.

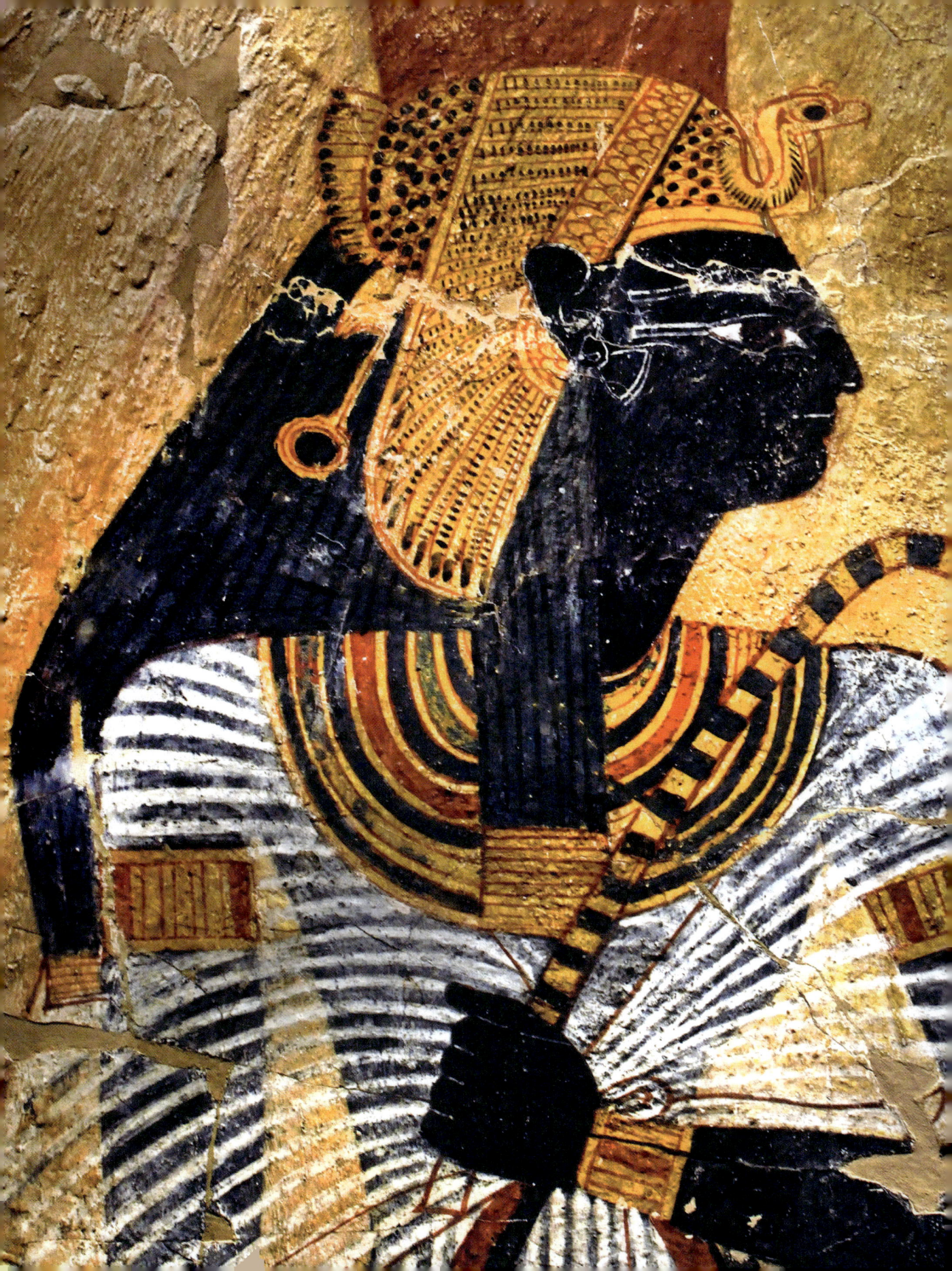

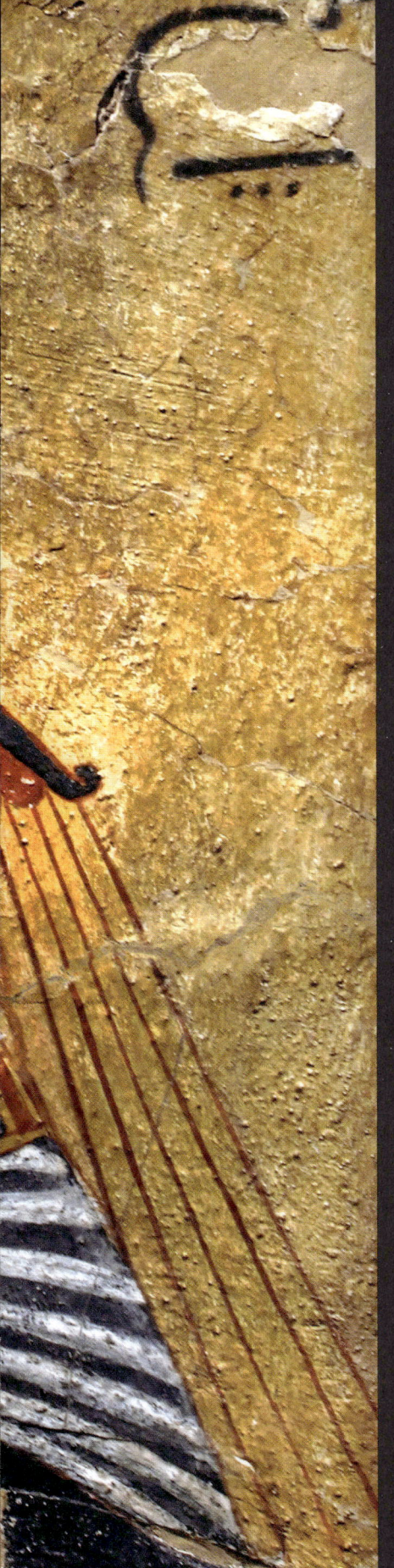

CHAPTER ONE

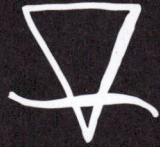

Black Earth

THE ORIGINS OF ALCHEMY

LEFT The deified Queen Ahmose Nefertari, in a detail from a wall painting from the tomb of Inherkau, c. 1152–1145 B.C., at Deir el-Medina, Thebes. Painting in ancient Egypt was typically a sacred art, invested with magical power and accompanied by rituals that gave it divine influence. As a result, the chemical production of pigments was awarded considerable social importance. Color use could be symbolic: the queen is often shown with black skin pigmentation, as black symbolized rebirth and regeneration, perhaps because of the fertile black soil of the Nile valley.

CHAPTER ONE

Alchemy began as an art more like the chemistry we know today: a matter of making new substances by transforming those we find around us in nature. People have conducted such transformations since time immemorial, often using the agency of fire or heat. Such practical crafts only truly became alchemy, however, when they merged with theories of the composition of matter, linked to metaphysical ideas about the fundamental nature of the world.

Alchemy is not, as is sometimes suggested, a superstitious and mystical practice that was supplanted by the modern scientific discipline of chemistry. Rather, before the seventeenth century, if you were concerned with transforming substances from one into another—making glass, metals, soap, charcoal, medicines—you would have been practicing alchemy. However, there were also philosophical, religious, and other contexts of alchemy that reach well beyond the boundaries of today's scientific discipline of chemistry—for patterns of thought in former times do not map neatly onto ours today.

Chemical transformations have always been as much a part of human making as tools and weapons, bread and wine, clothing, pots and pans, and ornaments for decoration or ritual. There is no obvious reason why, in changing from simple trial-and-error experimentation to the systematic and theoretically rigorous science of today, chemistry had to pass through a stage so imaginatively and symbolically rich, so elaborate and full of philosophical and theological implications, as alchemy. But that is what happened—and it is surely a testament to the scope of human inventiveness and ingenuity that people once found in the humble work of the artisan, allegories for the human condition and the mysterious nature of the cosmos.

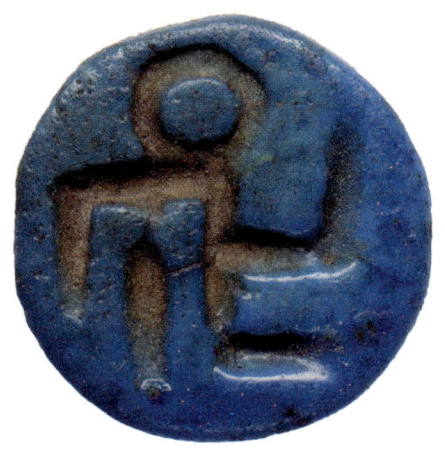

ABOVE Amulet featuring a motif of a squatting child, from Egypt, c. 2100–2080 B.C., coated in faience. This copper-based blue glaze was one of the most important products of ancient Egyptian chemical technology.

Bronze Age chemistry

According to the etymology asserted by the Roman writer Pliny the Elder, the word *chemistry* is derived from *khemia*, denoting the black soil of Egypt. (The *al* in *alchemy* is simply the standard Arabic prefix.)

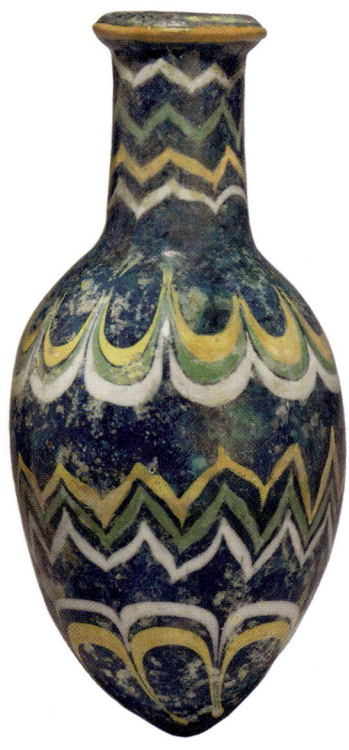

A more likely derivation is from the Greek *cheo*, meaning "to melt"; *chemeia* is the art of melting (and smelting) metals. But associating early chemistry with ancient Egypt makes sense, for there is abundant evidence of the importance of chemical transformations in Bronze Age Egyptian society, which used them to make dyes, glazes, ointments, cosmetics, and, indeed, bronze itself.

Glass was one of the most important and desirable products of ancient chemical technologies. It is made by melting sand (silicon dioxide or silica, as chemists would call it today) with soda or potash, alkalis containing sodium or potassium and typically extracted from plant ashes. The earliest-known glass dates from around 2500 B.C. in Mesopotamia, in present-day Iraq and Syria. The art of making it might have been discovered by accident during the production of glazed earthenware and ornaments—such as blue-glazed faience, which was produced in the lower Nile region from around 3000 B.C. and was later traded far and wide in the Middle East and Europe.

ABOVE This opaque, cobalt-blue glass bottle from Mesopotamia, mid- to late 15th century B.C., is among the earliest-known glass artifacts.

RIGHT Egyptians use foot bellows in the process of gold smelting in this scene from a wall painting in the tomb of the vizier Rehkmire, 1490–1436 B.C., at Sheikh Abd al-Qurna, Luxor, Thebes.

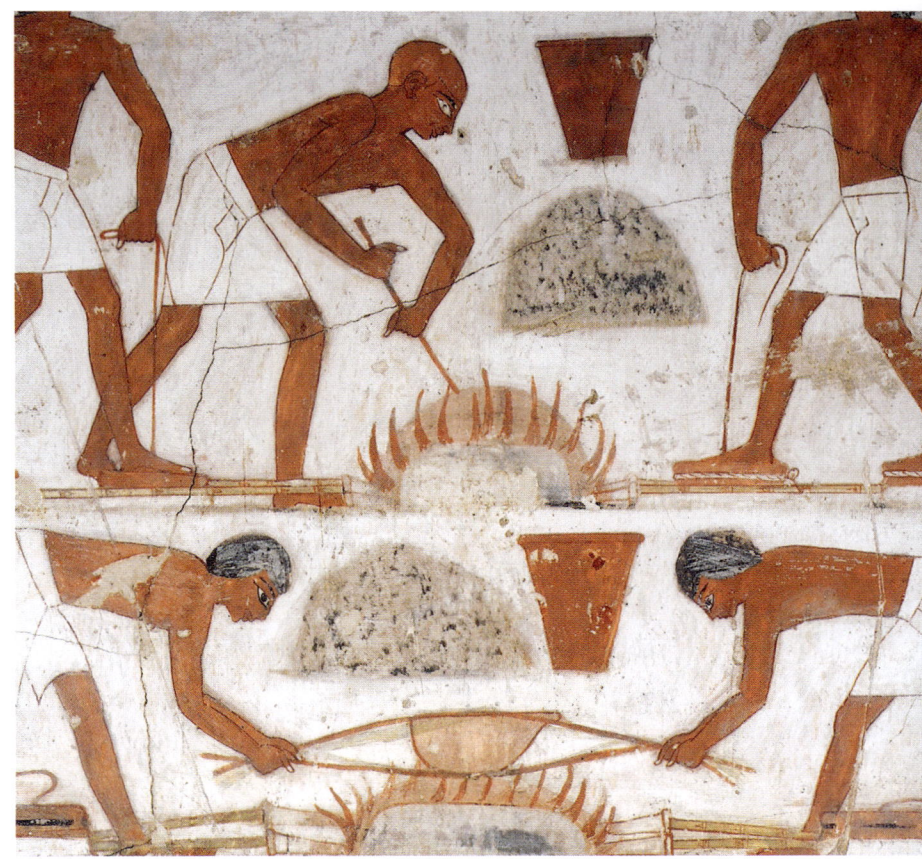

BLACK EARTH: THE ORIGINS OF ALCHEMY

BELOW This tattered and faded blue kerchief from Tutankhamun's embalming cache, c. 1336–1327 B.C., in Thebes, Egypt, displays an early use of indigo. This blue dye, as well as a rarer purple one, was produced from sea snails. Indigo was also found in southeast Asia, but there it was extracted from the plant *Indigofera tinctoria*.

Ancient glass could be improved by adding a little lime (calcium oxide), made by heating chalk. One recipe recorded on a clay tablet in the cuneiform script of Mesopotamia reads "Take sixty parts sand, a hundred and eighty parts ashes from sea plants, five parts chalk, heat them all together, and you will get glass"—an indication that the chemical technologies of the ancient world, while perhaps initiated by haphazard trial and error, were conducted with some quantitative precision.

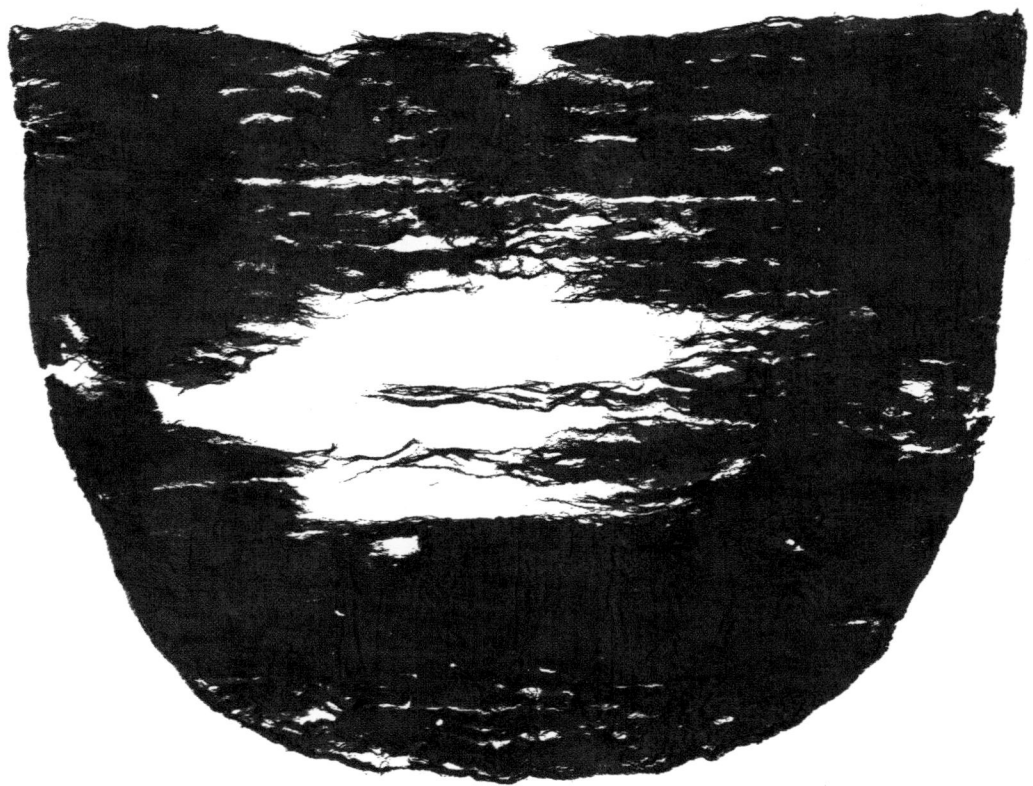

The chemical technologies of the ancient world, while perhaps initiated by haphazard trial and error, were conducted with some quantitative precision.

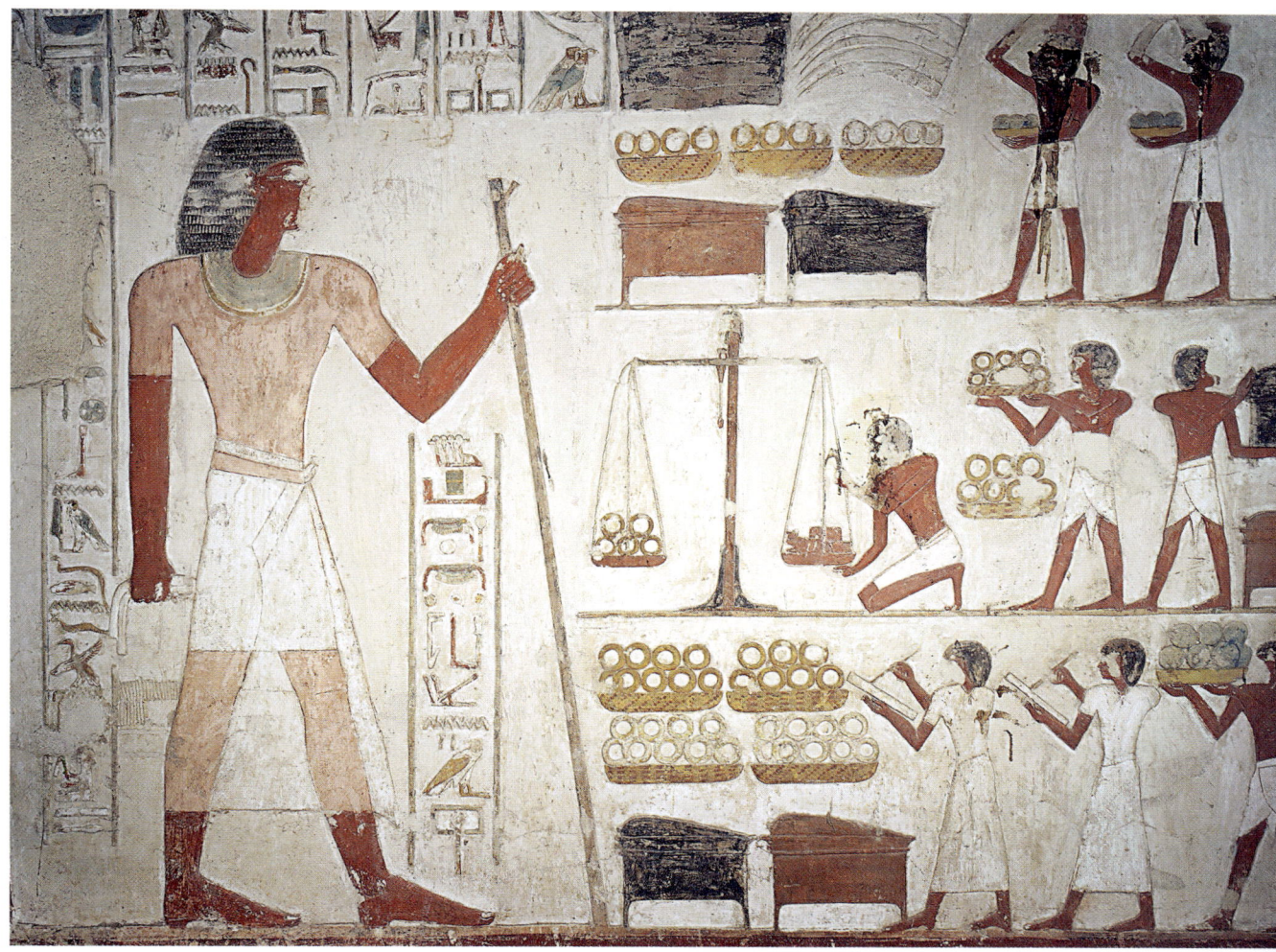

ABOVE Weighing copper during the reign of Amenhotep II in Egypt, depicted in a fresco in the tomb of Userhat, 1550–1295 B.C., at Sheikh Abd el-Qurna, Thebes. The earliest copper smelting dates back to around the 6th millennium B.C.

Using chemical techniques, the Egyptians also made pigments (such as the famous Egyptian blue, a kind of ground-up blue glass), cosmetics, dyes (typically extracts of plant or animal matter), ointments and medicines, and more. Their impressive knowhow in transforming the materials of nature into new and useful or attractive forms enriched ancient Greece and other cultures of the Mediterranean and Middle East.

The ancient chemical technology most closely associated with alchemy was metallurgy. Artisans in those times recognized several distinct metals. Silver and gold, which could be found naturally in their "native" forms, were relatively resistant to corrosion. The silvery liquid metal mercury was extracted from the red mineral cinnabar by simple heating; lead was drawn from its ores cerussite

BLACK EARTH: THE ORIGINS OF ALCHEMY

and galena by heating with charcoal; copper was sometimes found in its native state or in ores such as malachite; and tin could be obtained from cassiterite. Copper and tin together form bronze, which, being harder than ductile copper itself, was used to make tools, weapons, and armor from around 3300 B.C. in the Middle East. (This early bronze was made by heating a mixture of ores; pure tin was not produced until around 1800–1600 B.C.) Iron-working goes back at least to the early second millennium; iron implements from ancient Egypt and Central Anatolia dating from this period have been found, although it became more widely used after the demise of the Hittite empire around 1200 B.C.

All these metals were similarly dense and lustrous in nature. Ancient artisans, lacking today's understanding of different metals as irreducible chemical elements, naturally imagined that they had some kind of affinity—they were different varieties, perhaps, of the same fundamental substance, and therefore were potentially interconvertible.

Recipes

The diversity of chemical knowledge in the ancient world is revealed in two papyrus manuscripts believed to have been written in Egypt around 300 A.D. These documents were obtained—perhaps looted from a tomb—in 1828 by the Swedish vice-consul in Alexandria, who donated one to the Swedish Academy of Antiquities in Stockholm and sold the other to the Dutch government, which deposited it at the University of Leiden. Now known as the Stockholm and Leyden papyri, both are lists of recipes for making dyes and preparing gemstones, pearls, and metals. They may have been used in a workshop and were probably collected from older sources. Here is a recipe for "making silver" from the Stockholm papyrus:

Plunge Cyprian copper, which is well worked and shingled for use, into dyer's vinegar and alum and let soak for three days. Then for every mina of copper mix in 6 drachmas each of earth of Chios, salt of Cappadocia and lamellose alum, and cast. Cast skilfully, however, and it will prove to be regular silver. Place in it not more than 20 drachmas of good, unfalsified, proof silver, which the whole mixture retains and [this] will make it imperishable.

There are several things to note here. First, the instructions are rather straightforward, as well as precise, once you know what the

OPPOSITE Recipe for silver from the Greek manuscript known as the Stockholm papyrus (Papyrus Graecus Holmiensis), c. 300 A.D. This manuscript, believed to have been compiled in Egypt from older sources, is one of the most important windows on the chemical technologies of the ancient world, and includes recipes for making other metals that resemble gold.

ΑΡΓΥΡΟΥ ΠΟΙΗΣΙΣ
ΧΑΛΚΟΝ ΤΟΝ ΚΥΠΡΙΟΝ ΤΟΝ ΗΔΗ ΕΙΡΓΑΣΜΕΝΟΣ
ΚΑΙ ΕΚΤΑΣΙΝ ΕΧΟΝΤΑ ΤΗ ΧΡΗΣΕΙ ΚΑΤΑΒΑΠΤΟΝ
ΟΞΕΙ ΒΑΦΙΚΩΣ ΤΥΠ ΤΥΡΙΑ ΤΕ ΚΑΙ ΤΡΙΣΙΝ ΗΜ
ΕΑΥ ΒΡΕΧΕΣΘΑΙ ΤΟ ΤΕ ΔΗ ΧΩΝΕΥΕ ΤΕ ΤΗ ΤΟΥ
ΧΑΛΚΟΥ ΜΝΑ ΓΗΣ ΧΕΙΑΣ ΑΛΟΣ ΤΕ ΚΑΠΠΑΔΟΚΕΣ
ΚΑΙ ΣΤΥΠΤΗΡΙΑΣ ΣΧΙΣΤΗΣ ΕΚΑ ΔΡΑΧΜΩΝ Ϛ
ΑΝΑΜΕΙΞΑΣ ΕΜΠΕΙΡΩΣ ΔΕ ΧΩΝΕΥΕ ΚΑΙ ΕΣΤΑΙ
ΣΠΟΥΔΑΙΟΣ ΠΡΟΣΒΑΛΕ ΔΕ ΑΡΓΥΡΟΥ ΚΑΛΟΥ
ΚΑΙ ΔΟΚΙΜΟΥ ΤΟ ΥΑΠΛΟΥ ΜΗ ΠΛΕΓΩ ⳨ Κ
Ο ΔΙΑΦΥΛΑΞΕΙ ΤΗΝ ΣΗΝ ΠΑΣΑΝ ΜΕΙΞΙΝ
ΑΝΕΞΑΛΕΙΠΤΟΝ   ΑΛΛΟ
ΕΙΣ ΔΕ ΔΗΜΟΚΡ ΤΟΝ ΑΝΞΙΛΑΟΣ ΑΝΑΦΕΡΕΙ ΚΑΙ
ΤΟ ΔΕ ΤΟ ΥΣ ΚΟΙΝΟΥΣ ΑΛΑΣ ΑΜΑΣ ΤΥΠΤΗΡΙΑ
ΤΗΣ ΣΧΙΣΤΗΛΙΗΝΑΣ ΕΥ ΜΑΛΑ ΣΥΝΟΞΕΙ ΚΑΙ ΑΝΑ
ΠΑΣΑΣ ΚΟΛΛΟΥΡΙΑ ΤΑΥΤΑ ΕΠΙ ΤΡΙΣ ΗΜΕΡΑΣ
ΕΤΥΧΕΝ ΕΝ ΒΑΛΝΙΩ ΚΑ ΠΙ ΤΑ ΛΕΑΝΑΣ ΣΥΝ
ΕΧΩΝ ΕΥΕ ΤΟΝ ΧΑΛΚΟΝ ΕΠΙ ΤΡΙΣ ΚΑΙ ΥΔΑΤΙ
ΘΑΛΑΤΤΙΩ ΚΑΤΑΣΒΕΝΝΥΩΝ ΕΤΥΧΕΝ ΕΛΕ
ΞΕΙΤΟ ΑΠΟ ΒΗΣΟΜΕΝΟΝ Η ΠΕΙΡΑ
            ΑΛΛΟ
ΚΑΣΣΙΤΗΡΟΝ ΛΕΥΚΟΝ ΤΕ ΚΑΙ ΜΑΛΑΚΟΝ ΤΕ ΤΡΑ
ΚΙ ΚΑΘΗΡΑΣ ΑΚ ΤΟΥΔΕ Ϛ ΧΑΛΚΟΥ ΤΕ ΚΑΜΑ
ΤΙΚΟΥ ΛΕΥΚΟΥ ΜΝΑ Σ ΣΥΝΧΩΝΕΥΣΑΣ Ο ΜΗ ΚΕ
ΚΑΙ ΣΚΕΥΑΖΕΣΘΕ ΛΕΙΣ ΚΑΙ ΓΕΙΝΕΤΑΙ ΑΡΓΥΡΟΣ Ο
ΠΡΩΤΟΣ ΩΣ ΚΑΙ ΤΟΥΣ ΤΕΧΝΙΤΑΣ ΛΑΝΘΑΝΕΙΝ Ο ΤΗ
ΕΞ ΟΙΝΟΜΙΑΣ ΤΟΙΑΣ ΔΕ ΣΥΝΕΣΤΙ
      ΚΑΣΣΙΤΕΡΟΥ ΚΑΘΑΡΣΙΣ
Η ΔΕ ΤΟΥ ΚΑΣΣΙΤΕΡΟΥ ΚΑΘΑΡΣΙΣ ΤΟΥ ΧΩΡΟΥΝΤΟΣ
ΕΣΤΙΝ ΤΟΥ ΑΡΓΥΡΟΥ ΚΑΘΑΡΣΙΝ Η ΔΕ ΚΑΣΣΙΤΕΡΟΝ ΚΑ
ΒΑΡΟΝ ΕΑΤΥΓ ΝΑΙ ΚΑΙ ΑΛΕΙΤΑ ΣΕΛΕΩΤΕ
ΚΑΙ ΑΣΦΑΛΤΩ ΕΚΤΕ ΤΑΡ ΤΟΥ ΧΩΝΕΥΕ ΚΑΙ ΠΛΥΝΑΣ
ΑΠΟ ΘΟΥ ΚΑΘΑΡΙΣΩΣ ΠΡΟΣΒΑΛΛΕ ΤΟΙΣ ΤΕΤΤΑΡΣΙΝ ΤΟΥ
ΑΡΓΥΡΟΥ ΤΟΥ ΔΕ ΜΕΡΗ Ϛ ΚΑΙ ΧΑΛΚΟΥ ΤΟΥ ΜΑΤΙΚΟΥ
Ε Ξ ΚΑΙ ΛΗΣΕΤΑΙ ΠΡΟΚΕΙΜΕΝΟΝ ΩΣ ΑΡΓΥΡΩΜΑ
ΑΡΓΥΡΟΥ ΔΙΠΛΑΣΙΑΣΜΟΣ   ΟΙΚΟΝΟΜΙΑΣ ΓΕΙΝΕΤΑΙ
ΔΙΑΦΟΡΟΙΣ ΤΟΝ ΚΥΠΡΙΟΝ ΧΑΛΚΟΝ ΕΞΕΙΣ ΟΜΕΝΟΝ ΣΥΝ
ΑΛΟΣ ΑΧΝΗ ΕΞ ΑΚΙΧΩΝΕΥΣΑΣ ΕΠΙΒΑΛΕ ΤΟΝ ΑΡΓΥΡΟΝ
ΕΙΣ ΔΙΠΛΑΣΙΑΣΜΟΝ   ΑΛΛΟ
ΤΑ ΑΠΟΚΟΜΜΑΤΑ ΤΩΝ ΤΟΥ ΧΑΛΚΟΥ ΠΕΤΑΛΩΝ ΑΛΜΗ
ΚΑΤΑΒΑΠΤΕ ΚΑΙ ΣΤΥΠΤΗΡΙΑ ΠΛΑΒΕΙΤΕ ΑΝ Ϛ ΜΕ
ΝΗ ΓΛΥΚΕΙ ΥΔΑΤΙ ΑΠΟΒΡΕΧΕ Ϛ ΕΙΤΕ ΕΠΙΒΑΛΛΩΝ
ΗΡΕΜΑ ΤΟΝ ΑΡΓΥΡΟΝ ΧΩΝΕΥΕ

names mean—they could almost be from a chemistry textbook in early modern times. Second, this is a recipe for transmutation of metals: making silver from copper. So is it alchemical? In those times, the identity of some substances, especially metals, had to be judged largely from appearances—there were no reliable techniques for establishing this, nor the conceptual framework for making clear distinctions. Ancient metallurgists *did* know how to measure density—how much mass a given volume has—which was a clue to a metal's identity. They also knew how to purify some metals using a process called cupellation (see page 76). But if a metal had the bright, shiny appearance and hue of gold or silver, it seemed reasonable to suppose that it had at least some of the *nature* of those metals.

Another recipe explains how to "Giv[e] objects of copper the appearance of gold" by coating them with a preparation of powdered gold and lead. This account assures the reader that "neither touch nor rubbing against the touchstone will detect them . . . It is difficult to detect, because rubbing gives the mark of a gold object, and heat consumes the lead but not the gold." The implication is that there is intention here to deceive. Yet another recipe, for manufacturing *asem*, a natural alloy of gold and silver generally known as electrum, says that "the metal will be equal to true *asem*, so much so as to deceive even the artisans." At various points, these recipes appear to imply that one true metal is being made from another—or mimicking it well enough to be deceptive. Perhaps looking like gold and being like gold were just distinctions of degree.

These papyri come from precisely the moment when the practical chemical technologies of the ancient world were morphing into the discipline we now regard as alchemy. Their recipes are purely descriptive: they simply tell what ingredients and conditions are needed to attain a particular result. But alchemy proper arose when such artisanal knowledge became blended with theories about the composition of matter. That union seems to have taken place in Greco-Roman Egypt around the fourth century A.D.

At various points, these recipes appear to imply that one true metal is being made from another—or mimicking it well enough to be deceptive.

Mysteries

Evidently a great deal of useful chemical knowledge could be accrued without any theory to explain it. But it was surely inevitable that some of those performing or witnessing these transformations would wonder what they revealed about the nature of the world.

There is a hint of underlying theory in one of the earliest texts of the Greek-Egyptian era, called (in later Latin transcriptions) *Physica et mystica* (Natural and secret things). There is no original text of this work, and the author is unknown; we know it only thanks to its inclusion in an eleventh-century Byzantine anthology. Thought to date from the late first or the second century A.D., it was long attributed to the philosopher Bolos of Mendes in Egypt, but that idea has been debunked. The author named in the manuscript is Democritus, but it is surely not a work of the Greek philosopher of that name from the fifth century B.C. who postulated the idea of atoms. Like so many alchemical manuscripts of the ancient and medieval world, *Physica et mystica* seems to have been ascribed to a famous author to give it more authority.

The work is a recipe list like the Stockholm and Leyden papyri, with practical instructions for making gold, silver, gems, and dyes. But the title hints at something more. *Mystica* here does not imply anything mystical, but rather secret: this is one of the first suggestions that these arts are not intended to be seen or understood by everyone. After each trio of recipe-book passages comes the cryptic statement "Nature triumphs over nature. Nature rejoices in nature. Nature dominates nature." It is hard to know what this implies, but the rubric hints at some kind of scheme or principle that explains the transformations described.

One of the most important alchemical authors of this period is the Greco-Egyptian Zosimos of Panopolis, a city in Upper Egypt, in the early fourth century A.D. Rather little is known about him, and his works—he was said to have written twenty-eight books on alchemy—survive only patchily, interrupted here and there with those of other writers. But his text *Cheirokmeta* (Things made by hand) contains the hallmarks of alchemy proper: there are recipes aplenty, but also cryptic terminology, allusions to analogies between chemical processes and sexual reproduction, and hints at a unifying theory of these chemical transformations.

A modern reader might imagine that Zosimos sought to veil his accounts of ancient chemistry with metaphysical or mystical hocus-pocus—for example, in the way he speaks of his ingredients as having a "body" and a "spirit." But he was simply using the terms of reference familiar to him, based on the religious beliefs and practices that pervaded society, much as scientists today refer to, say, biological processes in the language of signal processing and information theory. When Zosimos accuses false alchemists of enlisting the aid of demons, he is again merely speaking the ubiquitous language of his times.

The Islamic alchemists

Alchemy first matured in Alexandrian Egypt around the third century A.D. When Alexandria and Upper Egypt were conquered in the mid-seventh century by the Islamic Rashidun caliphate, Arabic scholars inherited its intellectual riches. The works of Greek philosophers, physicians, mathematicians, and others, including Aristotle, Euclid, Hippocrates, and Galen, became widely translated into Arabic, especially in the great library of Baghdad, a city founded in 762 by the Abbasid caliph Abū Jaʿfar ʿAbd Allāh ibn Muḥammad al-Manṣūr.

RIGHT Marble bust reputedly depicting the alchemist Zosimos, 3rd century A.D. Zosimos was one of the first to describe the combination of chemical technology and philosophy that characterized true alchemy, and was later considered one of the foremost authorities on the art.

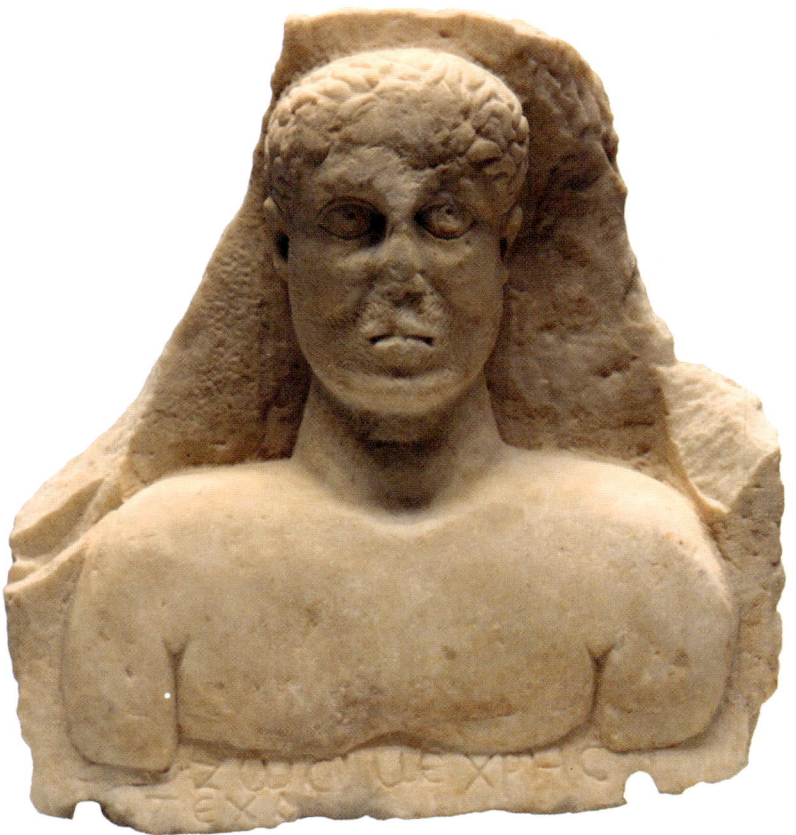

Cheirokmeta contains the hallmarks of alchemy proper: there are recipes aplenty, but also ... analogies between chemical processes and sexual reproduction, and hints at a unifying theory of these chemical transformations.

Islamic scholars did more than merely preserve and parrot what the Greeks had said. They extended, challenged, and corrected it, sometimes on the basis of practical investigations that warrant being called (with some caution) early experiments. Al-Khwārizmī introduced new elements of algebra (itself evidently an Arabic word); al-Rāzī pioneered medicine, not least by establishing a system of hospitals; and Ibn al-Haytham made pivotal discoveries in optics and astronomy. The eleventh-century scholar Ibn Sīnā was one of the most important philosophers and polymaths of the "golden age" of Islamic learning, whose works on physics and medicine were highly influential in the medieval West. (Islamic scholars

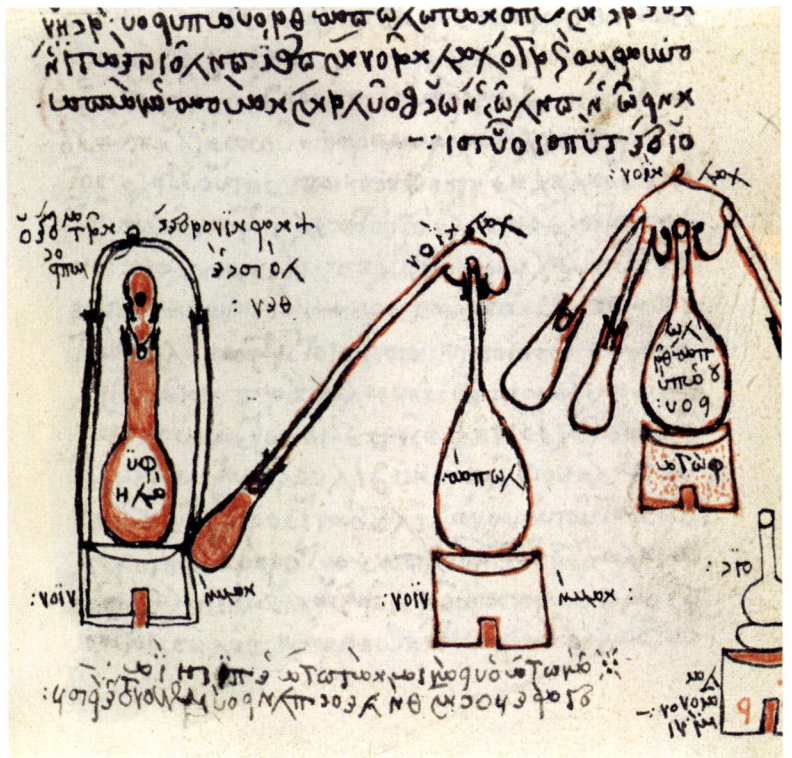

RIGHT Two alembics and their receivers, from a copy of a late third- or early 4th-century-A.D. treatise on alchemy attributed to Zosimos of Panoplis. Such illustrations testify to the practical nature of the discipline.

were often known in the West by Latinized names: al-Rāzī was known as Rhazes, al-Haytham as Alhazen, and Ibn Sīnā as Avicenna.)

Al-Rāzī wrote extensively on alchemy, but the most important Islamic alchemist was the late ninth-century scholar Jābir ibn Ḥayyān (see page 30). As is common for these early thinkers, the works attributed to Jābir were not necessarily all written by him; indeed, many more books have his name attached than one individual could possibly have written, perhaps three thousand in all—a body of work known as the Jābirian corpus.

In the writings of Jābir (that is, of the Jābirian authors) and al-Rāzī, alchemy begins to look like a systematic practical art

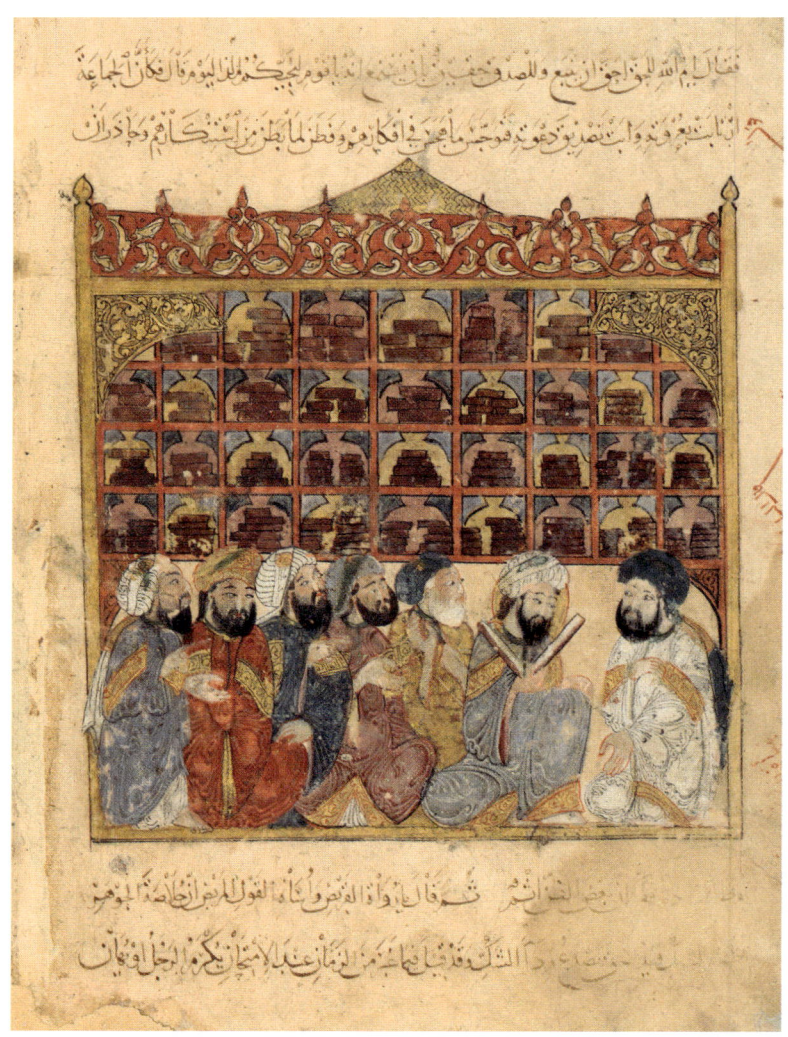

RIGHT Bayt al-Ḥikmah, the House of Wisdom, or Grand Library of Baghdad, founded in 762 A.D., from the *Maqāmāt of al-Ḥarīrī*, illustrated by Yaḥyā al-Wāsiṭī (1237). The library clearly played an important role during the "golden age" of Islamic scholarship until it was destroyed by the Mongols in the 13th century. However, its reputation as a kind of proto-university is now widely considered by historians to be something of an exaggeration.

that embraces a phenomenal range of substances, from metals to minerals, salts, vapors, and acids, supported by both careful empirical study and rational theory. But Jābir didn't merely collect recipes, nor was he content to augment them with a few grand and cryptic axioms. Rather, he undertook a systematic program of constructing theoretical principles that could account for the nature and transformations of matter that he observed, both in nature and in the workshop. It is not too much of a stretch to regard this mission as comparable to that of modern science: to seek unifying but typically "hidden" rules that can explain a wide range of observations.

One of the foundational texts of the Greco-Egyptian alchemical writings is the *Secret of Creation*, attributed—again, almost certainly spuriously—to the first-century-A.D. Greek philosopher

BELOW Page from a later copy (1700–1899) of *Sirr al-Asrār* (Secret book of secrets) by the 10th-century Persian polymath Abū Bakr Muḥammad ibn Zakariyyā' al-Rāzī. The book is in three parts: the first ("On Simples") describes the raw materials of alchemy, and the second ("On Instruments") the apparatus. The third and major part ("On Methods") discusses procedures for the chemical manipulation of materials such as arsenic, sal ammoniac, and sulfur.

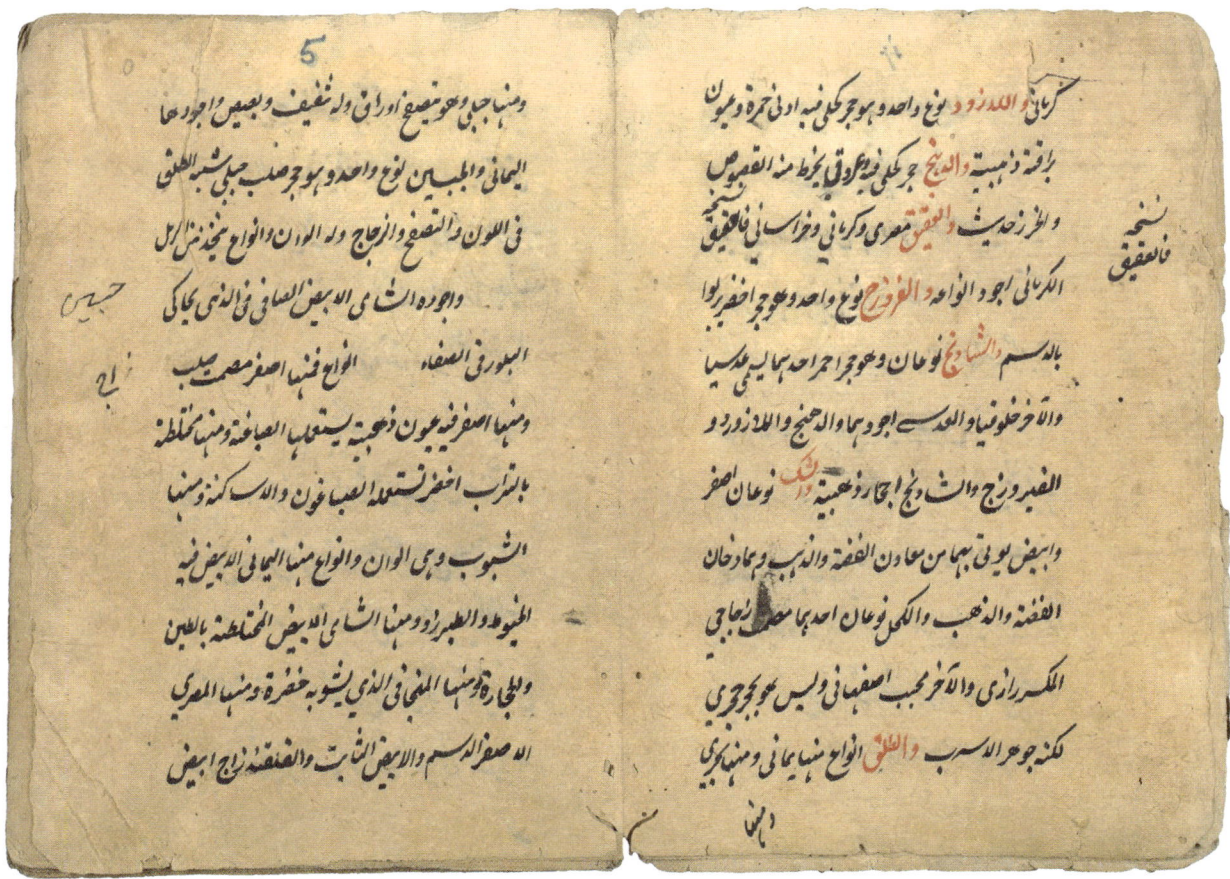

BLACK EARTH: THE ORIGINS OF ALCHEMY

Apollonius of Tyana. We know of the work only through Arabic writers of the seventh or eighth centuries, who refer to the author as Balinus. Within the *Secret of Creation* sits a short text called the *Emerald Tablet* that has acquired almost mythic status.

Balinus says this passage was itself written by Hermes Trismegistus, the "thrice-great Hermes," who may or may not have had any connection to the Greek god Hermes, deemed to be the divine originator of the alchemical (or "Hermetic") art. Others say that Hermes Trismegistus was a magician who lived at the time of Moses, and that his *Emerald Tablet* was discovered by Sarah, the wife of Abraham. When the passage was translated into

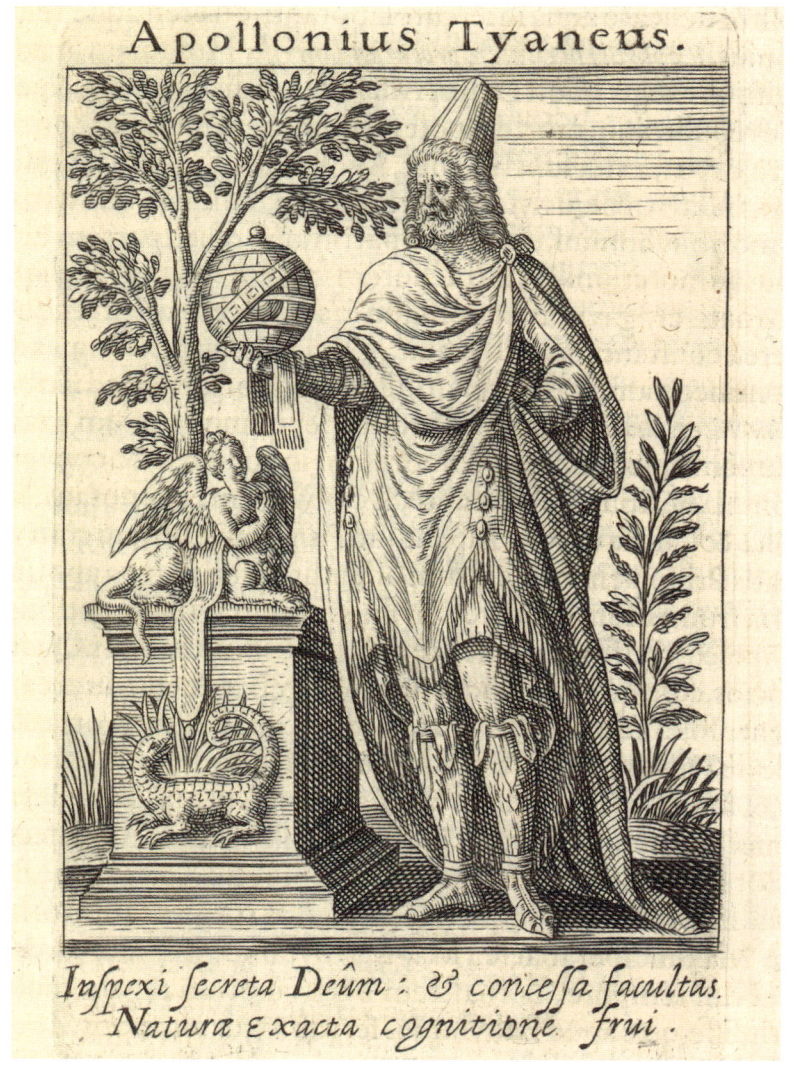

RIGHT Imaginary portrait of the 1st-century Greek philosopher Apollonius of Tyana, supposed author of the Greco-Egyptian *Secret of Creation*, from Jean Jacques Boissard's *Tractatus posthumus . . . de divinatione & magicis praestigiis* (A posthumous treatise. . . of divination and magical tricks), engraved by Theodor de Bry (Oppenheim, *c.* 1615). Apollonius was a mystic, rumored to possess magical powers and said to have been executed for conspiring against the Roman emperor. But in what has been written about him it is hard to distinguish fact from myth, and there is no indication that he actually had any connection with early alchemy.

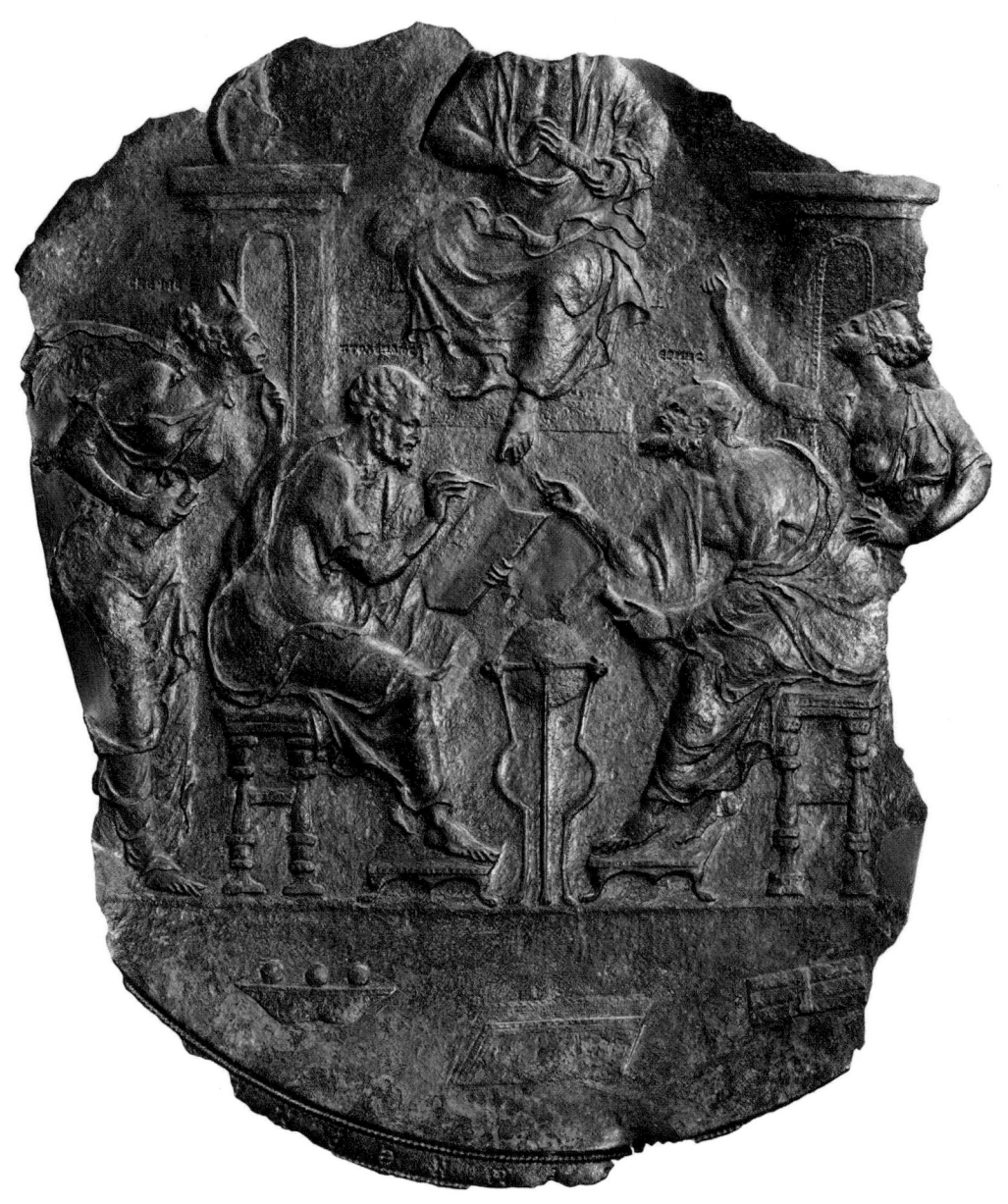

ABOVE Hermes Trismegistus, the "thrice-great Hermes," teaches Ptolemy the World System on this Byzantine silver plate with relief from the eastern Mediterranean, c. 500–600. Hermes Trismegistus was considered a legendary founder of the art of alchemy.

medieval Latin in 1200, it excited copious commentary. The key to the *Emerald Tablet* appears in its first phrase: "What is below is like what is above, and what is above is like what is below, to accomplish the miracles of one thing." This notion—"As above, so below"—became a central theme in the Neoplatonic philosophy of the early Christian era and a core concept in early and medieval alchemy.

PROFILE

Jābir ibn Ḥayyān

(c. 721–c. 815)

Despite being the most celebrated of the Islamic alchemists, Jābir ibn Ḥayyān might never have existed at all. That's an extreme interpretation, but many of the three thousand or so works in the corpus attributed to him were surely composed by other authors—most of whom lived later than the eighth century, when Jābir is said to have worked in the court of the Abbasid caliph Hārūn al-Rashīd in Baghdad.

The reign of Hārūn—his epithet al-Rashīd translates as "the just" or "the upright"—is considered to have been the start of the golden age of learning in the Islamic world. Hārūn is said to have established the great library known as the House of Wisdom in Baghdad, and, despite conflicts with the Eastern Roman Empire of Byzantium, the caliph forged a friendly relationship with the Frankish emperor Charlemagne to the west.

Jābir was allegedly a physician from the city of Kufa, south of Baghdad, where his family moved from eastern Persia. He was said to have learned alchemy from various adepts, including a Christian monk named Marianos, before entering the caliph's court. Toward the end of his life he fell from favor for political reasons and spent his days under house arrest, or at least in seclusion, at his home in Kufa.

Perhaps Jābir's greatest contribution to alchemy as early chemistry was to stress the importance of experimentation. "He who performs not practical work nor makes experiments will never attain to the least degree of mastery," he wrote. In his *Kitāb al-Raḥma al-kabīr* (Great book of mercy; which historian Paul Kraus considers the oldest work in the Jābirian corpus), Jābir seems skeptical of transmutation, saying that those who give themselves over to the search for gold are of two types: "the deceivers and the deceived." That, however, is not the same as saying that gold-making is impossible.

Some of the Jābirian texts date from the ninth century, and Kraus suggests that this is when the real Jābir, whoever he was, actually lived. The works have little of the mysticism, allegory, and cryptic character of later alchemy, reading much more like straightforward accounts of laboratory practice. Some are concerned with eminently practical chemistry: making steel, dyeing cloth and leather, and preparing varnishes, inks, and colored glass. The corpus is not limited to alchemy, touching also on medicine, philosophy, mathematics, logic, and religion.

Although the Jābirian works contain the influential notion that all metals are composed of the two principles sulfur and mercury

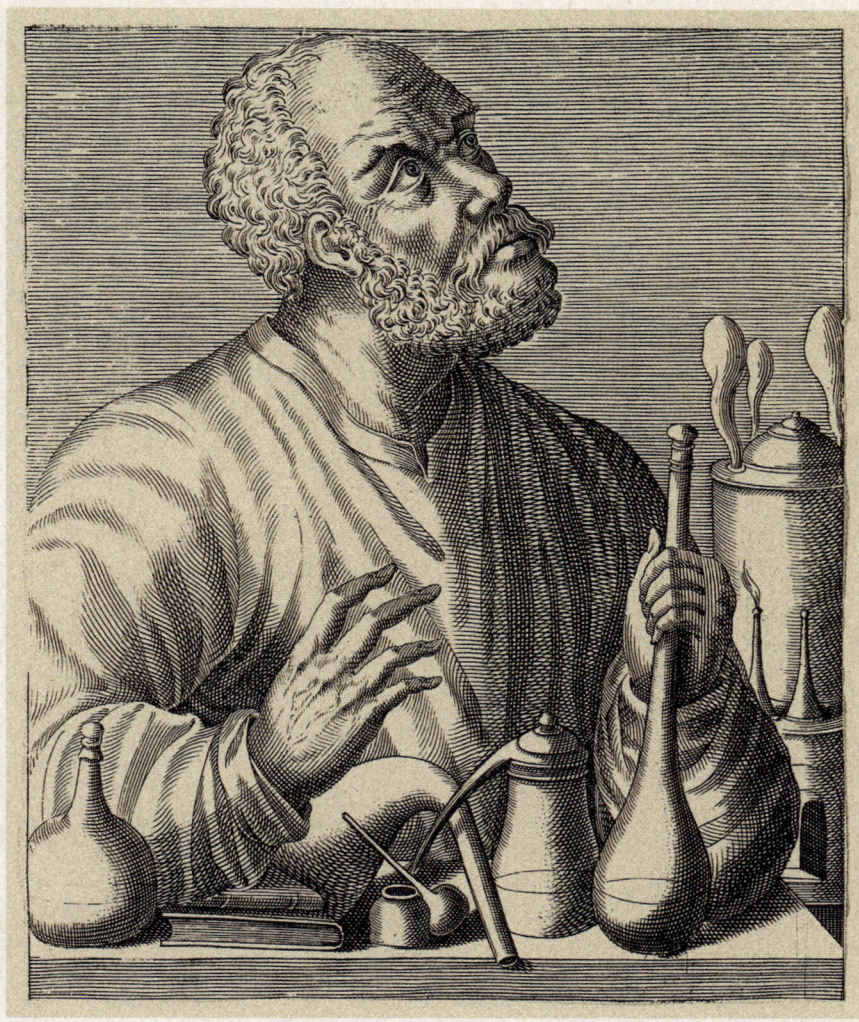

RIGHT Engraved portrait of the 8th-century Arabic polymath Jābir ibn Ḥayyān, known in the West as Geber, from André Thevet's *Les vrais pourtraits et vies des hommes illustres grecz, latins et payens* (The true portraits and lives of illustrious Greek, Latin, and pagan men; Paris, 1584).

(see page 67), they also observe the Aristotelian system in which all substances contain the complementary principles of hot/cold and dry/moist, in varying proportions. Jābir considers these to be actual materials that can be extracted, and he writes that "cold" is "a white and pure substance which, when it is touched by the smallest degree of moisture, dissolves and is again transformed into water."

In the Middle Ages this corpus was widely considered (although there were skeptics even then) to be the work of a single author, known by the Westernized name Geber. (To confuse matters further, one of the most famous Jābirian texts was certainly written by a medieval European author under that name—see page 69.) Despite the notable clarity of the writing in much of the Jābirian corpus, Samuel Johnson argued in his 1755 English dictionary that the word "gibberish" derived from the alchemical "jargon of Geber and his tribe." The jibe reflects the Enlightenment contempt for alchemy, though in this case it is not only probably etymologically unsound but also unjustly directed.

BLACK EARTH: THE ORIGINS OF ALCHEMY

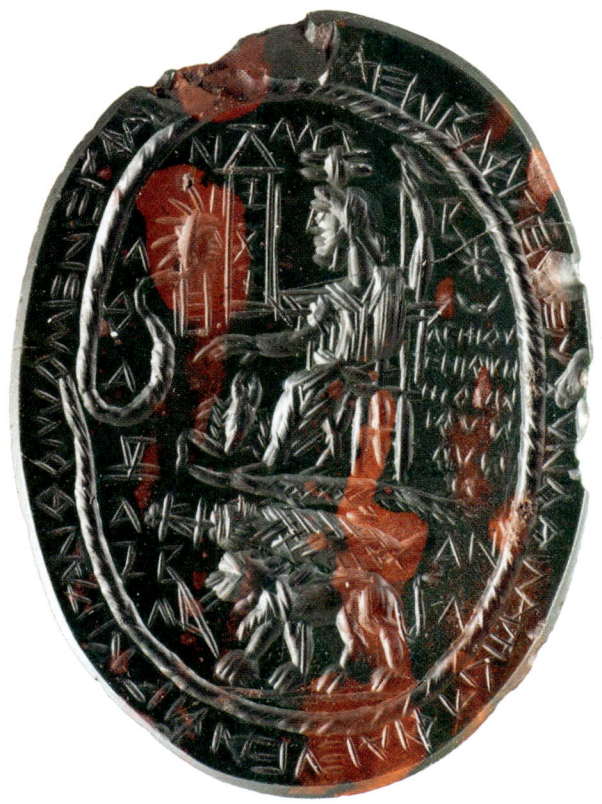

ABOVE "Magical gem" amulet from the Roman period, showing the Egyptian god Horus on a lotus, surrounded by an Ouroboros (serpent eating its own tail) and incantations. The Ouroboros was a common symbol in ancient Egyptian iconography, and became widely used in later alchemical imagery as a representation of the Great Work of goldmaking and the cycle of creation and destruction.

OPPOSITE Page depicting a stela (upright monument) dedicated to the Egyptian king Amenemhat II, from an Arabic treatise on alchemy, *Kitāb al-aqālīm al-ṣabʿah* (Book of the seven climes), incorporating Egyptian and Arabic motifs. The book is an 18th-century copy of a 13th-century text by Abū al-Qāsim al-ʿIrāqī, which itself reproduced alchemical illustrations from earlier Arabic texts, including some with Egyptian sources.

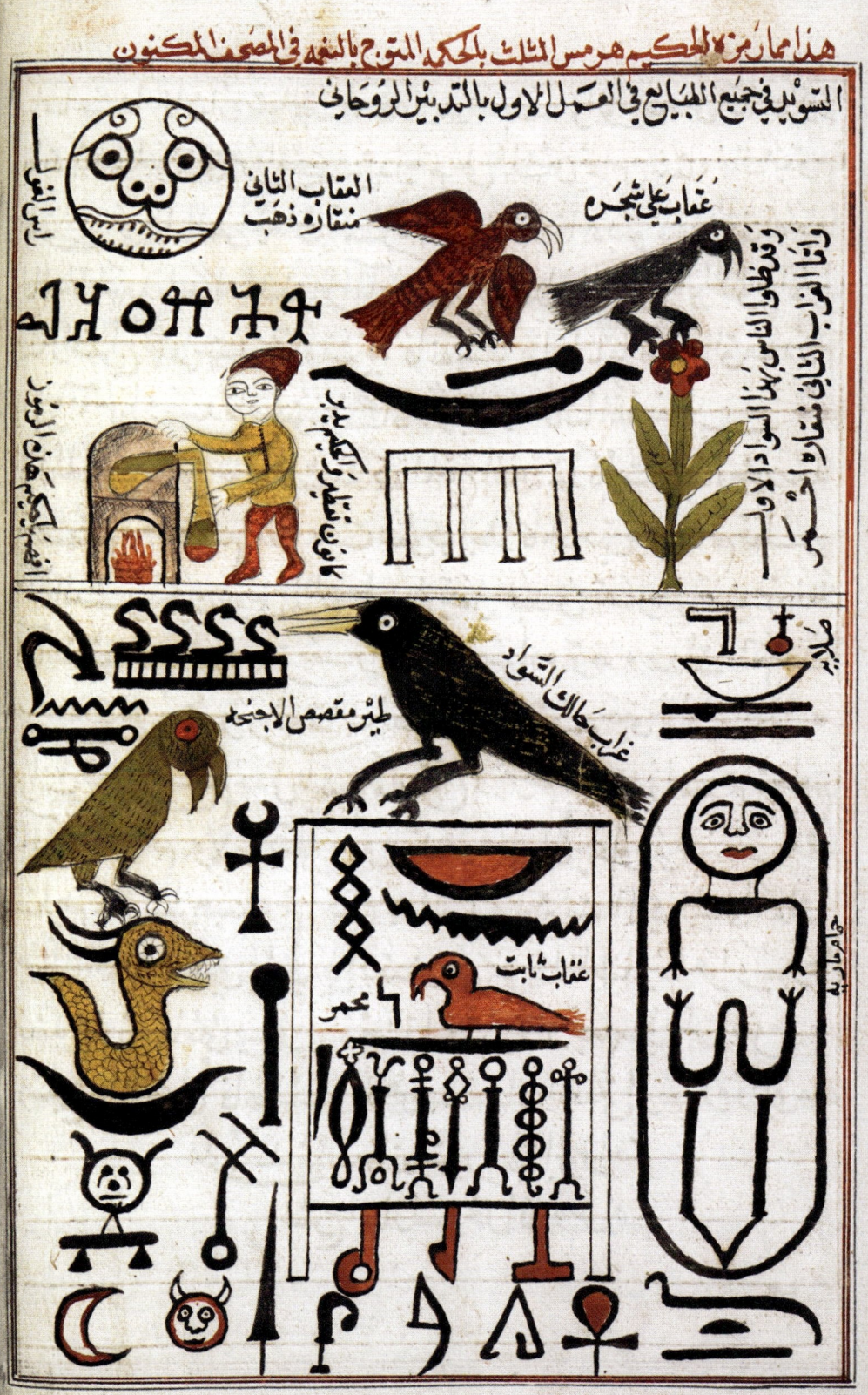

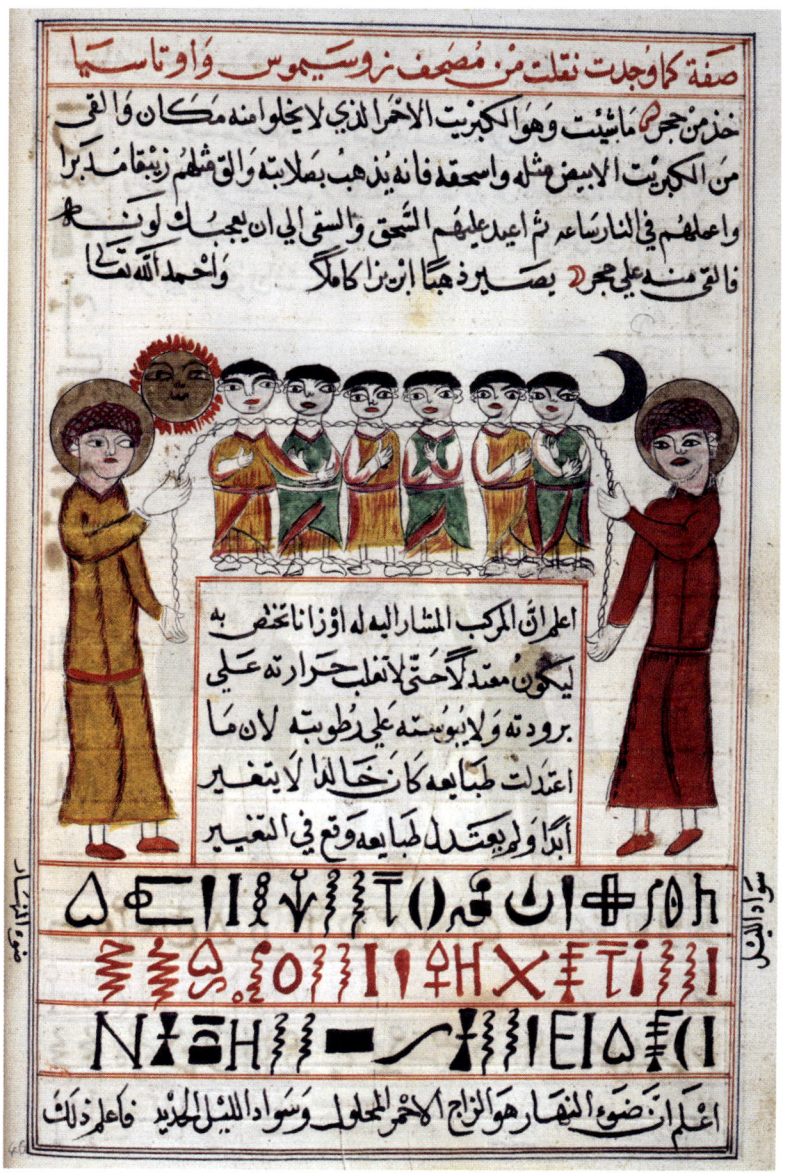

ABOVE The Greco-Egyptian alchemist Zosimos of Panopolis (left, with the sun over his head), and Theosebeia, a Deaconess (right, with the moon), from an 18th-century Arabic alchemical manuscript by Abū al-Qāsim al-ʿIrāqī, *Kitāb al-aqālīm al-sabʿah* (Book of the seven climes). It is based on an earlier illustration in *Muṣḥaf al-ṣuwar* (Book of pictures), a 13th-century alchemical emblem-book purported to be by Zosimos.

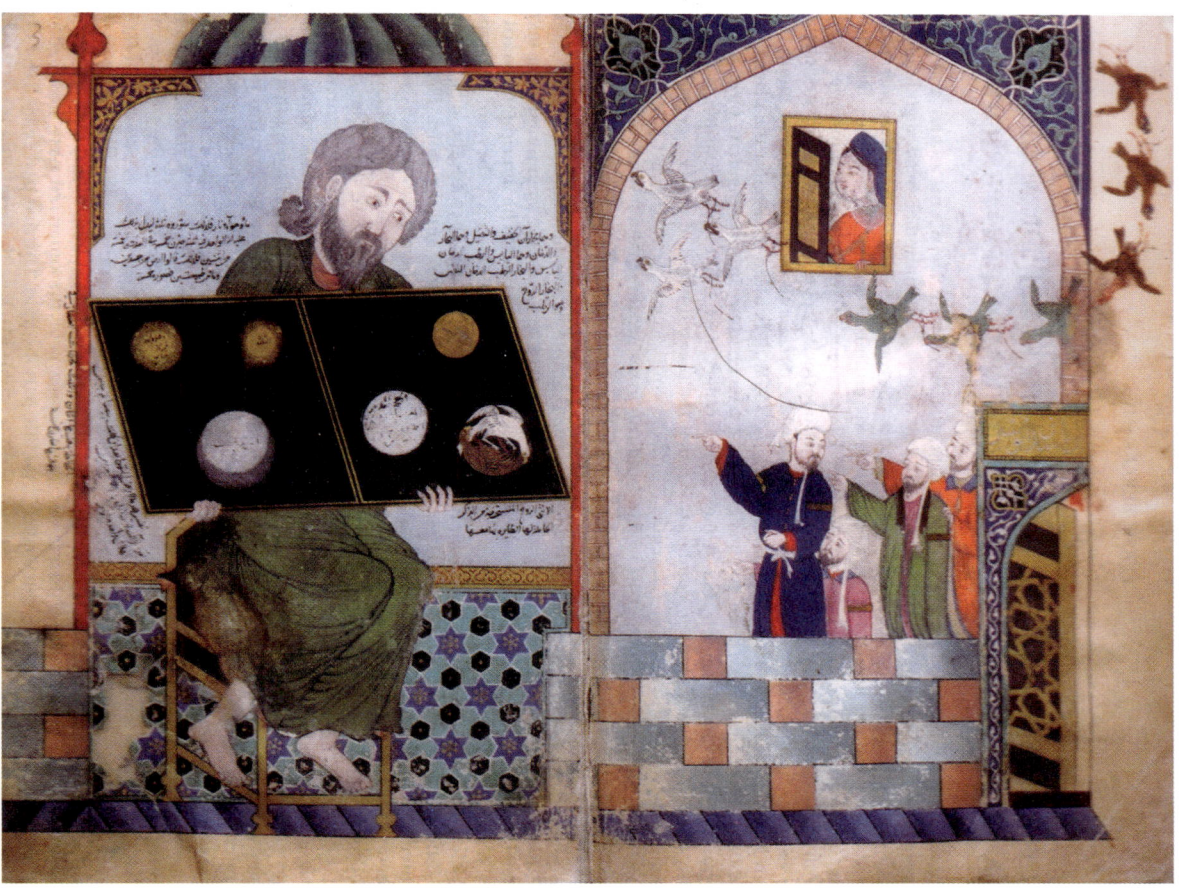

ABOVE Hermes Trismegistus holding the *Emerald Tablet*, illustrated in a transcript of *Al-māʾ al-waraqī* (The silvery water) by the Arabic scholar Umail al-Tamīmī, produced in Baghdad in 1339. Many of the notes written around the tablet, called the "Letter from the Sun to the Moon," are mathematical relationships between the hieroglyphs. The *Emerald Tablet*, which first features in an early 9th-century work ascribed to the Greek writer Apollonius of Tyana, is a short set of rather gnomic dictums later believed to represent the core principles of alchemy, although it is unclear whether they bear any true relation to alchemy at all.

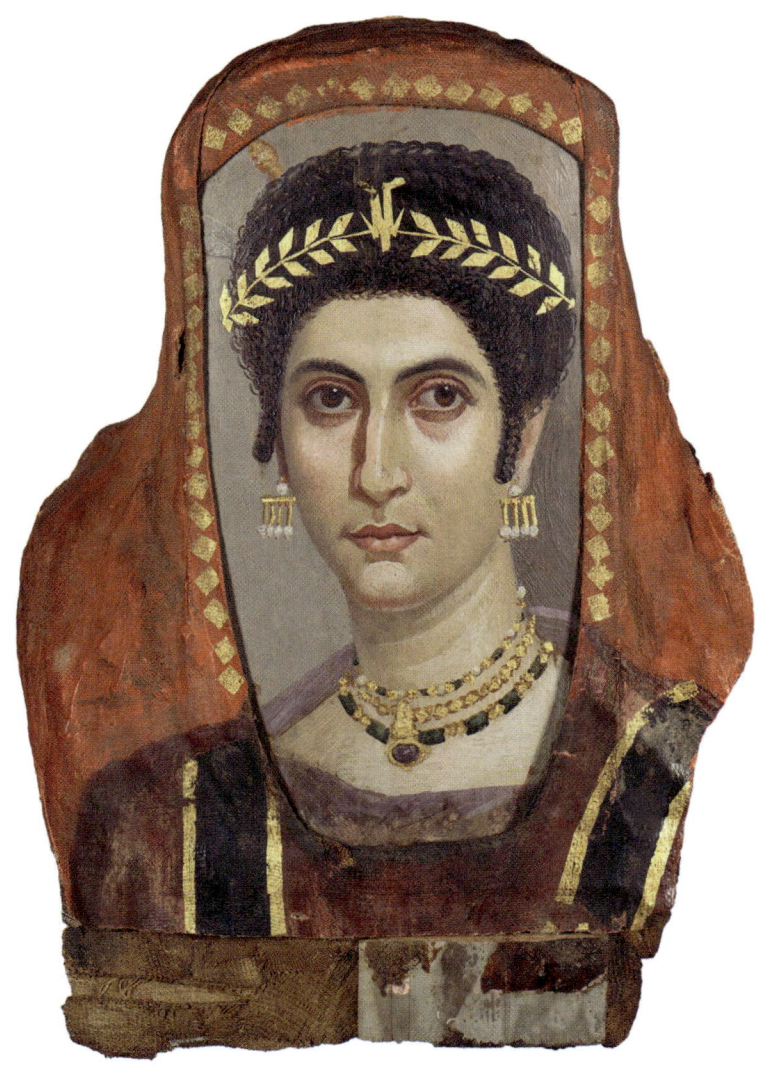

ABOVE This Roman-Egyptian portrait on a mummy casing, depicting a woman apparently called Isidora, dates from around A.D. 100, and is rendered in wax-based encaustic on linden wood. The pigments used here include red lead (lead tetroxide), the iron mineral jarosite (yellow-brown), madder (pinks), copper mineral green, and lead white. Red lead was sometimes applied to Roman-Egyptian mummy wrappings themselves, perhaps symbolically to ward off evil spirits. No more is known about the identity of the woman, but she was evidently a person of wealth and social status.

OPPOSITE Page from a 14th-century copy of *Kitāb al-Qānūn fī l-ṭibb* (The canon of medicine) by the physician and polymath Avicenna (Ibn Sīnā; 980–1037). Avicenna was considered in the Middle Ages to be an authority on medicine, but he was skeptical about alchemy and especially about the possibility of the transmutation of metals. He felt that human art always fell short of what God has created in nature: "human industry is not the same as what nature does," he wrote; even if alchemists could make artificial gold, it would not be of the same quality as natural gold but a mere pale imitation.

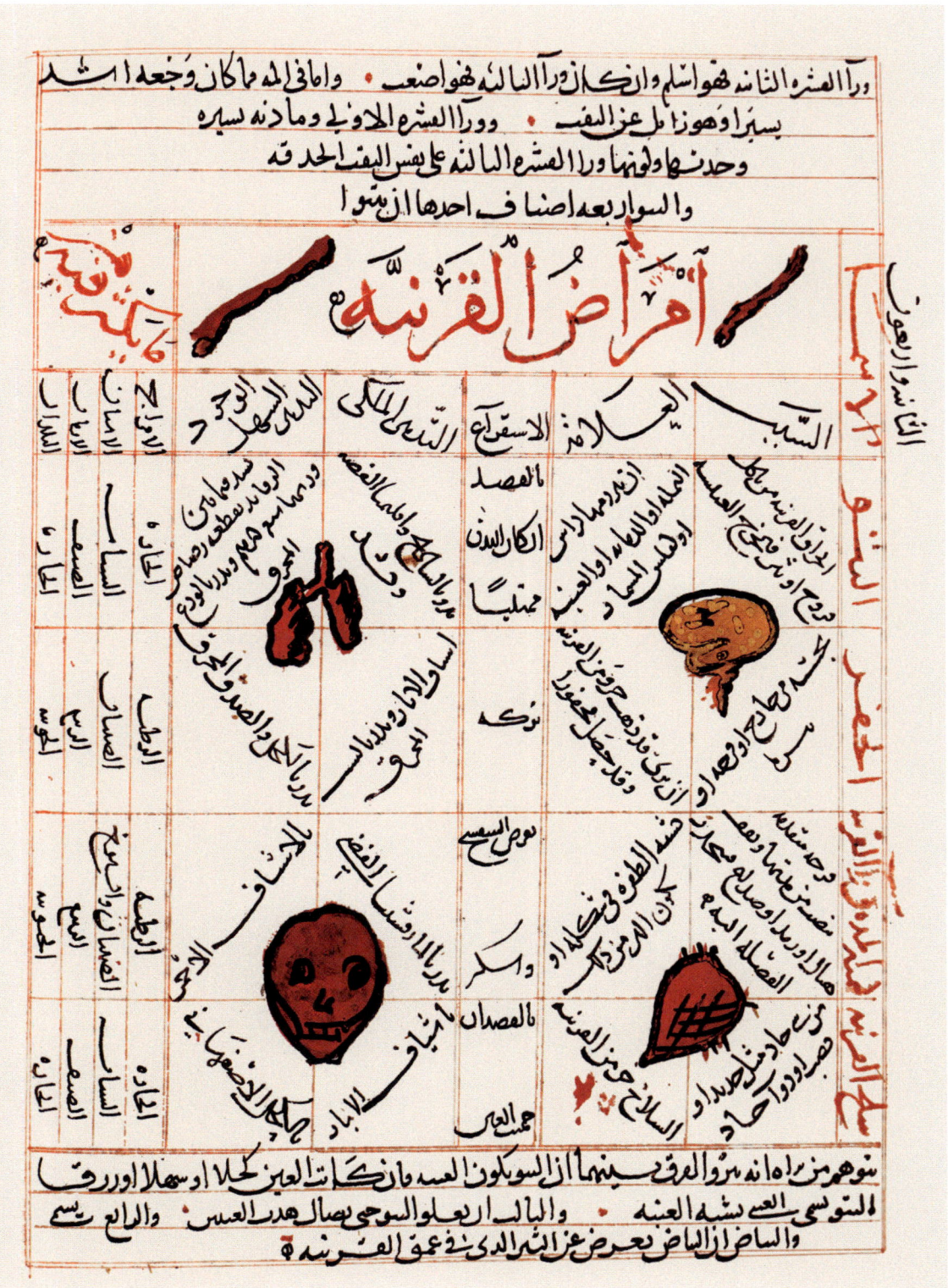

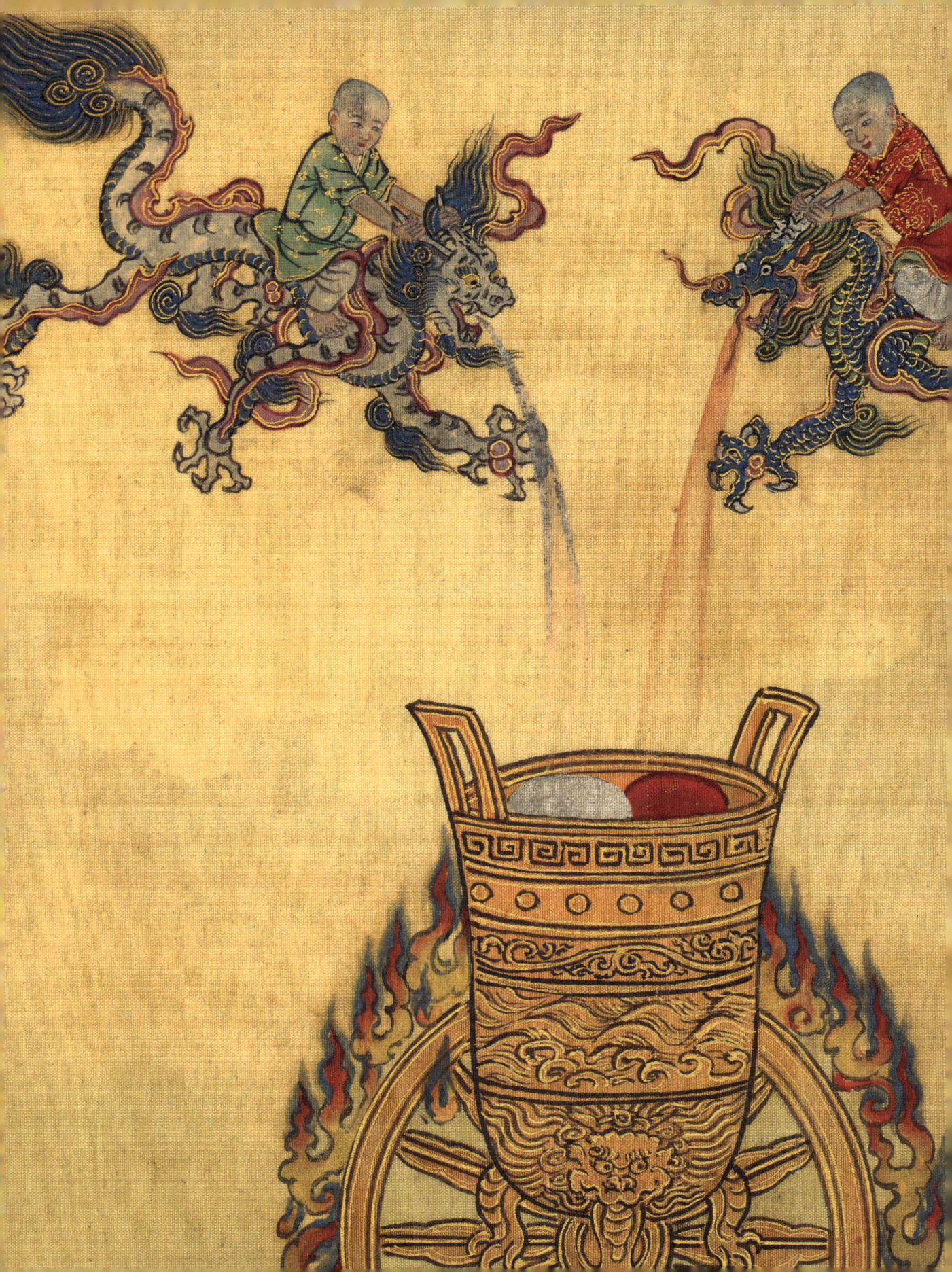

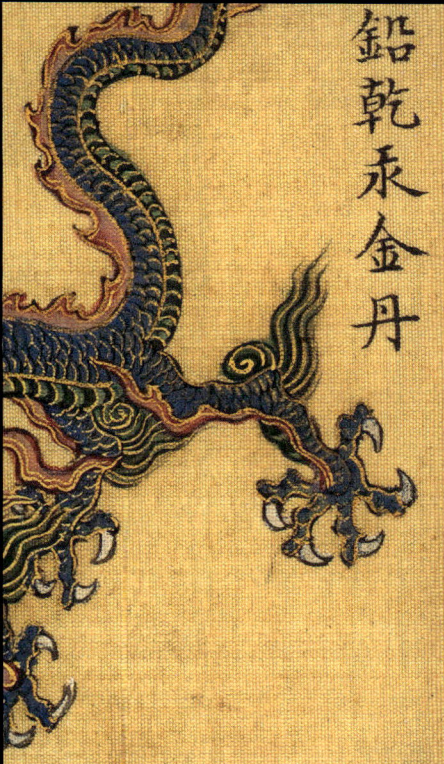

CHAPTER TWO

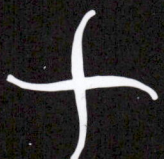

Elixirs
ALCHEMY IN THE EAST

LEFT Detail from a 17th-century Qing-dynasty Chinese scroll painting depicting a tiger and a dragon above a heated crucible containing an elixir. In the Daoist *Waidan* tradition ("outer" chemical alchemy), the tiger represents mercury and the dragon represents lead. In the tradition of *Neidan* (the "inner" alchemy of the human body), they symbolize respectively the complementary principles of *yin* and *yang*, which are brought together in the body through meditation practices.

CHAPTER TWO

Alchemy was long practiced in the East, particularly in China and India, although to what extent those traditions were connected to alchemy in the West, if at all, remains unclear. Sinologist and historian Nathan Sivin claimed in 1968: "It is impossible to believe that two traditions which shared so many materials, methods, and goals could have remained mutually exclusive over two millennia of unceasing cultural contact." But historians today are more cautious. "Eastern and Western alchemy," says Lawrence Principe, "are embedded in such widely divergent cultural and philosophical contexts that trying to squeeze them into a single narrative damages the uniqueness of each one." With that proviso in mind, let's see what alchemy meant east of the Nile.

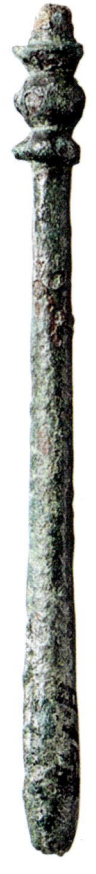

In his three-volume tome on the origins of alchemy published between 1919 and 1954, the German chemist Edmund Oscar von Lippmann asserted that it could never have originated in China because "the Chinese possessed no characteristic chemical methods of their own, nor any apparatus of original design." Meanwhile, the American sinologist Homer Dubs argued in the mid-twentieth century that, precisely because of the alleged backwardness of ancient Chinese chemistry, claims that gold could be made from other substances were able to take root in China in a way they could not in the Middle East, since no one could reliably dispute them.

These statements say more about the prejudices of Western historians in former times than they do about Chinese metallurgy and chemical technology—which was, since ancient times, sophisticated and complex. So much so that it can be hard to figure out what the rich vocabulary of metals in ancient China really means—what, in other words, these artisans were making. "Gold" was a somewhat loose concept before early modern times: often a metal need only have a bright yellowish sheen, showing merely a superficial resemblance to the natural metal, to qualify as a kind of gold. Chinese metalworkers could, for example, make various forms of brass, an alloy of copper and zinc with a golden color. Several

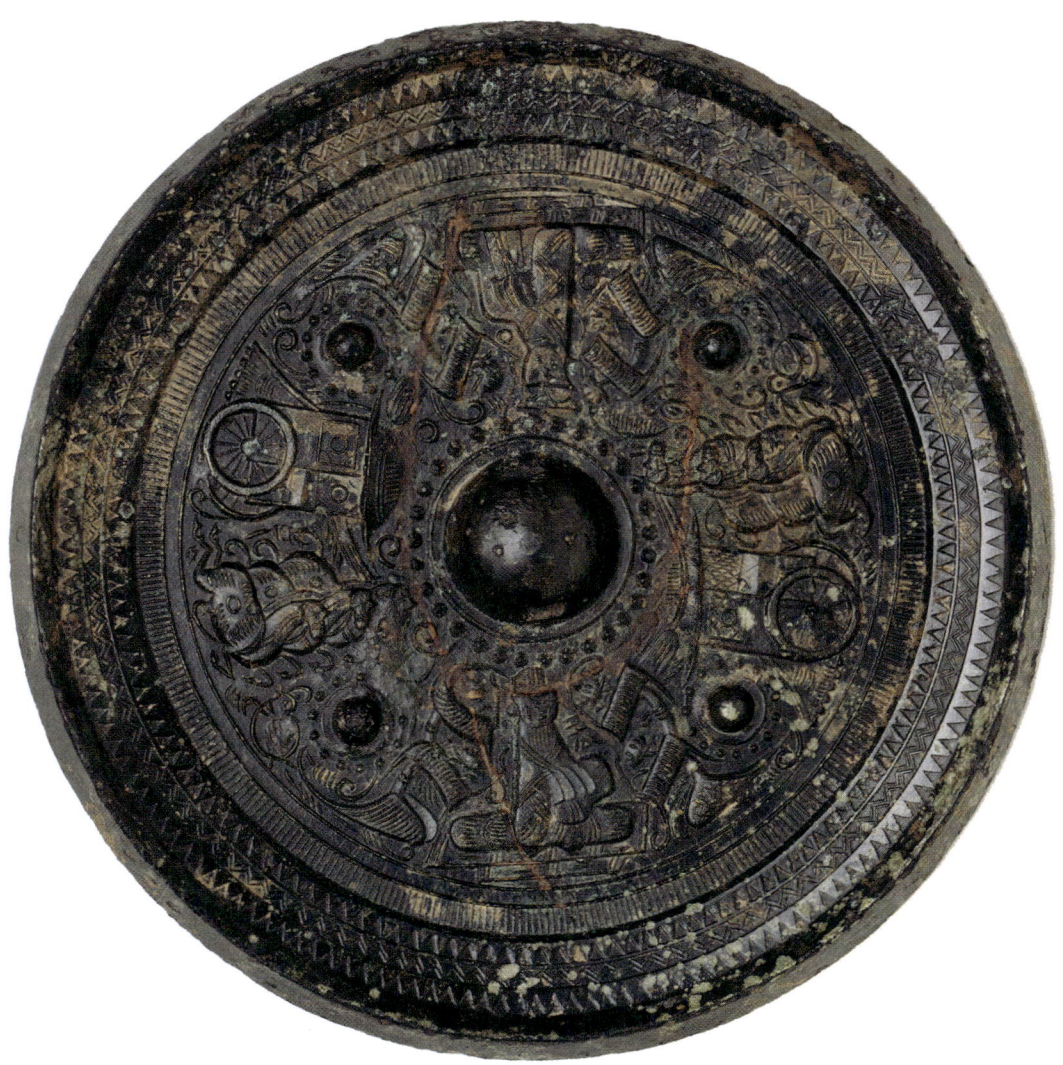

ABOVE A bronze mirror back with reliefs, probably from Zhejiang province, Eastern Han dynasty, 25–220 A.D., featuring the Queen Mother of the West (Xi Wang Mu) and the King Father of the East (Dong Wang Kong) attended by Immortals. These Daoist deities figured prominently in the quest for eternal life, and for life-preserving elixirs, in Han-era China.

OPPOSITE An ancient pin made from an alloy of copper and zinc (sometimes known in the West as Indian tin) from the Temple of Ishtar, in the Royal Palace of Mari, eastern Syria, 2500–2340 B.C.

ancient Chinese texts speak of a "yellow silver" (*huang yin*) that is probably brass, and the sinologist Joseph Needham claims that "many of the artificial golds produced by the Chinese aurifictors [his neat name for gold-makers] and alchemists were brasses of suitable composition."

Although zinc was not recognized as a distinct metal until much later, zinc bracelets have been found in Greek ruins dating from before 500 B.C. In the sixteenth century, it seems to have been largely imported from India and the Far East, and Europeans often called it "Indian tin." In China, it seems to have gone by the name *wo qian*, meaning "poor lead." However, zinc doesn't need to be

ELIXIRS: ALCHEMY IN THE EAST 41

extracted and purified from its ores to make alloys like brass: it appears as an impurity of lead and tin.

Bronze (an alloy of copper and tin) was made in China since the start of the second millennium B.C. within the Shang culture of the Yellow River basin. Vessels made from this alloy were very costly, available only to rulers and nobles and often associated with religious rites. With its metallic brown color, bronze itself would hardly be mistaken for gold. But arsenic mixed with copper or bronze imparts a silvery or golden sheen, as well as making the alloy more ductile and easier to work. This was a common process throughout the Middle East and Asia during the Bronze Age, although it's unclear whether arsenic was added deliberately or was naturally present as an impurity of copper.

The ancient Chinese prowess in making and transforming metals raises the question of whether we should think of it as alchemy—trying to make gold—at all, or rather as straightforward artisanship. The answer is most probably that the distinctions would not have been clear at the time.

BELOW This woodblock illustration, taken from the encyclopedic *Tiangong kaiwu* (Exploitation of the works of nature) compiled by Song Yingxing in 1637, shows the process of lead compounds being removed from impure silver, in a crucible containing saltpeter.

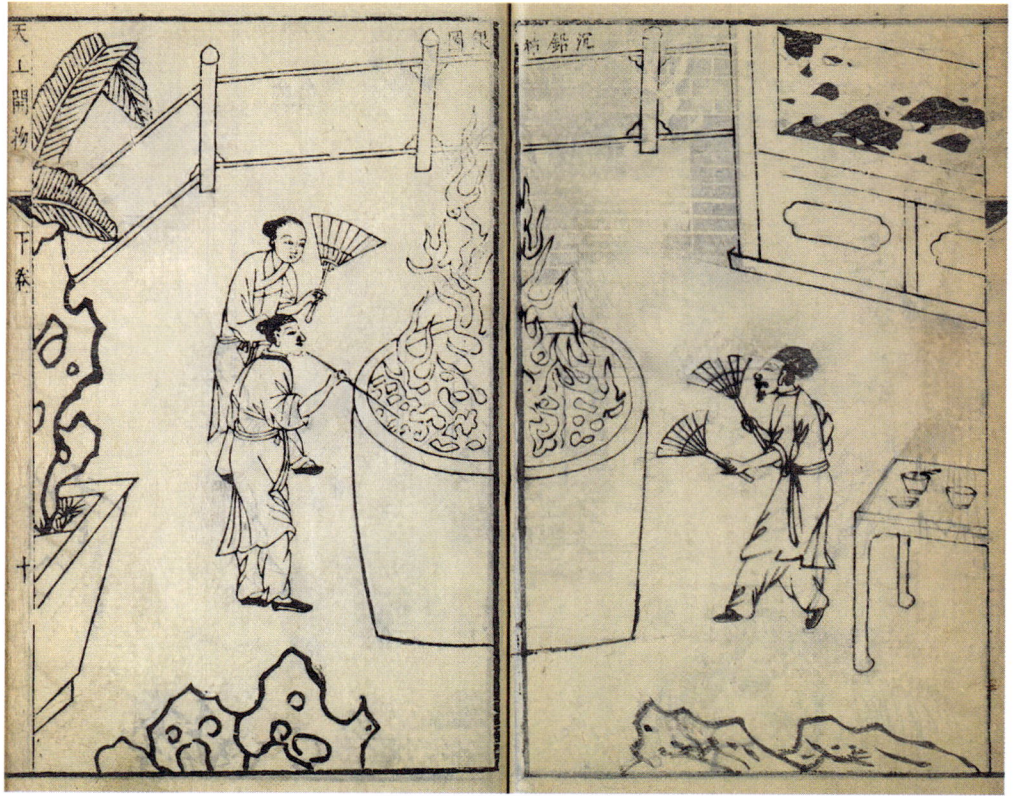

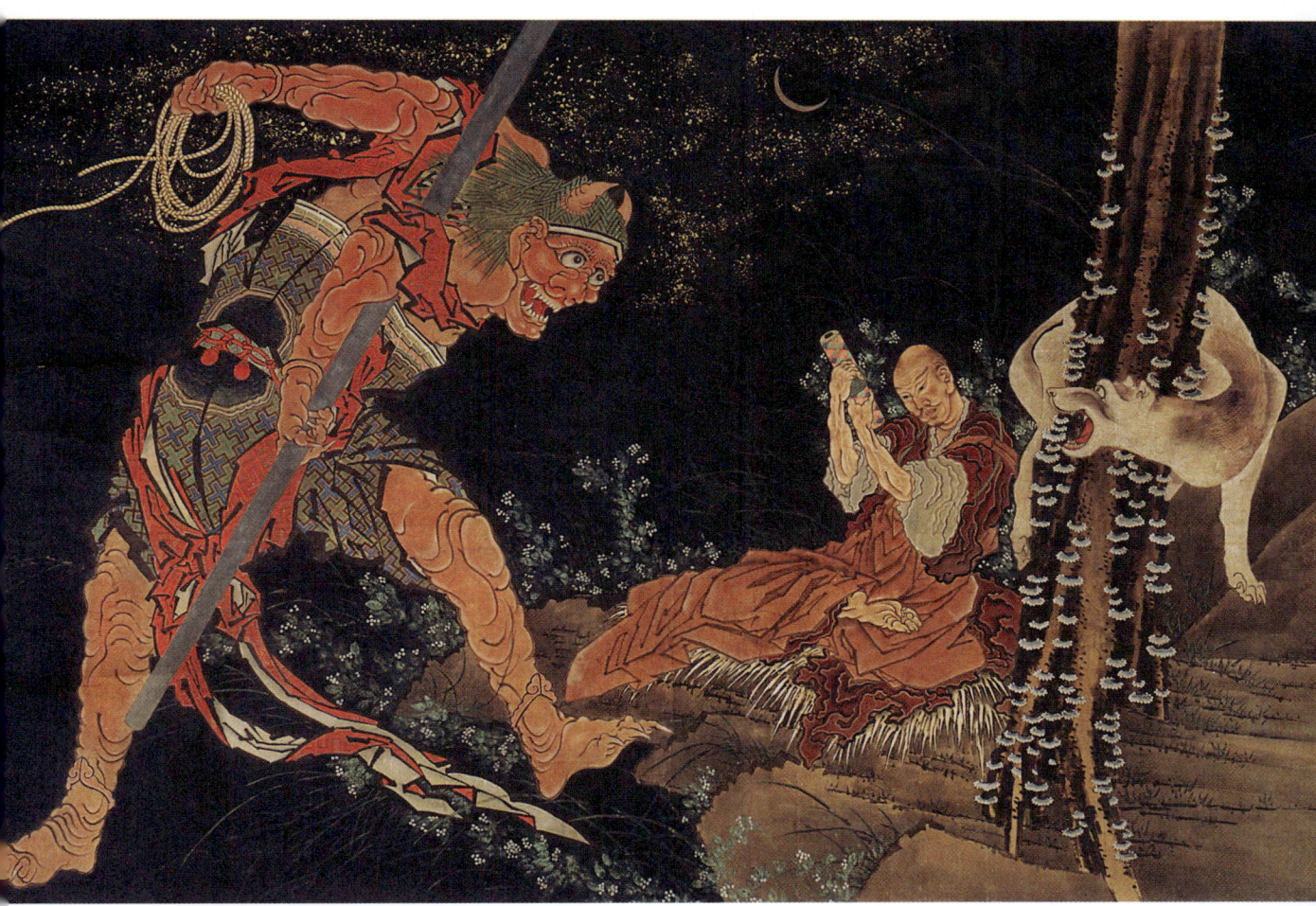

ABOVE A Japanese priest (*kōbō daishi*) practices the spiritual discipline called the Tantra, with a demon before him and a wolf behind, in this drawing by Katsushika Hokusai, before 1849. The Tantric tradition originated in India, and was incorporated into both Hinduism and Buddhism. It spread into China and also into Japan—for example, in the Shingon tradition of Buddhism. Tantrism provided a conceptual foundation for the alchemical tradition in India.

Varieties of gold

Chinese philosophers shared with their Western counterparts the conviction that transmutation of metals was possible and that metals and minerals were slowly transformed from one to another within the earth. In the fourth century A.D., Ge Hong, arguably the most influential of all Chinese alchemists, wrote that "transformation is something spontaneous in nature." Consider, he said, how caterpillars become moths, or—rather blunting his argument from today's perspective—how men may turn into women, snakes into dragons, and oysters into frogs. "Why then," he asked, "should we demur to the possibility of making gold and silver from other things?" Ge Hong argued that artificial gold might even be superior to the natural kind, because it acquired all the essences and virtues of its various ingredients.

RIGHT Artificial or mosaic gold (the compound tin disulfide) is evident in the sleeve detail of Andrea del Verrocchio's *Virgin and Child with Two Angels*, c. 1476–78. The detail is also shown below. As well as real gold, mosaic gold was commonly used in manuscript illumination, possibly chosen for its subdued shimmer rather than merely as a cheap substitute for the costly metal.

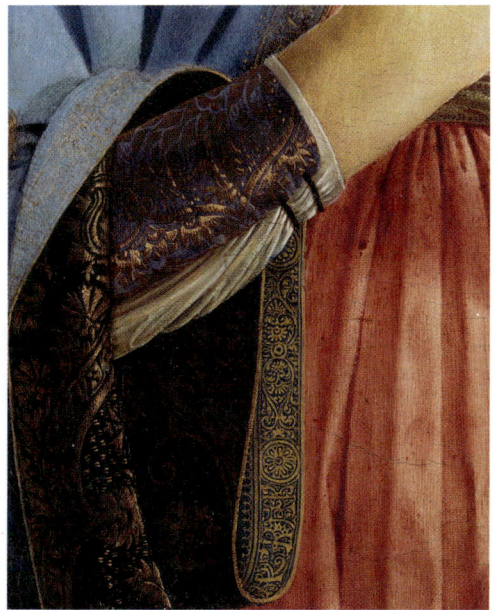

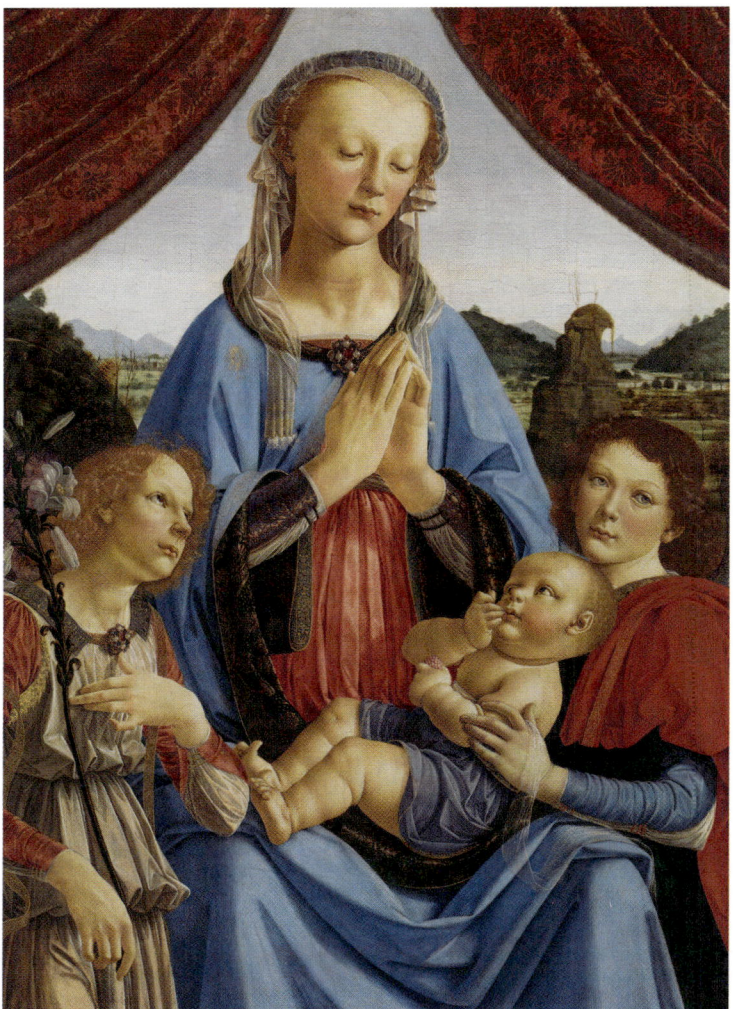

How did one facilitate that transformation artificially? Ge Hong says that a type of gold might be made from cinnabar. "That is why gold is generally found in the mountains below cinnabar deposits," he explains. The alchemist and polymath Tao Hongjing of the fifth and sixth centuries A.D. says that copper can be turned into gold using realgar (the sulfide ore of arsenic). There was evidently some latitude in what passed as gold: Ge Hong says that some artificial golds could be softened with wine, or made miscible in water, or turned into a kind of paste. These transformations, says Needham, are all consistent with the idea that such "artificial gold" was the orange-yellow compound tin disulfide, often called mosaic gold in medieval Europe.

Artificial gold was often classified according to tint: an eighth-century Chinese manuscript mentions various hues progressively produced in alchemical transformations, from pale greenish to yellow, red, and finally purple. Ge Hong refers to a "purple sheen gold of superior hue," while a fourteenth-century text tells of an artisan of Jiangsu province who became rich and famous for his jewelry made from a reddish purple "full-color" gold. In Japan, *shakudō*, an alloy of copper with a little gold, has been popular since at least the twelfth century; it acquires a deep purplish or indigo patina when subjected to a chemical process called *niiro*, which involves a mixture of copper compounds. One would hardly mistake *shakudō* for gold—but it's an indication of how Eastern gold-making alchemy was bound up with other metal-working crafts and technologies.

There were also recipes for making silver. The eleventh-century scholar He Wei claimed to have seen an adept make it from copper using saltpeter and a "white, lustrous" reagent. The real product here was possibly an alloy of copper and nickel, like that used for "silver" coins today: nickel removes the redness from copper very effectively.

Metal objects could also be gilded with gold and silver by dissolving the precious metals in mercury to form amalgams. The butter-like amalgam was rubbed on a surface and heated to evaporate the mercury, leaving behind a gleaming coat. This process might have been developed around the fourth century B.C. both in the East and in Greece.

By the end of the Tang dynasty in the tenth century there was an extensive classification of different types or grades of gold, distinguished by their origin. A document from around 918 lists twenty varieties, from natural forms found in alluvial deposits or mined from quartz veins to various artificial golds such as "realgar gold," "sulfur gold," and "brass gold." Needham argues that such matter-of-fact distinctions of composition made by the Tang metallurgists mark a "decisive step forward in scientific thinking."

Some of these gold-making (or perhaps gold-faking) recipes may have been discovered in the second century B.C. in the court of a prince named Liu An in Huainan in north-central China, one of the kingdoms of the Han empire. Liu An was the grandson of the dynasty's founder Liu Bang, and he welcomed alchemists and magicians to his court; the prince was himself an eminent scholar and philosopher and one of the authors of an important Daoist text called the *Huainanzi*.

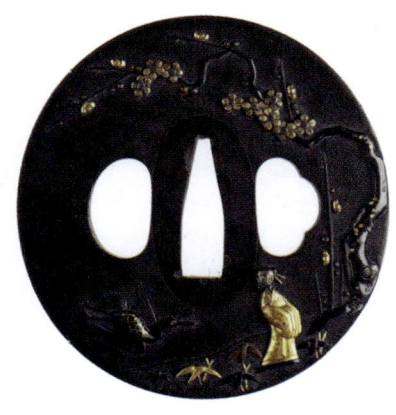

BELOW A Gotō-style sword guard (*tsuba*), from Japan, date unknown, showing a sage under a blossoming plum tree and a flying crane. The item is an example of the metallurgical art of *shakudō*, an alloy of copper and gold turned an indigo color by patination, here inlaid with gold and silver.

Alchemy in ancient China was itself closely allied to Daoist tradition. Gold-making procedures were deemed to require a ritualistic, spiritual element, and Ge Hong says that the gold-maker should seek seclusion and engage in bodily purification by fasting and avoiding pungent foods. Needham has speculated that some alchemical discoveries might have been made not in the artisan's furnace but in Daoist incense burners, in which all manner of minerals and plant or animal extracts were burned.

Eternal life

The connection to Daoist belief is surely also a reason why the emphasis of Chinese alchemy was less on making gold than it was on making medicines—specifically, elixirs (to use the anachronistic, Arabic-derived term) for sustaining life and ultimately conferring immortality.

In the Christian context of the Hellenistic and medieval worlds, attempting to attain immortality could have carried a taint of impiety, as though one were trying to thwart God's will. But there was no such obstacle in China, for the Daoist tradition focused on the material and spiritual pleasures of this world. It was widely believed that some individuals did attain immortality: semi-divine Immortals feature in many Chinese legends, where they constitute a kind of heavenly bureaucracy. "The world-view of ancient China," claims Needham, "was the only milieu capable of crystallizing

RIGHT Ge Hong moving house, his wicker baskets packed with a variety of alchemical equipment, in a detail from a 16th-century Ming-dynasty scroll painting. Ge Hong is one of the most important authors on Chinese alchemy. He was also an official, a linguist, and a philosopher in the Western Jin dynasty during the 4th century.

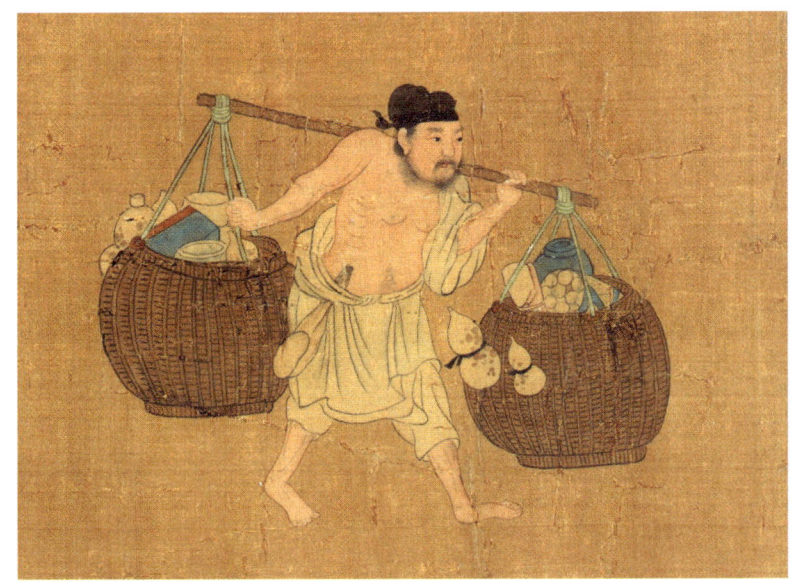

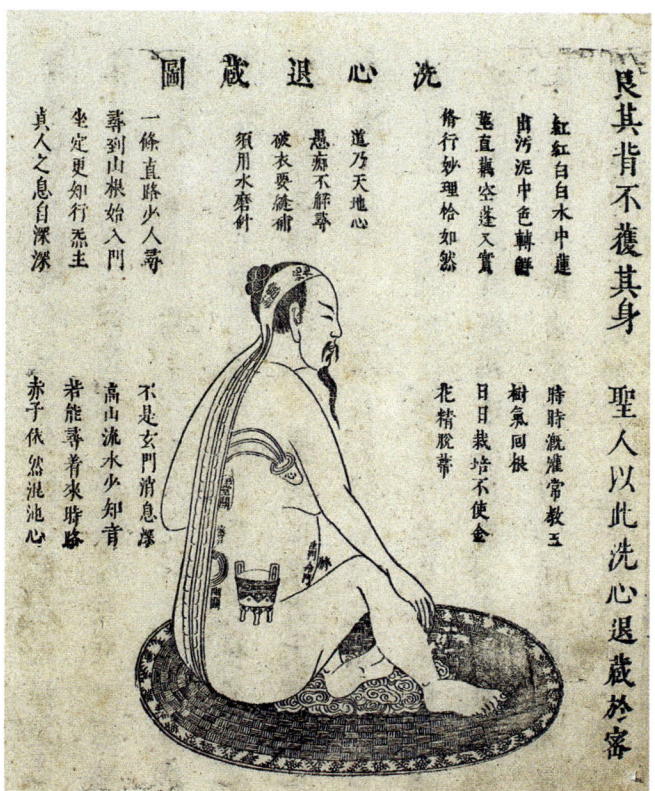

RIGHT Bodily purification was advocated by Ge Hong and promoted in Daoist texts such as Yi Zhenren's *Xingming guizhi* (Pointers on spiritual nature and bodily life), a classic of Chinese internal alchemy published in 1615. The woodcut illustrates the practice called *Xixin tuicang*, meaning to wash the heart and retire to a hidden place.

belief in an elixir, good against death, as the supreme achievement of the alchemist." Historians today might debate that statement, but it's certainly conceivable that the elixir tradition in Islamic alchemy came originally from the East.

Great and perhaps indefinite longevity was thought to be possible through Daoist techniques of meditation and purification. But this required immense commitment and discipline; an elixir offered a shortcut. There was nothing especially mysterious or magical about such agents—they were considered no different from any medicine that maintained and prolonged good health. Some, like many traditional cures (even today in Chinese medicine), might be extracts of plants. It has been suggested that the oldest elixirs may have been made from the juice of the psychotropic and toxic fly agaric mushroom, a technique brought to China from the Vedic tradition of India. However, the fact that gold does not tarnish suggested that it, too, is imbued with a kind of incorruptibility that might be transferred to humans: eating with gold utensils was thought to lengthen life.

Tao Hongjing avers that "natural unrefined gold wards off evil influences."

The *Atharvaveda* from around the tenth century B.C., one of the foundational texts of Hinduism and a repository of magical and religious rites and spells, comments that "gold that was born from Fire and is immortal has been deposited with mortal creatures." Tao Hongjing avers that "natural unrefined gold wards off evil influences"—although he warns that it contains poison, implying that alchemical manipulation is needed to make it into an elixir.

Many formulas for making elixirs are recorded in a text attributed to the Tang physician Sun Simiao called the *Danjing yaojue* (Essential formulas of alchemical classics), which Sivin calls "as close to a modern laboratory handbook as anything we are likely to find in ancient literature." The glorious names of these potions, such as "Scarlet snow and flowing pearl elixir," are already enough to make them sound enticing.

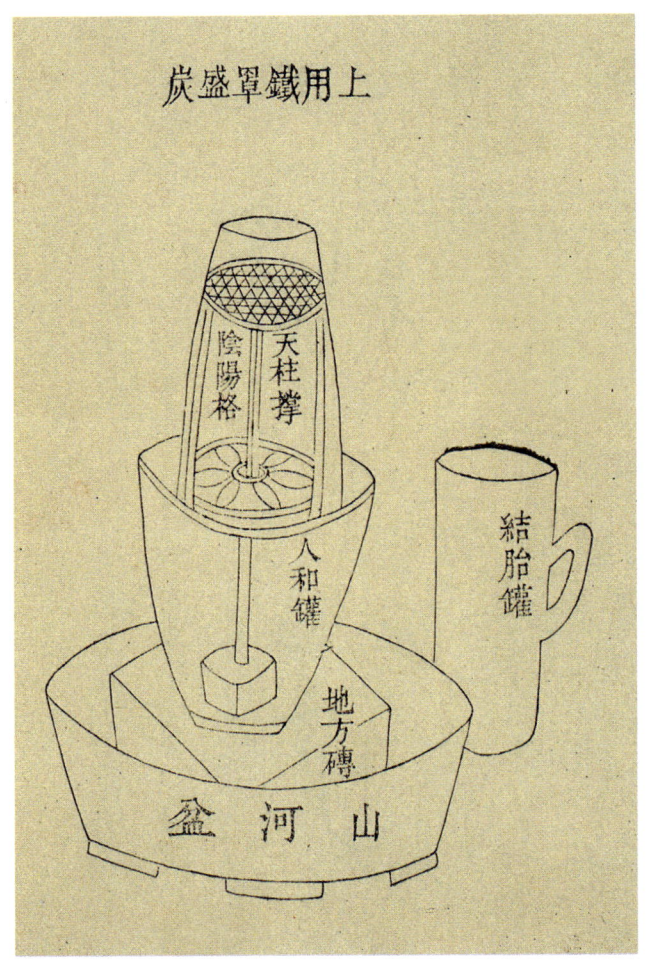

RIGHT A woodblock illustration of a Chinese alchemical furnace, showing its internal structure, from the 19th-century *Waike tushuo* (Pictorial manual of external medicine). The illustration shows implements for creating the *bagua da jiangdan* (Eight-trigram grand descending elixir). Used to treat various ailments, the elixir was a compound obtained by mixing mercuric chloride ($HgCl_2$) and mercurous chloride (Hg_2Cl_2).

OPPOSITE A woodblock portrait of Sun Simiao, author of the *Danjing yaojue* (Essential formulas of alchemical classics), attributed to the Tang-dynasty (618–907) artist Gan Bozong, from a 17th-century edition of *Bencao mengquan* (Introduction to the pharmacopoeia). Sun was best known as a physician, but his classic work on alchemy provides some of the most detailed accounts of the methods and materials of the art at that time.

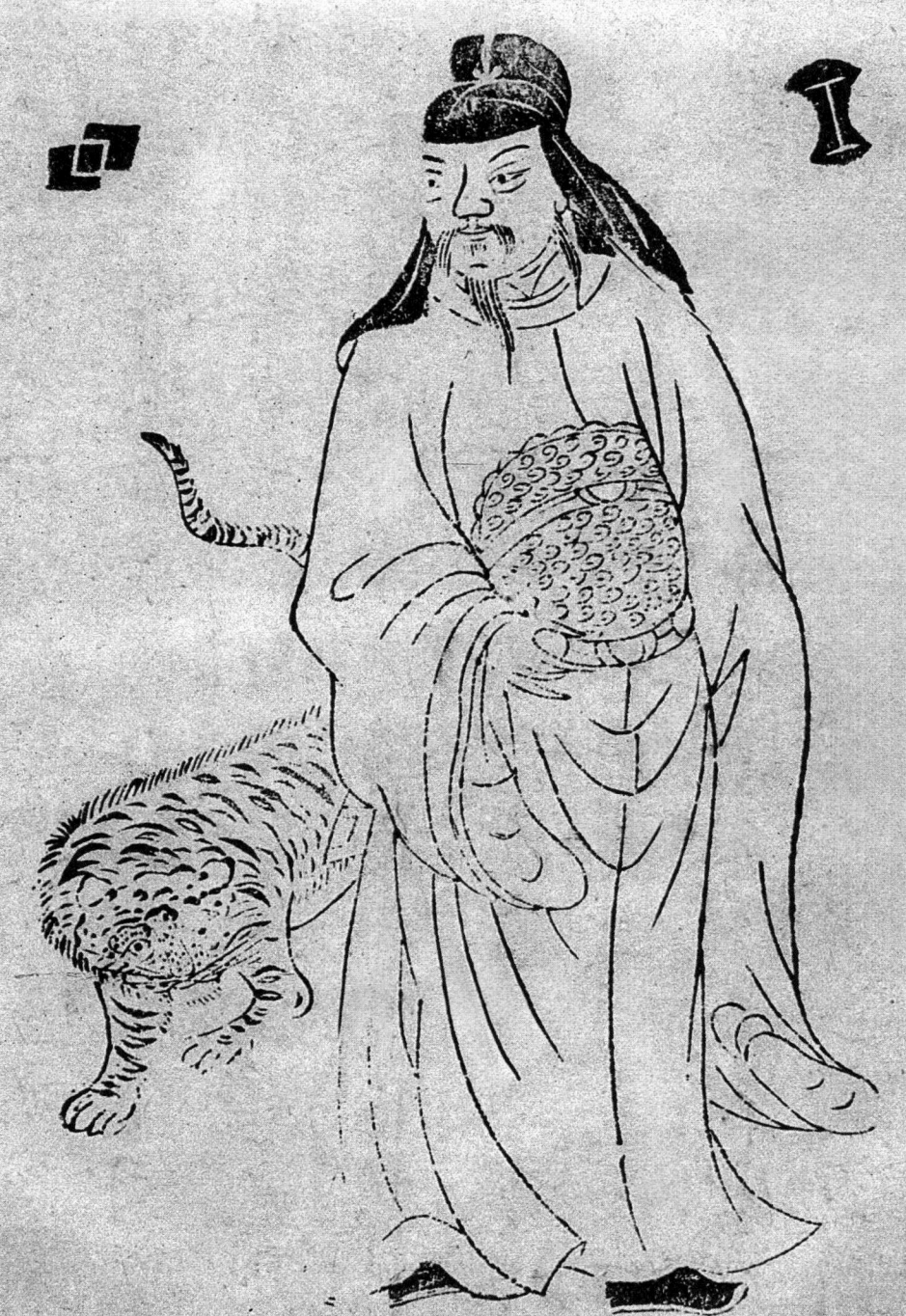

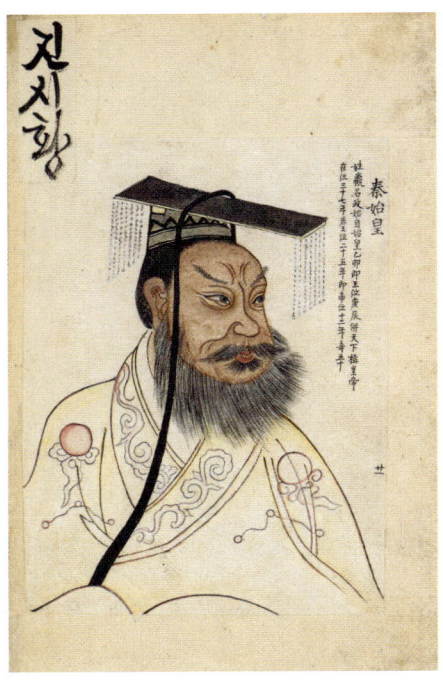

ABOVE The first emperor of China, Qin Shi Huang (260–210 B.C.), head of the Qin dynasty in the 3rd century B.C., shown in a Qing-dynasty (1644–1911) portrait. The emperor craved elixirs to prolong his life and perhaps even make him immortal. In pursuit of this goal he may well have consumed considerable quantities of mercury compounds. His tomb, in present-day Shaanxi province, was said to contain a scale model of China with rivers of mercury.

Whether they did any good is another matter. It seems likely that many if not most were harmful, as they contained toxic elements such as mercury, lead, and arsenic. How then did the popularity of such potions, not merely useless but hazardous, persist for so long? The same question can be directed at any early medicine and is partly answered by the evident difficulties of attributing cause and effect in health and illness, especially when life was so precarious. Those eager to experience, for example, the alleged aphrodisiac properties of sulfur (imbued with male *yang*) and arsenic compounds like realgar and orpiment might have been prepared to put up with their ill effects. What's more, some elixir-makers would employ the timeless gambit of quacks who claimed that a patient feeling bad was evidence the medicine was working. As a sixth-century Chinese text says,

After taking the elixir, if your face and body itch as though insects were crawling over them, if your hands and feet swell dropsically, … if you feel as though you were going to be sick most of the time, … if you have to go often to the latrine, or if your head or stomach violently ache—do not be alarmed or disturbed. All these facts are merely proof that the elixir you are taking is successfully dispelling your latent disorders.

This ambivalent attitude whereby the hazards of elixirs are interpreted as a sign of their potency is attested in the story told about the semi-legendary alchemist Wei Boyang in the kingdom of Wu in the second century A.D. (Wei was probably a real person, although the texts attributed to him might have been compiled collectively; one of them records the first known recipe for gunpowder.) In the presence of his assistants, Wei gave a potion he had prepared to his dog, which promptly collapsed, dead. Undeterred, the alchemist and one of his pupils took the potion themselves, with the same result. The other two disciples thought better of the matter and headed off to arrange for their master's burial. In the meantime, Wei, his student, and his dog all regained life and discovered they had become immortal.

A salutary example of the dangers of the alchemical quest for immortality is provided by Emperor Qin Shi Huang, the first Qin-dynasty emperor of the third century B.C. Desperate for everlasting life, the emperor allegedly consumed great quantities of herbal and alchemical elixirs, and even sent a minister on a voyage over the eastern seas in a fruitless search for a mythical immortality potion.

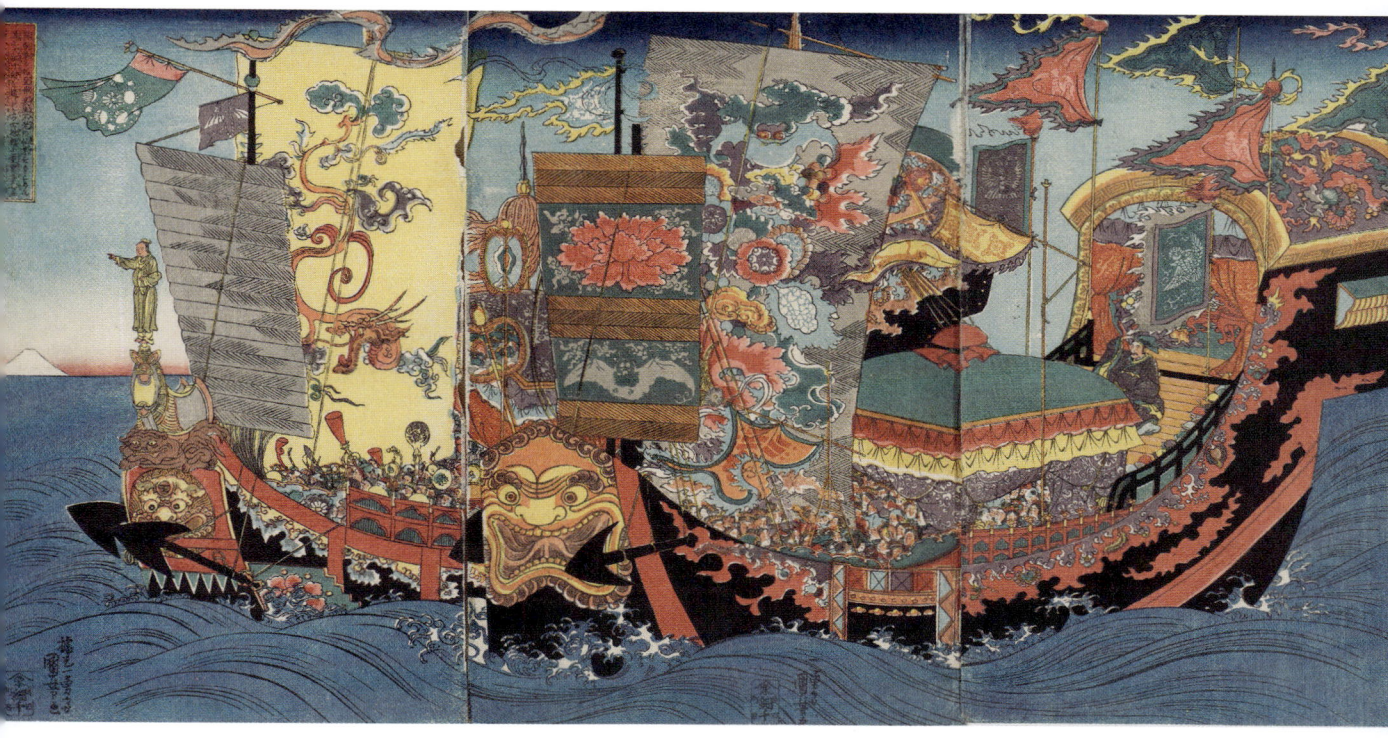

ABOVE A Japanese woodblock print of the *ukiyo-e* genre by Kuniyoshi, c. 1839–40, depicting the great expedition sent by the Chinese emperor Qin Shi Huang, led by Wu Xu (together with three thousand virgin boys and girls), to retrieve the elixir of immortality from the Immortals on their legendary home on Mount Penglai in c. 219 B.C.

Even before his conquests that ended the Warring States period and made him the first ruler of a united Chinese empire, the Qin emperor ordered the construction of a vast and elaborate mausoleum in present-day Shaanxi province from which, in the event of his death, he could continue his cosmic rule from the spirit world. According to the Han-dynasty historian Sima Qian, 700,000 men worked on the tomb, creating entire palaces, towers, and scenic landscapes underground through which the emperor's spirit might roam. Sima Qian's description sounded rather fanciful until the remains of the mausoleum were discovered in the 1970s. Archeological excavations have now unearthed an army of thousands of life-size clay soldiers, along with horses and chariots and figures of officials, acrobats, and laborers.

Those eager to experience the alleged aphrodisiac properties of sulfur and arsenic compounds might have been prepared to put up with their ill effects.

Sima Qian also records that vast quantities of mercury were used for the rivers of a scale model of China built inside the tomb—an idea supported by modern archeological studies that have revealed high levels of mercury in the overlying soil. (The burial chamber itself has not been excavated and has probably collapsed.) While the amount of mercury implied by Sima Qian's account seems unrealistic, extracting mercury from cinnabar was surely an important business in the Qin era, and one-fifth of the country's current reserves lie in Shaanxi. At that time, mercury was a common ingredient of medicines for treating infected sores, scabies, and ringworm, and as a sedative for alleviating mania and insomnia.

The Daoist concept of *yin* and *yang*, the two fundamental and complementary principles of life, encouraged an idea that cold, watery (*yin*) mercury and bright, fiery (*yang*) gold might be blended in ideal proportions to sustain vitality. Chinese legend tells of one Huang An, who prolonged his life for at least ten thousand years by eating cinnabar, perhaps mixed with wine and honey as in the elixirs allegedly prepared for Emperor Qin Shi Huang.

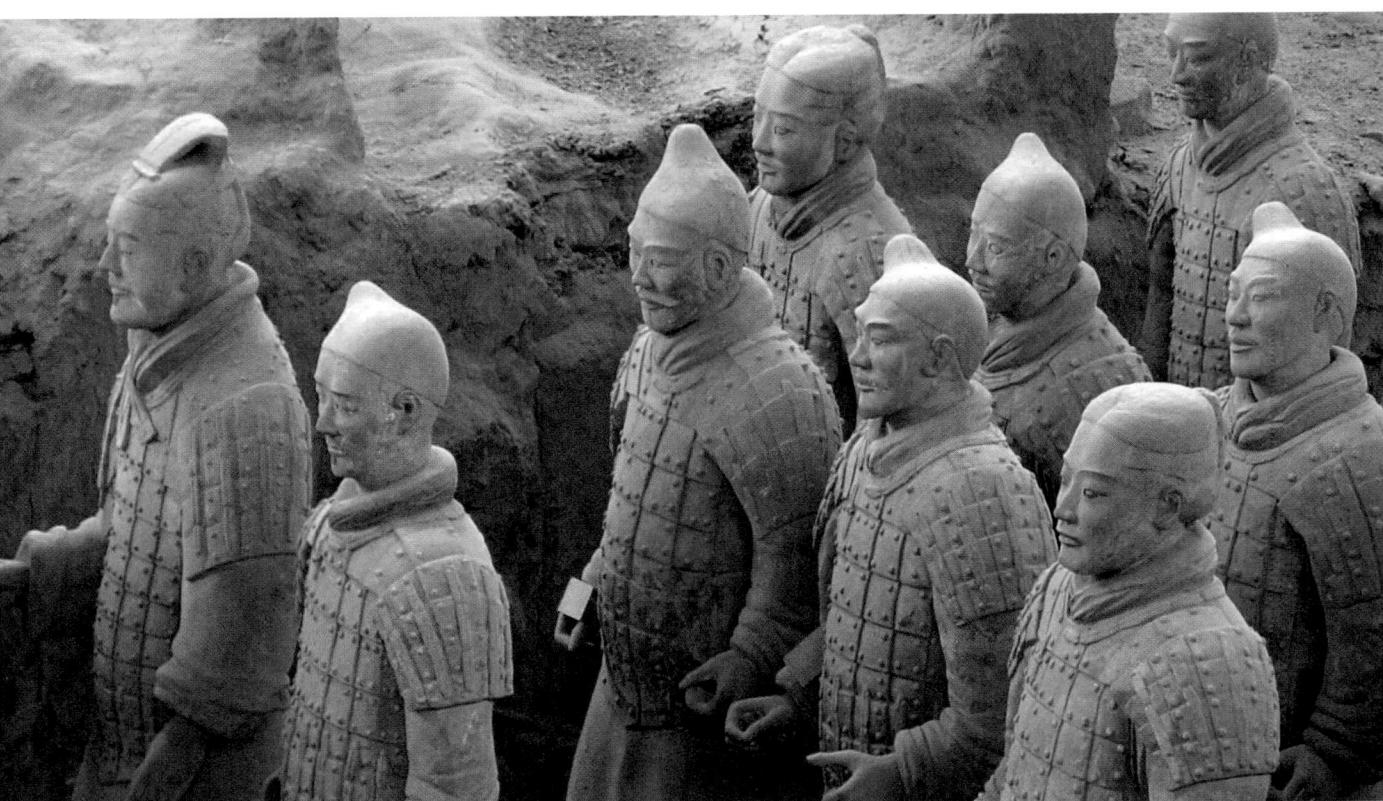

BELOW A close view of the vast terracotta army of Emperor Qin Shi Huang in his 3rd-century-B.C. tomb near present-day Xian, Shaanxi province, China. These figures were part of a retinue of soldiers, officials, and servants that would serve the emperor in the afterlife. They were originally painted, and some have retained their pigments, analyses of which offer another window on ancient Chinese chemical technology.

Indian alchemy

In India, too, life-preserving elixirs were the dominant aspect of alchemy—although these were more in the nature of regular medicines than agents of immortality. Archeological studies of the Indus Valley, the locus of the ancient Harrapan culture in the fourth and third millennia B.C., show that chemical manipulations were already being conducted in that prehistorical period. Indian alchemy as such did not develop until much later, however, flourishing around the tenth century A.D.

Just as Chinese alchemy was strongly influenced by Daoism, so Indian alchemy drew on the Tantric spiritual tradition. Originating around the fourth century A.D., Tantrism emphasized ways to sustain the body through meditation, yoga, and spiritual practices, and attributed cosmic significance to the sexual union of man and woman.

The preparation of elixirs of longevity in India is again accompanied by many tales of poisoning and death. That's no surprise given that mercury plays a central role here too: the Sanskrit word designating alchemy, *rasayana*, means "the way of mercury," and the workshops in which the art was practiced were often called temples of mercury. The word used for mercury, *rasa*, could denote any kind of vital fluid; the silvery liquid metal was itself sometimes identified with the semen of the Hindu god Shiva.

A text called the *Rasarnava* (Ocean of mercury) from the eleventh or twelfth century speaks of a mercury-containing pill that can transmute into gold a hundred times its mass of base metals and which, when held in the mouth, will convey near-immortality. Much of the mercury used in Indian alchemy was imported from China, with which trade had existed at least from the first millennium A.D. No doubt ideas and theories were transmitted too, and from India these passed on to Central Asia and Persia, and farther west. We may never know to what extent the alchemy of East and West originated independently—but the dreams of longevity, health, wealth, and secret knowledge that these traditions fed were surely universal.

PROFILE

Ge Hong

(c. 283–343 a.d.)

The most famous and revered of the ancient Chinese alchemists, Ge Hong (spelled Ko Hung in the older Wade-Giles system of romanization) was born in what is now Jiangsu province around 283 A.D., during a time of ongoing social unrest and rebellion. He spent his early working life as a public and military official in the Eastern Jin dynasty but seems to have craved the quiet life of a philosopher, taking on the pseudonym Baopu, loosely meaning "The master who embraces simplicity." Integrating Confucian and Daoist beliefs, he once remarked that "Confucianism is difficulty in the midst of facility; Daoism is facility in the midst of difficulties."

After retreating from public duties to study philosophy, Ge Hong learnt alchemy from the Daoist Bao Liang, who allegedly lived in the sacred Songshan mountain range of Henan province. He married Bao Liang's daughter Bao Gu, one of the first female physicians in recorded Chinese history.

Although Ge Hong had many alchemical works attributed to him, the only one for which the authorship can be assigned with any confidence is the *Baopuzi*, which ranges from discourses on elixirs, immortality, and alchemy to politics and literature. Here Ge Hong portrays himself as the archetypal humble alchemist:

I am an unsophisticated person; dull by nature, and a stammerer. My physical frame is unpleasant to look at; and I am not competent enough to boast of myself and gloss over the defects. My hat and shoes are dirty; my clothes sometimes the worse for wear or patched; but this does not always bother me ... My speech is frank and sincere; I engage in no banter. If I do not come upon the right person, I can spend the day in silence.

In Joseph Needham's estimation, there is much in the *Baopuzi* "which is wild, fanciful and superstitious"—but also passages that are "scientifically as sound as anything in Aristotle." Among the recipes in the book are ones for making artificial golds, such as "mosaic gold" (a yellow powder used as a pigment) and lime water (a solution of the alkali calcium hydroxide), as well as a description of what seems to be tin foil. The book is widely seen as the first systematic account of Chinese alchemy.

Ge Hong was also a physician. Anticipating later medically oriented alchemists, he sought remedies that all people could afford, eschewing expensive ingredients available only to the rich. While he believed in an elixir of immortality, he disapproved of those who sought it only for themselves and asserted that a long life

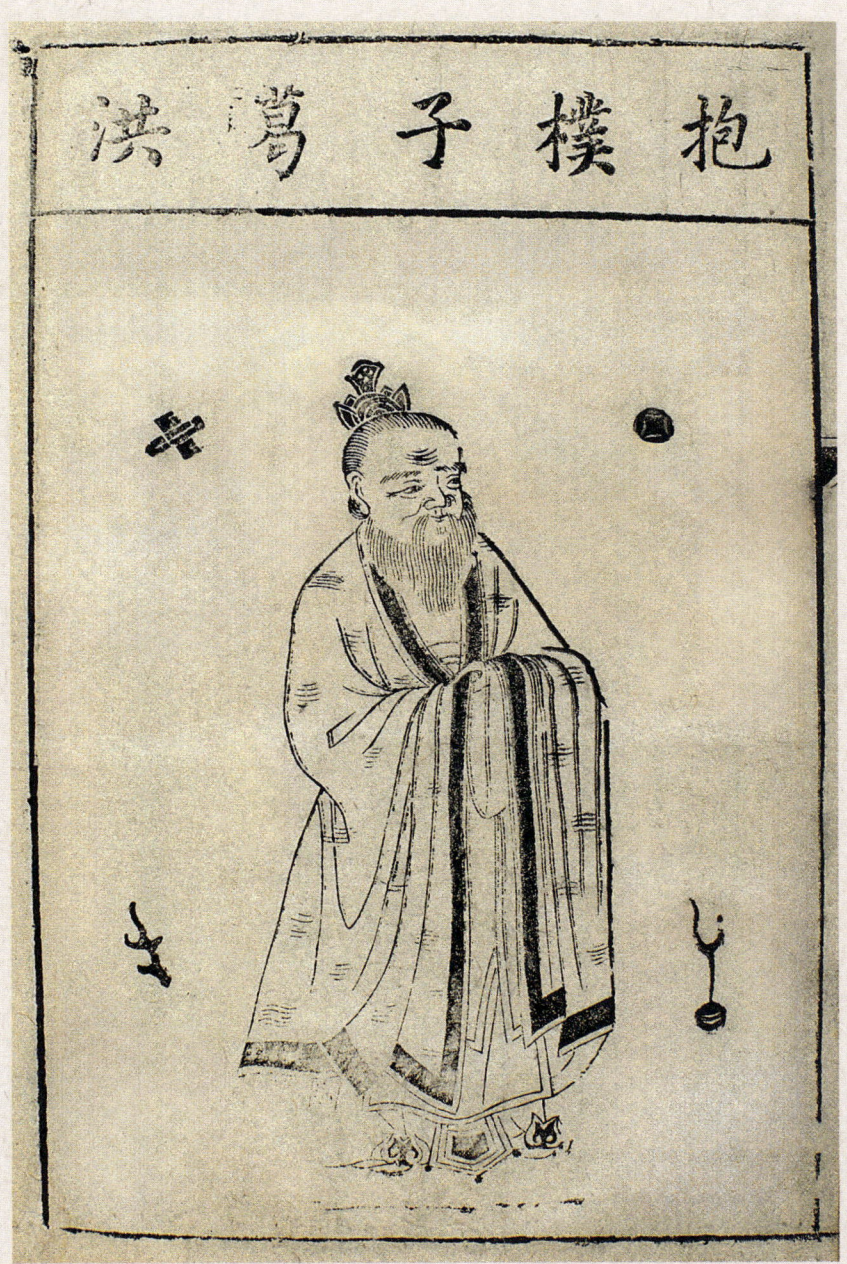

RIGHT Ge Hong, known as Ge Xianweng (Ge the Immortal Sage), depicted in a woodcut from a series of legendary founders of Chinese medicine from a 17th-century edition of *Bencao mengquan* (Introduction to the pharmacopoeia), after the Tang-dynasty (618–907) artist Gan Bozong.

demanded also breathing and meditative practices as well as personal virtues such as loyalty and friendliness. Ge Hong said that the condition we now call malaria could be treated with an extract of the sweet wormwood shrub (*Artemisia annua*), common in Asia—a suggestion that encouraged the use of the plant as a folk remedy and which ultimately inspired the Nobel Prize–winning discovery by Chinese chemist Tu Youyou of the antimalarial compound artemisinin.

ELIXIRS: ALCHEMY IN THE EAST

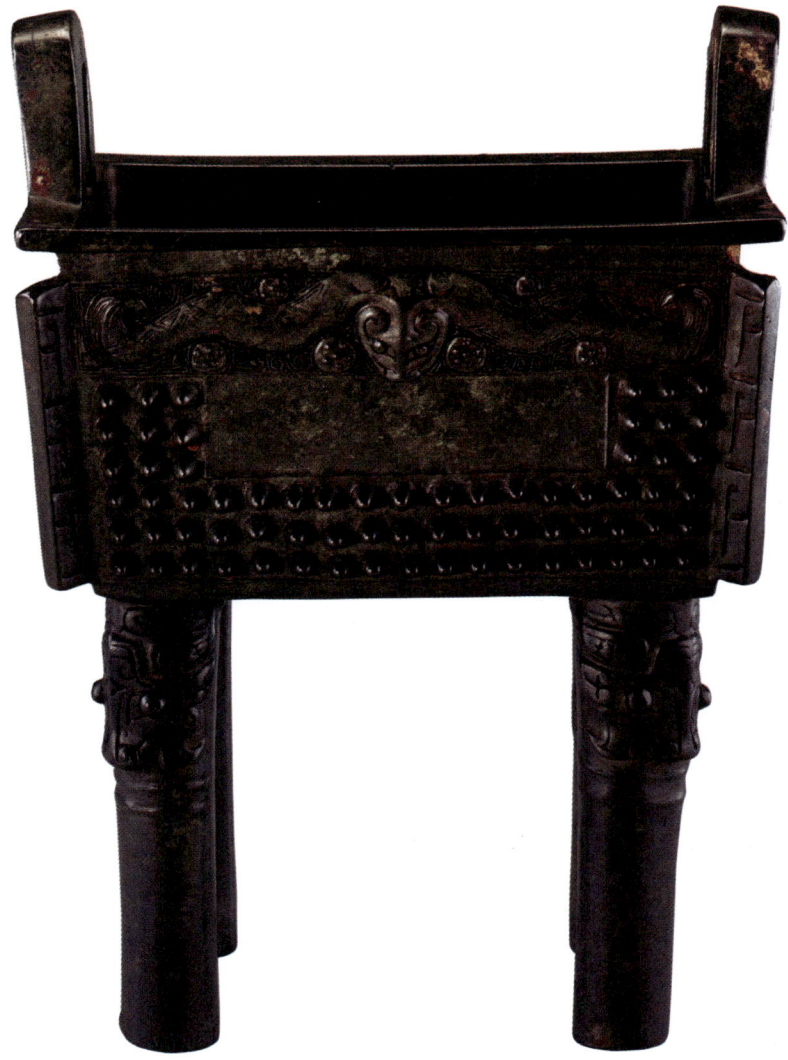

ABOVE A bronze ritual cauldron (*fangding*) with masks (*taotie*) and snakes, from the late Shang dynasty, c. 1100–1050 B.C., found probably in Anyang, Henan province, China. Vessels like these were typically used for holding ritual offerings to the gods or ancestors in ceremonies. *Taotie* are symmetrical face-masks that might have been based on masks worn by shamans. The cauldrons were cast in clay molds and testify to the metallurgical prowess of the Shang culture of the Yellow River valley, generally regarded as crucible of Chinese civilization.

OPPOSITE Wang Xizhi watching geese, in a detail from a handscroll painting by the 13th-century artist Qian Xuan, c. 1295. Like Ge Hong (see page 54), this legendary alchemist and master of calligraphy was said to derive great inspiration from natural forms. He lived during the Jin dynasty in the 4th century, serving as an administrator and politician. Wang was reputed to be a master of Daoist alchemy and sought elixirs of immortality. According to one story, he died from poison; others say he was executed by the emperor for disobedience.

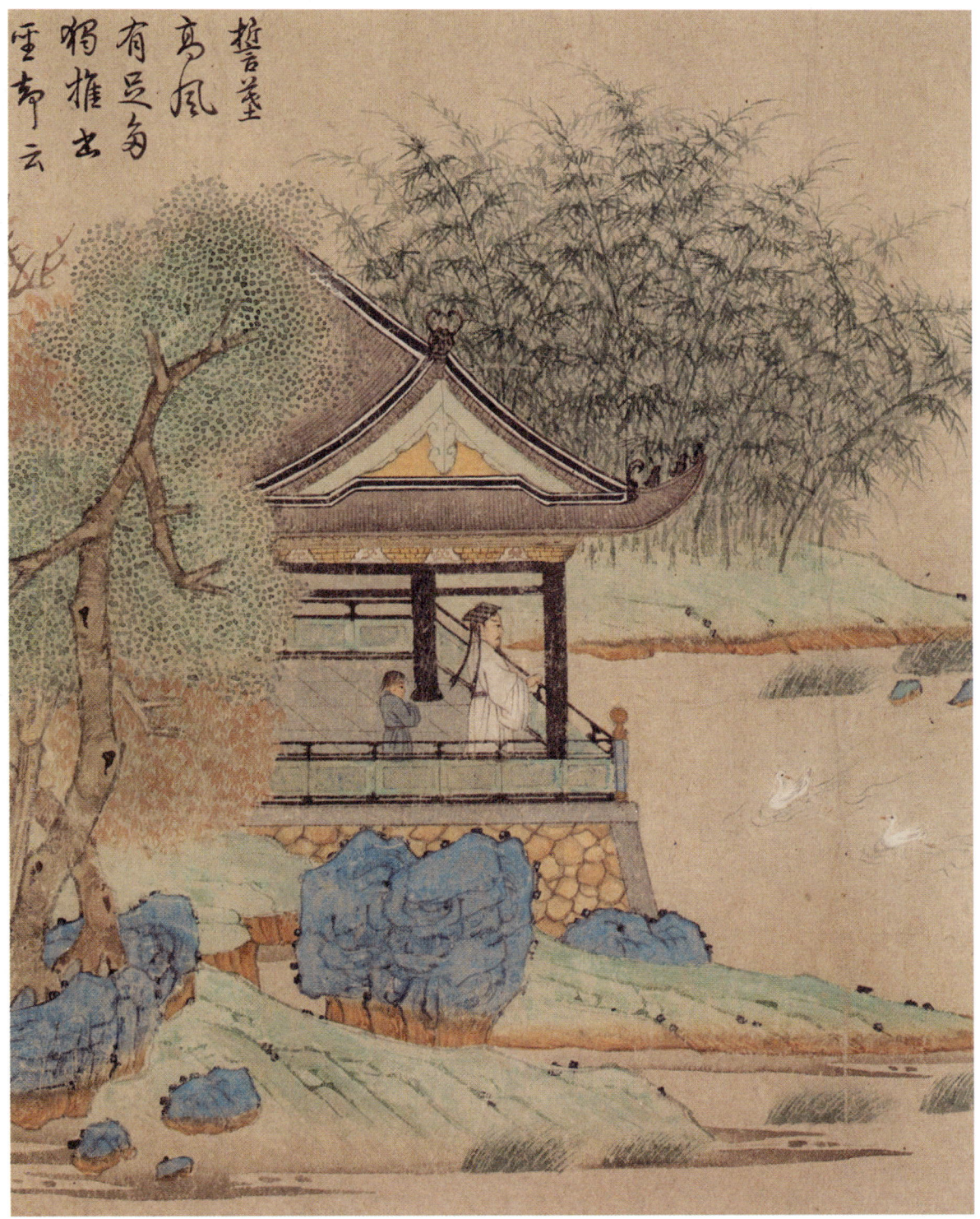

內經圖

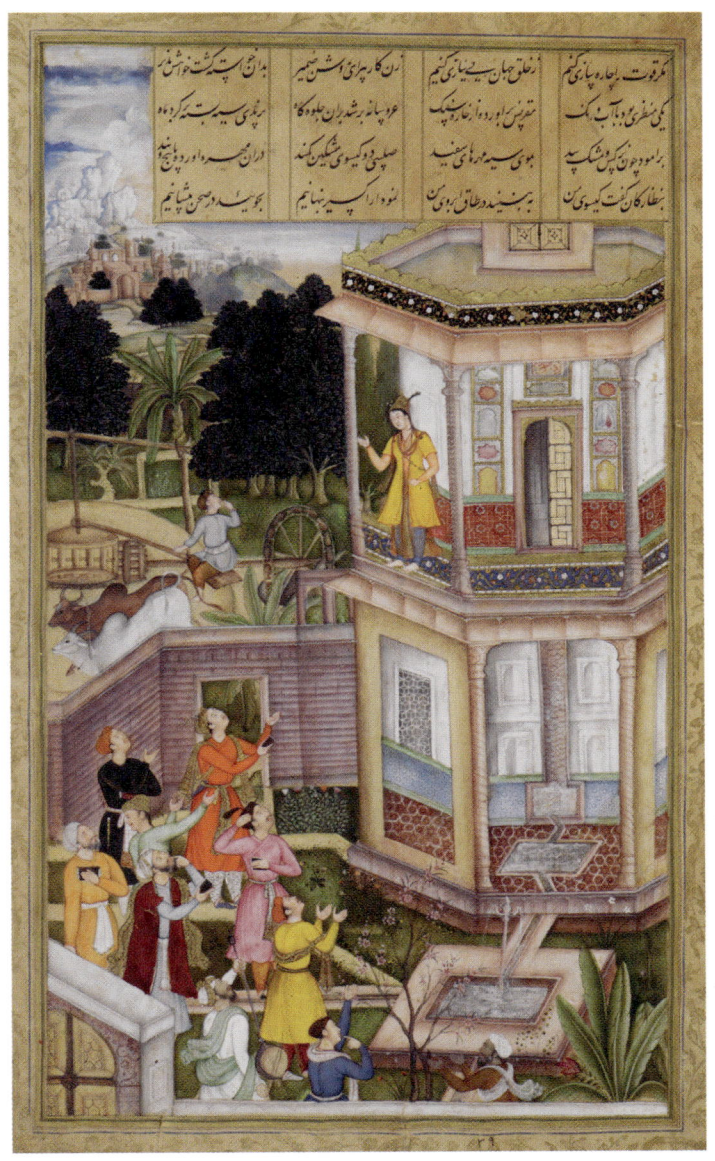

OPPOSITE Illustration showing *Neijing tu* or the "inner landscape" of the human body, a concept of the tradition called *Neidan* (Daoist internal alchemy). *Neidan* (also known as *Jindan*, "golden elixir") involved various physical, mental, and spiritual practices aimed at prolonging life and ultimately immortality. The tradition seems to have originated during the late Tang dynasty (9th century) but become popular in the Song dynasty that succeeded it. The practices treated the human body as a kind of alchemical cauldron in which the vital forces of life, such as *qi* (breath), were refined.

ABOVE Miryam, a Christian woman versed in the art of alchemy, is consulted by other alchemists in this miniature from the *Khamsa* (Five poems) of the Persian poet Nizami, from India, 1595. The poems, composed in the 12th century, became a popular text for illuminations in Mughal India in the 16th and 17th centuries. One describes how Miryam, driven from her native land in present-day Syria after the death of its ruler, her father, went to the court of Alexander the Great and gained knowledge of making gold. After she was restored as ruler of her land, other alchemists visited her to plead for the secret of chrysopoeia.

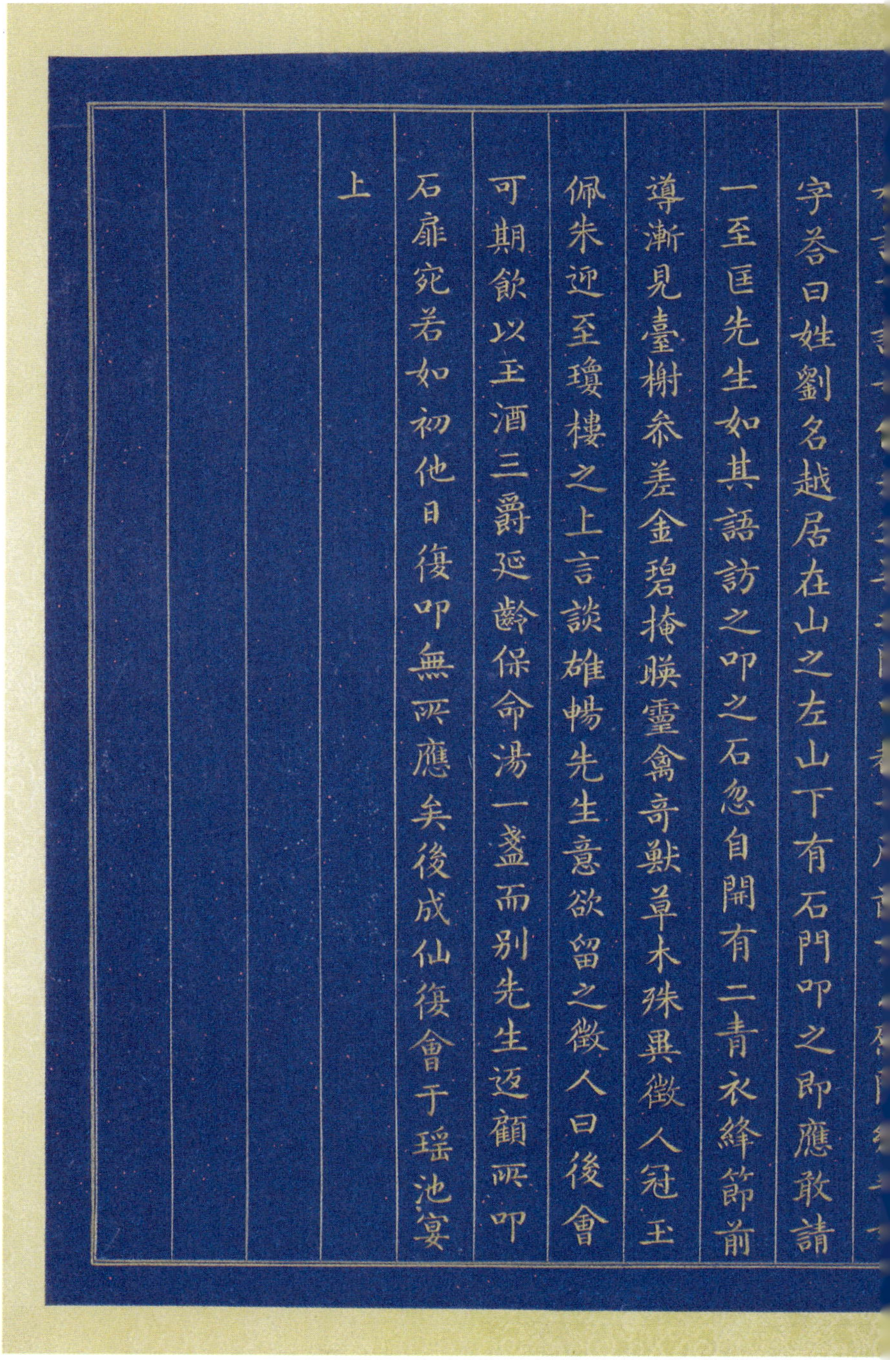

字荅曰姓劉名越居在山之左山下有石門叩之即應敢請
一至匪先生如其語訪之叩之石忽自開有二青衣絳節前
導漸見臺榭参差金碧掩暎靈禽奇獸草木珠異徵人冠玉
佩朱迎至瓊樓之上言談雄暢先生意欲留之徵人曰後會
可期飲以玉酒三爵延齡保命湯一盞而別先生返顧兩叩
石扉宛若如初他日復叩無兩應矣後成仙復會于瑤池宴
上

RIGHT This painting, from the 18th-century album *A Keepsake from the Cloud Gallery/Yuntai Xianrui*, which illustrates stories of Daoist adepts, depicts the Chinese quest for immortality. Here, Liu Yue, who lived on the Mountain of the Southern Peak, seeks everlasting life by drinking a special potion.

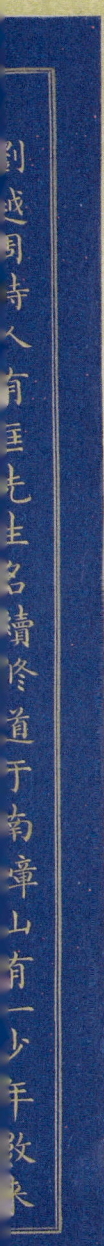

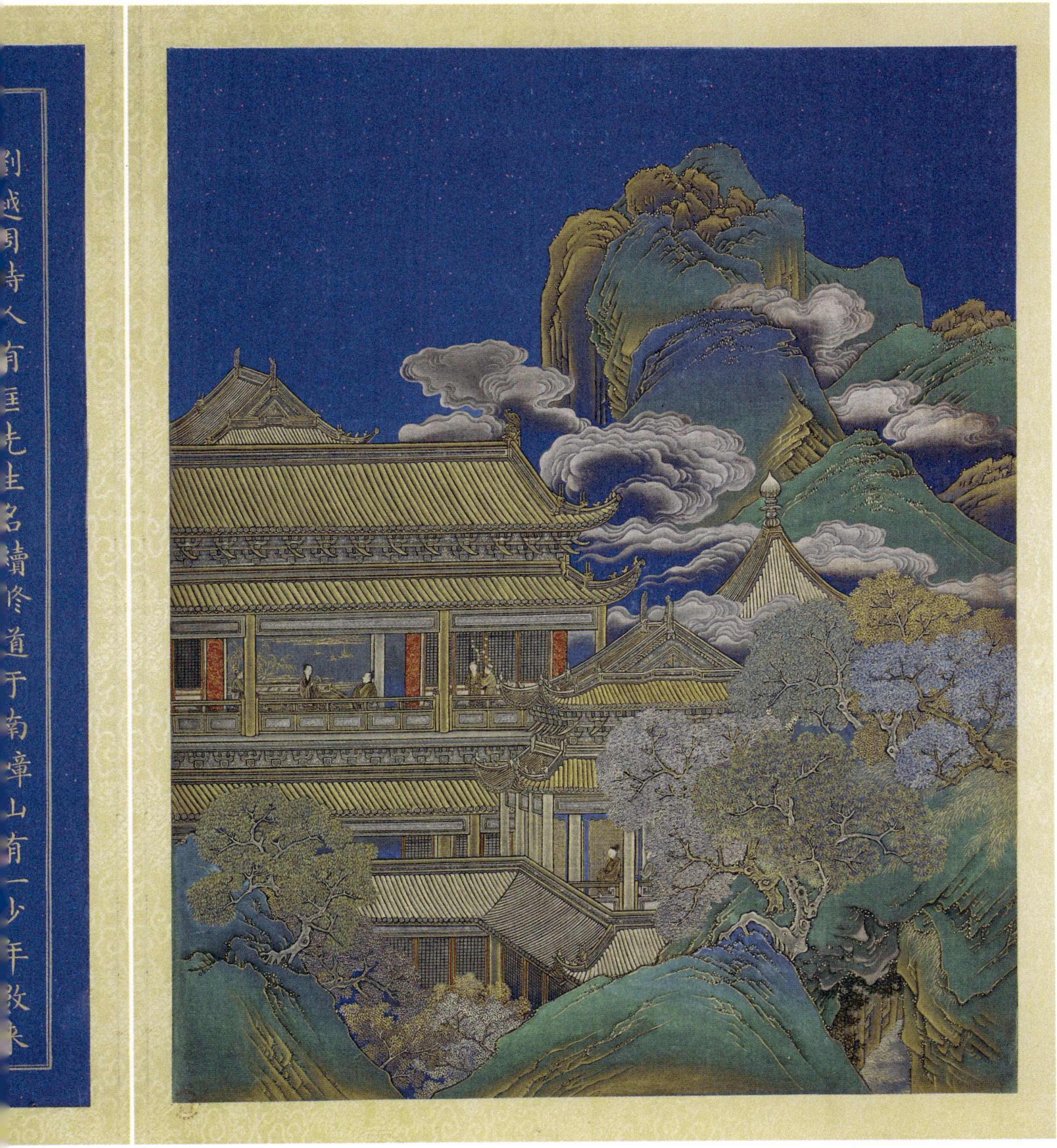

ELIXIRS: ALCHEMY IN THE EAST

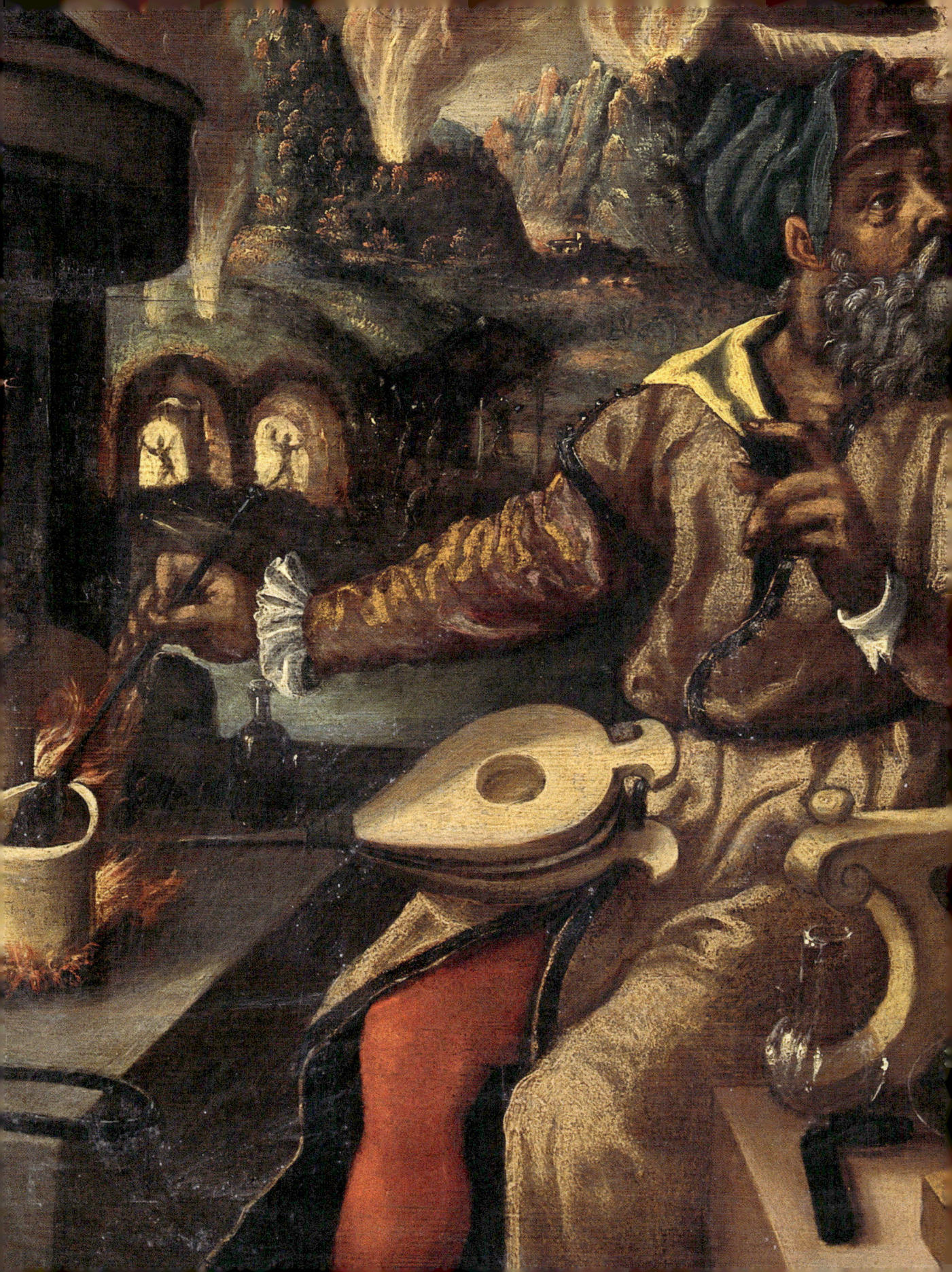

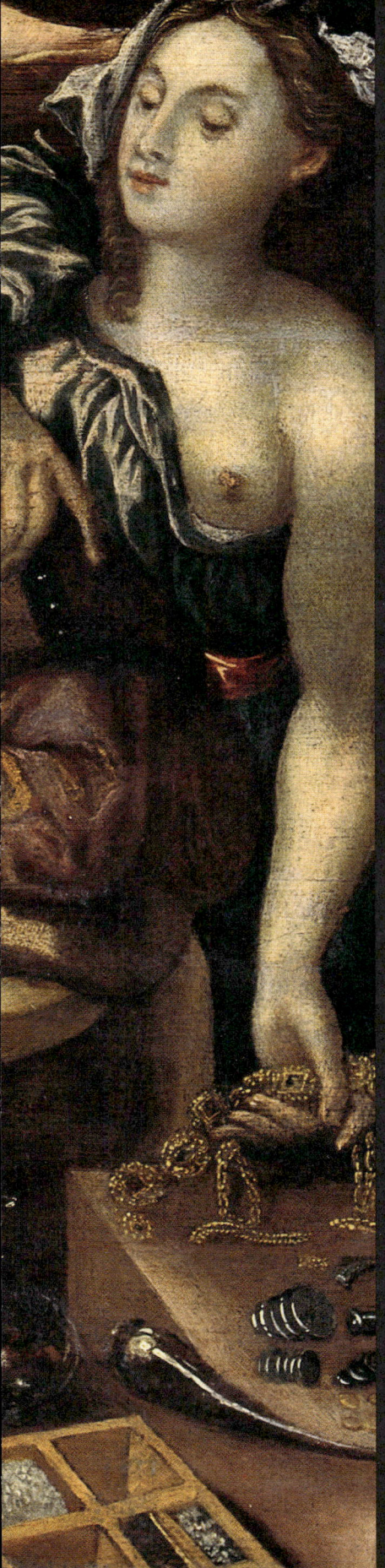

CHAPTER THREE

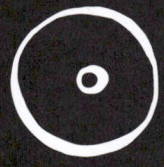

Chrysopoeia
THE QUEST FOR GOLD

LEFT An alchemist, tempted by Luxuria (or perhaps warned by Prudentia), risks the dangers of the "fallacious art of alchemy" and its quest for gold. These dangers ranged from charges of impiety and necromancy to accusations of fraud and impoverishment, but if one were successful in making gold, the rewards would be beyond measure. This 16th-century painting was made after a print by the Flemish artist Marten de Vos, known for his allegorical compositions.

CHAPTER THREE

If there is one thing everyone knows about alchemy, it is that its practitioners sought to transform base metals such as dull, heavy lead into gold. Gold-making seems clearly futile and even delusional from today's perspective, when we know that no chemical manipulation can change one element to another. But for early alchemists there was no obvious reason why it should be impossible, and many genuinely believed they could do it.

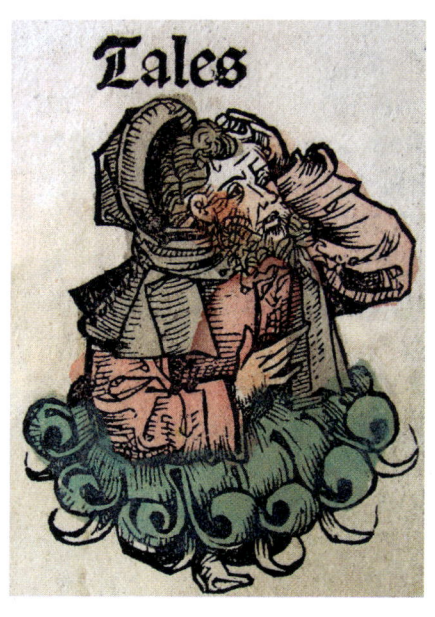

ABOVE A fanciful portrait of Thales of Miletus by Michel Wolgemut or Wilhelm Pleydenwurff. Thales was an ancient Greek philosopher who argued that all substances are derived from one fundamental element: water. This wood engraving is from a colored copy of Hartmann Schedel's Weltchronik (World chronicle), known as the Nuremberg Chronicle after the city in which it was published in 1493.

According to the current understanding of the nature of matter, it is impossible to convert one metal to another using any chemical process. To transmute lead into gold we would need to alter the composition of the very atoms themselves. Modern physics permits that, but only using enormous energies way beyond what could be achieved in a medieval furnace.

It's not surprising, then, that the so-called Hermetic art has long been deemed a fool's quest. But judging people in the past by the standards and the knowledge we have today gets us no closer to genuinely understanding what they did, and why. Three hundred and more years ago, there was no obvious reason for a rational person to think that making gold—which modern historians commonly describe using the Greek-derived term *chrysopoeia*—was impossible. It was widely believed that the conversion of other metals to gold happens naturally in the depths of the earth—not an unreasonable idea, seeing as gold was often found in nature mixed with other metals such as silver or copper. This geological transformation was thought to be very slow, and the alchemist sought to induce it at a faster pace in the laboratory.

What is more, reports of gold-making were widely attested, often by reliable witnesses. We can be confident that no one ever really *did* witness a transmutation, but such seemingly credible accounts kept the dream of chrysopoeia alive for centuries.

It works in theory
To understand why chrysopoeia was thought feasible, we need to appreciate what was once believed about the composition of matter.

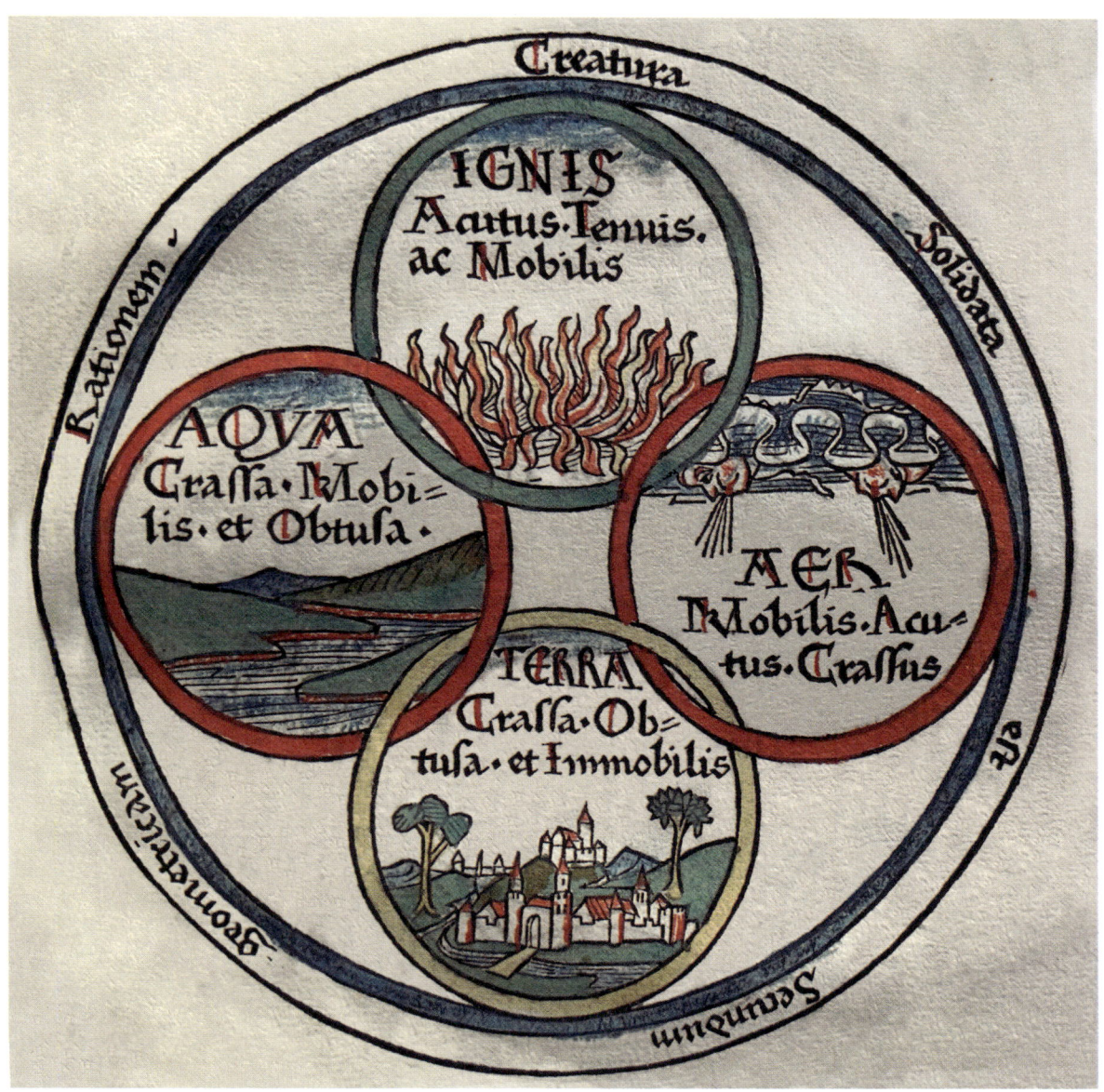

ABOVE Hand-colored diagram of the Four Elements of Empedocles, from Isidore of Seville's *De responsione mundi et de astrorum ordinatione* (On the response of the world and the arrangement of the stars; Augsburg, 1472). Isidore, a theologian and bishop of the 7th century, was one of the most important early Christian writers. His most celebrated work, *Etymologiae*, was a compendium of classical texts that is credited with saving them from vanishing altogether.

Various theories were posited by ancient Greek philosophers, all predicated on the idea that the vast diversity of substances we find in the world is made up of just a few fundamental constituents: the *elements*. For example, Thales of Miletus in Asia Minor in the sixth century B.C. argued that there was but a single element: water. Empedocles of Akragas on Sicily around the late fifth century B.C. recognized four elements—earth, air, fire, and water—in a theory that was adopted by the two most important philosophers of classical Greece, Plato and Aristotle, in the fourth century B.C. and which went on to have enormous influence. (A fifth element, called

CHRYSOPOEIA: THE QUEST FOR GOLD 65

aether, which comprised the imperishable heavens, was later added to the Aristotelian system.)

Crucially, these elements weren't themselves immutable. That, indeed, was the point. Metals, possessing the characteristics of solid earth, could be melted so that they flowed like water. (Mercury already possessed that attribute at everyday temperatures.) Water itself could be frozen to form solid ice, or it could be heated to change it into an invisible vapor, like air. In such ways, the Empedoclean/Aristotelian elements offered a system for thinking about the changes that occur in matter.

According to Aristotle, transformations of the elements happened through an alteration of four qualities: hot, cold, dry, and wet. Each element had two of these: water was wet and cold, fire hot and dry, and so on. So water becomes earth (cold and dry) by changing wetness to dryness, and becomes air (hot and wet) by changing cold to hot through heating.

RIGHT An illustration of Philosophical Mercury, one of the two "principles" considered to be the constituents of all metals, from a 15th-century Italian alchemical manuscript. Red, fiery sulfur is associated with the sun, and cool, silvery mercury with the moon. Mercury was considered the "feminine" principle, and the union of the two in the Great Work of alchemy was often depicted as a conjoining of male and female to produce an androgynous or hermaphroditic entity.

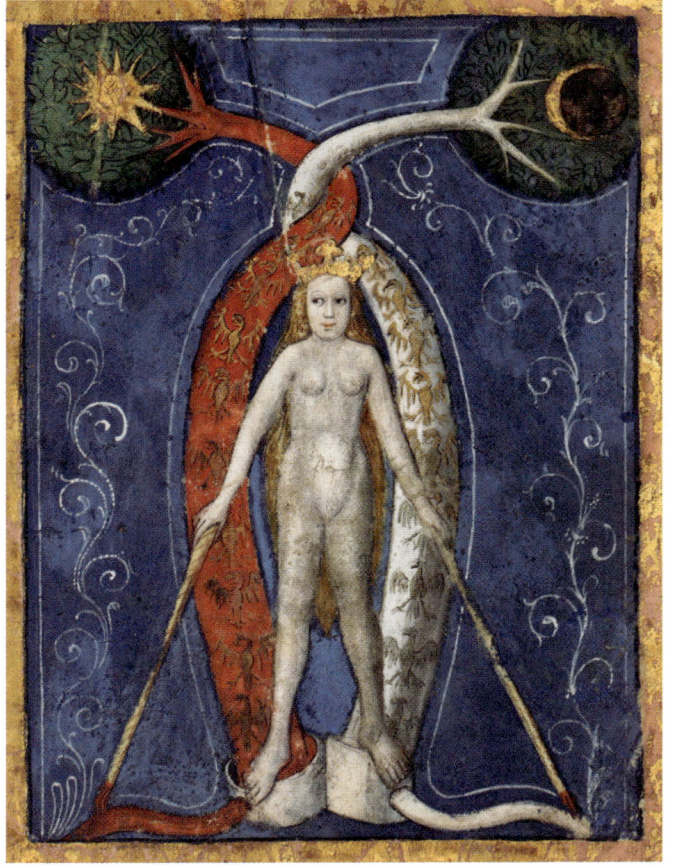

Given that a solid and earthy substance could be changed to something tenuous and airy, it would hardly seem unlikely that one metal could be transformed into another; their differences were apparently far smaller. When this transformation happened within the earth, it was believed to amount to a kind of maturation, rather like the aging of a fine wine.

These philosophical ideas about the composition of matter began to impinge on the practical chemical knowledge of artisans around the third century A.D. in Alexandrian Egypt. It was then that the idea of using chemical transformations to *imitate* gold began to morph into the notion of *making* gold. Zosimos of Panopolis talks of the "tingeing" of metals, as though the artisan is merely applying some superficial coloration or patina—but he regards this as a more profound transformation, a transfer of the "spirit" (*pneuma*) of one metal to the "body" (*soma*) of another.

It was with the Arabic alchemists, especially Jābir ibn Ḥayyān (see page 30), that the chrysopoeian art acquired a firm theoretical basis. Jābir established a notion foreshadowed in the works of Zosimos: that all metals are composed of the same two ingredients, sulfur and mercury, mixed in different proportions. In this theory, sulfur and mercury were not elements like those in the Aristotelian quartet (which Jābir accepted), but rather "principles"—closer, perhaps, to properties or essences. Quite how they were thought to be related

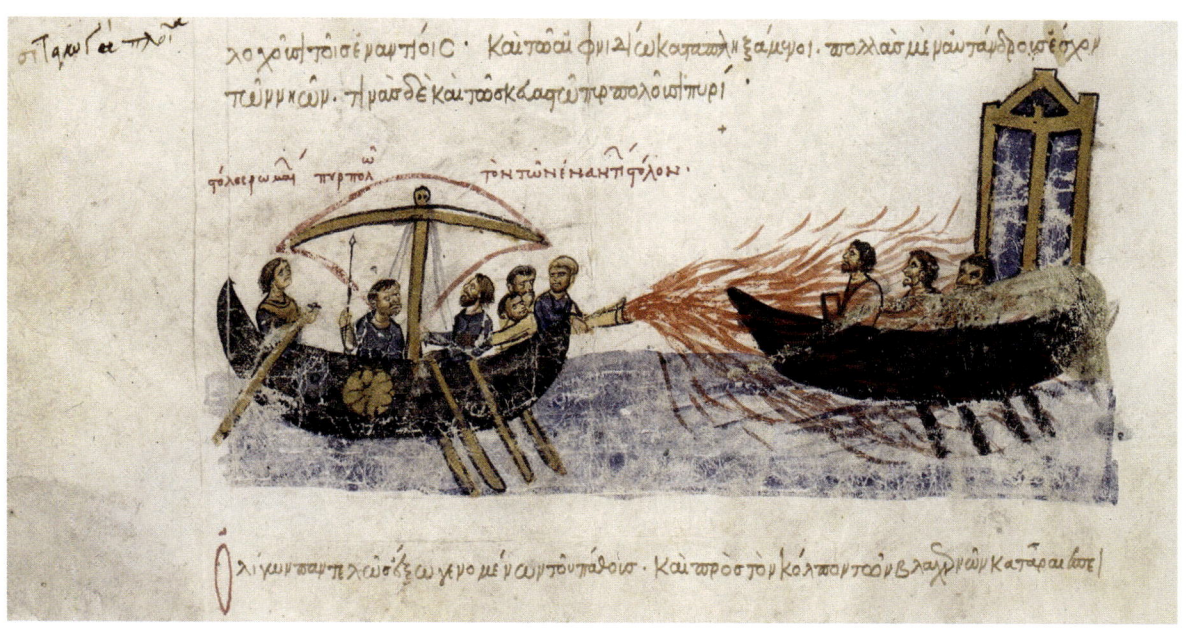

BELOW So-called "Greek fire," using sulfur as an incendiary agent, being used at sea in an illustration from the *Skylitzes matritensis (Madrid Skylitzes)*, a Greek manuscript of the *Synopsis of Histories* by Byzantine historian John Skylitzes produced in Palermo in the 12th–13th century and now housed in Madrid.

CHRYSOPOEIA: THE QUEST FOR GOLD

to the ancient Greek elements was never spelled out explicitly, except to say that sulfur—an inflammable substance, used in the legendary incendiary agent called Greek fire and later in gunpowder—was deemed to be imbued with the qualities of hotness and dryness, while mercury was cold and wet.

To confuse matters further, natural philosophers knew of sulfur and mercury as real, physical substances. (Deposits of yellow, pungent sulfur occur naturally in volcanic regions.) So how did these tangible substances relate to the alleged constituents of metals? Historians of chemistry still debate that issue. Some say they are one and the same; others claim that the principles were abstract and non-material—which is why they are commonly denoted Philosophical Sulfur and Philosophical Mercury.

RIGHT The nature of transmutation is symbolized by this illustration demonstrating the purification of gold from *The Book of Lambspring* contained in Georg Hellmerich's *Musaeum hermeticum* (Hermetic museum; 1692). The text accompanying the illustration states: "[As] The salamander lives in fire, and the fire changes it into the best colour, [so] The repetition, gradation, and improvement of the tincture, or the sharpening of the philosopher's stone, is to be understood."

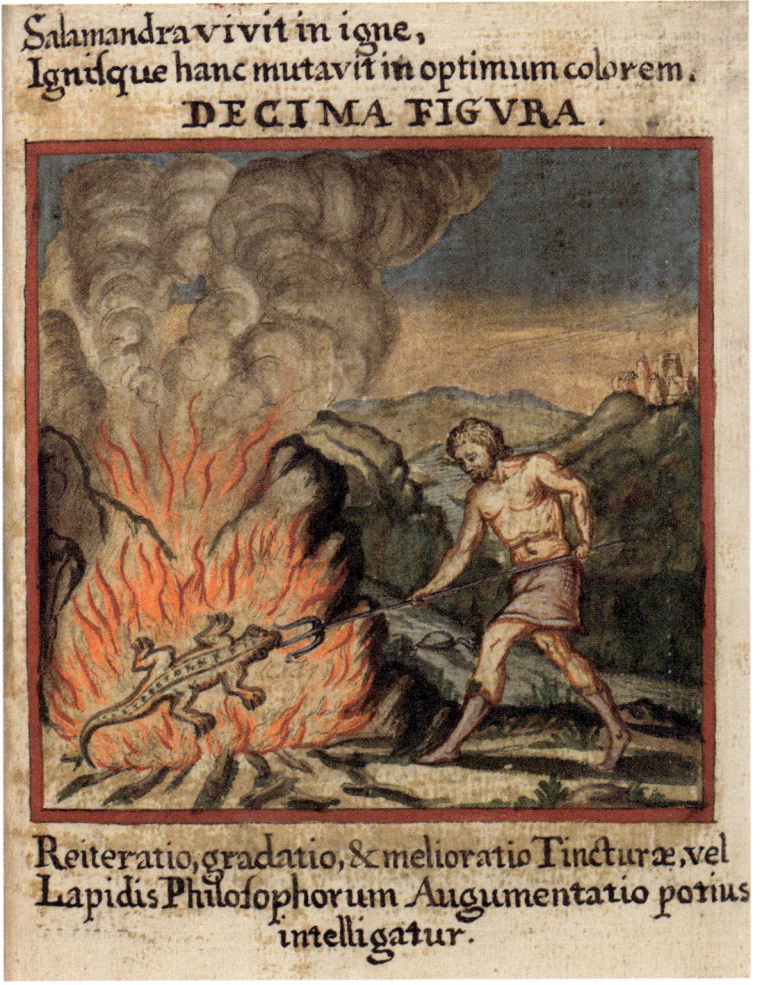

To achieve transmutation, one needed to alter the proportions of this Philosophical Sulfur and Philosophical Mercury to achieve the "perfect" ratio that occurs in gold. But how was that done?

The Stockholm papyrus (see page 20) contains various recipes for how to "make" gold by giving a metal object the appearance of that metal, sometimes in a manner that can evade detection. But Zosimos also refers to a "medicine of metals," the *xerion*, that could seemingly "cure" their imperfection and raise them to the status of true gold. The Greek word connotes a dry powder and was later rendered in Arabic as *al-iksir*: an elixir.

For Jābir, transmuting metals meant adjusting their Aristotelian qualities to bring them closer to the hot, wet character of gold. It was important to know exactly how much of a given quality was needed, and Jābir employs a seven-point graded scale, although these proportions were deduced not empirically but by logical deduction using numerological principles—a reminder that what sometimes sounds like science in former times could coexist with ideas that now seem anything but.

An elixir was the agent that could bring about this adjustment in qualities. Exactly what kind of elixir you needed depended on what you were starting with—there were elixirs that would only work, say, for "cold, dry" lead. Some thought that only minerals could be used to make elixirs, others advocated using substances extracted from animals or plants. The purest grade of elixir could transmute *any* metal into gold.

This marvelous substance came to be known to medieval alchemists as the *lapis philosophorum*, the philosopher's stone. The term was popularized by a medieval text called the *Summa perfectionis* (The height of perfection), which was attributed to Jābir (in the Latinized form Geber) but was most probably written around the start of the fourteenth century. Several other works by this "Geber" began to appear at that time, and it is suspected they were produced either by an individual or a group of scholars using the name of the illustrious Arabic alchemist; their author is now denoted Pseudo-Geber, and one suggestion is that he was an Italian Franciscan named Paul of Taranto. Pseudo-Geber says that while some agents of transmutation produce an appearance of gold, only the true philosopher's stone guarantees to make the real thing.

Even during the times of the Crusades, European scholars did not deny their intellectual debt to the "infidels." Alchemy entered the medieval Western world via Latin translations of the Arabic texts of Jābir, al-Rāzī, Ibn Sīnā (Avicenna), and others, beginning with the translation in 1144 of *De compositione alchemiae* (On the composition of alchemy; the original author is unknown) by the English scholar Robert of Chester.

"Geber" was by no means alone in having works posthumously attached to him. Many celebrated scholars of the Middle Ages "became" authors of alchemical texts, such as the German bishop Albertus Magnus (see page 78) and the thirteenth-century Catalan mathematician and preacher Ramon Lull (now commonly regarded as a pioneer of cryptography and code-making). A key opus attributed to Lull was the *Liber de secretis naturae seu de quinta essentia* (Book of the secrets of nature or of the quintessence), a compilation of more than one hundred texts that drew heavily on the works of the radical Franciscan friar John of Rupescissa

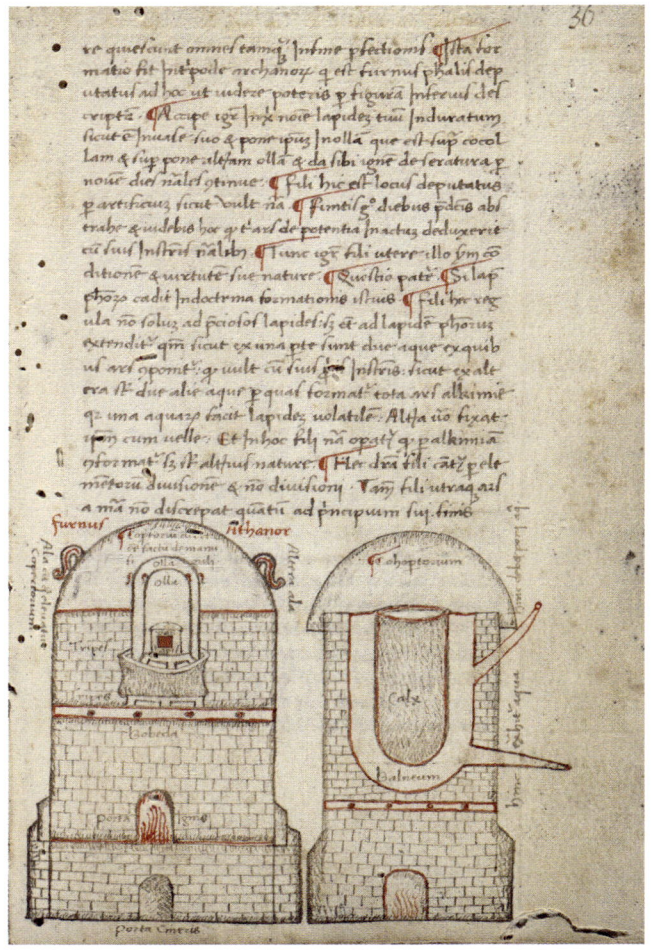

RIGHT Alchemical furnaces illustrated in *Liber de secretis naturae seu de quinta essentia* (Book of the secrets of nature or of the quintessence; 1498), attributed to Ramon Lull, drawing on the works of John of Rupescissa. Lull was a Catalan theologian and philosopher who in fact wrote none of the alchemical works widely attributed to him; the author of these works is generally designated today as Pseudo-Lull, although it might not be a single person.

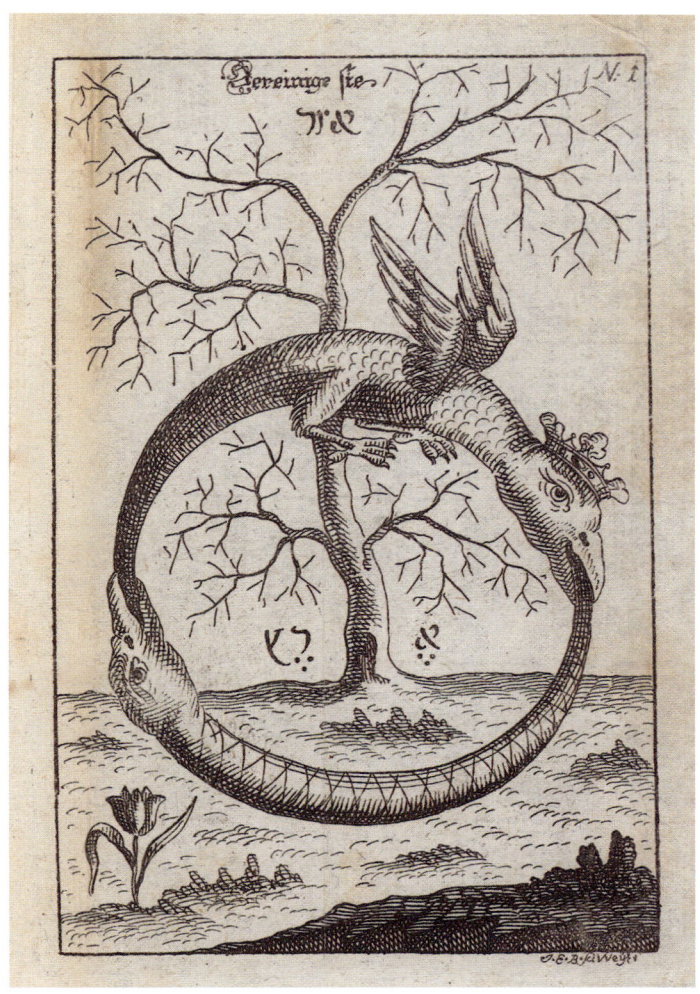

RIGHT The Ouroboros: a serpent and a crowned dragon form a circle—a common alchemical symbol of the unification of primal matter with a "universal spirit." This illustration is from a work on alchemy by the pseudonymous Abraham Eleazar, who claimed to be recording the work of a rabbi named Samuel Baruch, *R. Abrahami Uraltes chymisches Werk* (R. Abrahami Eleazari's ancient chymical work; Leipzig, 1760).

(see page 130). Another eminence to whom several important alchemical works were falsely attributed was the thirteenth-century Catalan physician Arnald of Villanova, who acted as doctor to the popes Boniface VIII and Clement V. As with Lull, the real Arnald probably wrote nothing on alchemy.

One such work of the alchemical "Pseudo-Arnald" is the *Rosarium philosophorum* (Rosary of the philosophers). A later (sixteenth-century) version was one of the first alchemical texts to include illustrations, which expressed the concepts and processes of the Hermetic art in cryptic symbolism. A king and queen make love and merge into a winged hermaphrodite, for example, and a lion devours the sun. Both are typical allegories for the chemical manipulations thought to be involved in transmutation using the philosopher's stone.

RIGHT In Cornelis Bega's 1663 painting *The Alchemist*, the alchemist is about to weigh a red powder representing the philosopher's stone. At his feet a charcoal brazier holds a small crucible containing a molten metal, to which the powder will be added to transmute the metal into gold. Beside the brazier lies a "monk," an instrument used to form moistened bone (usually deer) ash into a cupel to be used for testing the quality of gold.

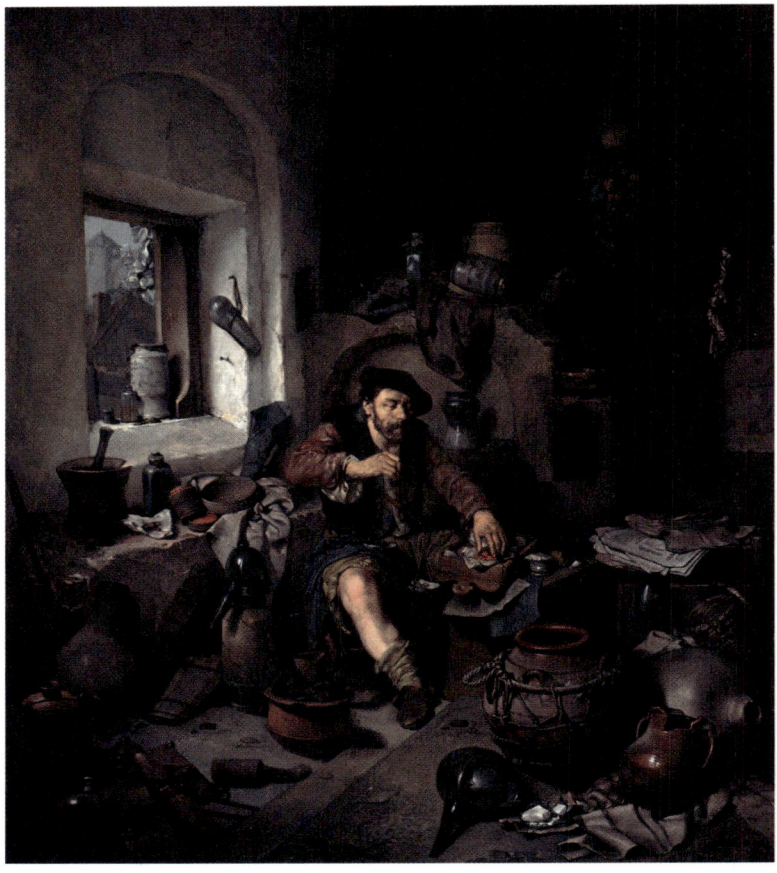

Gold-making in the Middle Ages

It's often said that the philosopher's stone is like the modern chemical concept of a catalyst: a substance that brings about a chemical change without being altered itself. This is not quite right, for the stone was indeed consumed in a transmutation. But you only needed a little of it, which would be mixed into the base metal after it was melted in the furnace. A good-quality sample of the philosopher's stone might transform many thousands of times its own weight of base metal into gold. Some transmutations were said to need a little gold to begin with, which was then "multiplied." The cost of failure could thus be rather high: having transformed gold in some way, the alchemist might fail not only to multiply it but even to recover what had been there at the outset.

Many alchemists agreed that the philosopher's stone was a red powder, and in the Middle Ages there was a common belief that it was made by ushering the ingredients through a series of

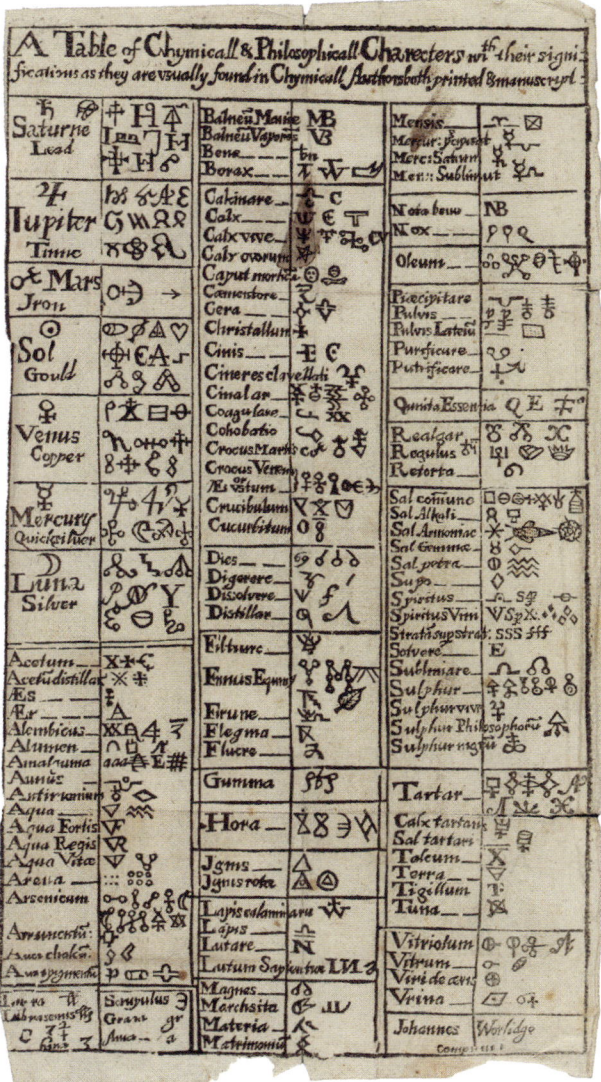

RIGHT "Chymicall & Philosophicall Charecters," a table of alchemical symbols from the *Last Will and Testament* of the pseudonymous alchemist Basil Valentine (London, 1671). This list of chymical ingredients and operations anticipates the systematic tabulation of elements in later textbooks of chemistry.

characteristic color changes. The common sequence began with blackening (*nigredo*), followed by whitening (*albedo*), then eliciting a yellow tint before finally the substance turned red, or perhaps purplish. The color changes recorded by some alchemists offer hints at what was really happening in their apparatus: for example, lead might turn black, yellow, or red when heated in air as it forms its colored oxides. ("Red lead," or lead tetroxide, was widely used as a pigment since antiquity.)

But such recipes can be misleading, because the names given to ingredients can't always be taken at face value. Sometimes this

> Some believed that there could be different *kinds* or grades of gold, and gold made by alchemy was suspected by some to be inferior to that found in nature.

would be a deliberate attempt to hide the precious knowledge: alchemists might use "code names," or what became known by the German term *Decknamen*, which only true adepts would understand. This tradition began in Arabic alchemy, where, for example, sulfur might be called *al-wāqid*, "the burning;" *al-zuhār*, "the moaning;" *al-'alam*, "the sign;" or many other variants. Alternatively, ambiguities in naming might simply reflect the rudimentary chemical knowledge of the times, and substances we now recognize as distinct chemical compounds may not have been so clearly distinguished at the time: "vitriol," for example, generally referred to sulfate minerals but could occasionally mean other metal ores.

Some believed that the agent of transmutation could be extracted from gold itself—a kind of "essence of gold," or, as some called it, a Tincture or Sulfur of Gold. On one thing, however, there was widespread agreement: the philosopher's stone was a combination of two central ingredients. This union was often described in symbolic terms as a conjoining of male and female (such as the amorous king and queen), or the coming together of the sun (symbolizing hot, yellow sulfur) and the moon (cool, silvery mercury).

The key thing to remember amid all this puzzling encryption is that the philosopher's stone, and the means of making it, were deemed to be entirely within the natural order of things. Whether the process was in any sense "magical," as romantic tradition often implies today, is another matter, and a complicated one, as we'll see.

Not all authorities believed transmutation was possible at all. Ibn Sīnā felt that the best we humans could do was to imitate gold, for "human industry is not the same as what nature does." Imitations—metals that have been "dyed" or "tinged"—might be excellent, he said, but in them "the essential nature remains unchanged." That idea harks back to an old debate, joined by Plato and Aristotle, about the nature of human "art" and its relation to the products of nature.

Yet imitation did not necessarily entail crude deception, like covering a metal object with gold leaf. Some believed that there could be different *kinds* or grades of gold, and gold made by alchemy was suspected by some to be inferior to that found in nature. The thirteenth-century English friar Roger Bacon, a pioneer in studies of optics who seems to have conducted alchemical experiments, challenged this idea, asserting that, on the contrary, alchemical gold might be even better than the natural variety.

OPPOSITE A three-headed dragon or *Mercurius*, symbolizing mercury, with the conjoined heads of the sun (hot, yellow sulfur) and the moon (cool, silvery mercury), from an alchemical and Rosicrucian compendium produced in the Lower Rhineland, c. 1760.

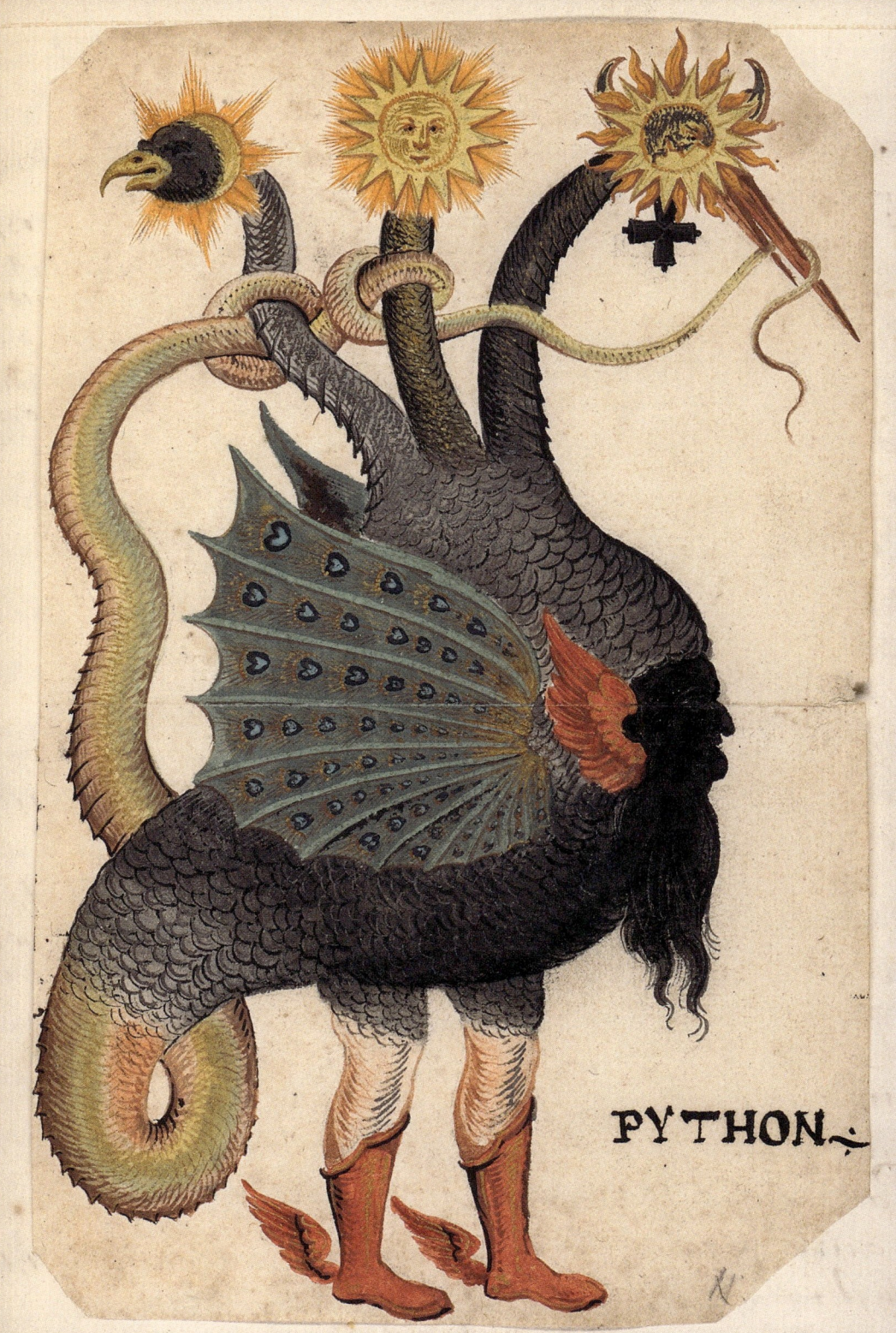

What really happened?

Since we can be sure that no alchemist ever conducted a successful transmutation, one has to wonder at the number of "successful" accounts of chrysopoeia in historical records. Why were such demonstrations believed? One reason is that methods for distinguishing one metal from another in ancient and medieval times were somewhat crude and relied on specialist knowledge. Gold was the densest of the known metals, but the main method metalworkers had for verifying the identity (or the purity) of gold was a chemical technique called cupellation that separated it from other metals. If putative gold weighed the same after cupellation as it did before, it was the genuine article.

Cupellation was an ancient process: there is evidence that gold was being refined and purified this way in the Middle East and Egypt since at least the second millennium B.C. In the fourth century A.D.—the era when alchemy proper starts to emerge in Hellenistic culture—Themistius of Byzantium, a writer and adviser to emperors, avers that wise people seek out methods like this for assaying gold so as not to be tricked in the marketplace. But cupellation can't be done without destroying (by melting) at least some of the supposedly golden artifact—and besides, it was expert knowledge, even a sort of trade secret. At any rate, the

RIGHT Separating metals by the process of cupellation in a furnace, in a woodcut from Georgius Agricola's *De re metallica* (On the nature of metals; Basel, 1556). Agricola's treatise on mining and metals, one of the pre-eminent sources of information on 16th-century metallurgy, eschews the cryptic style of alchemical works in favor of a practical presentation of mineralogy and techniques for extracting and preparing metals. The author was skeptical of alchemy, largely on the grounds that its practitioners were apt to make false and inflated claims.

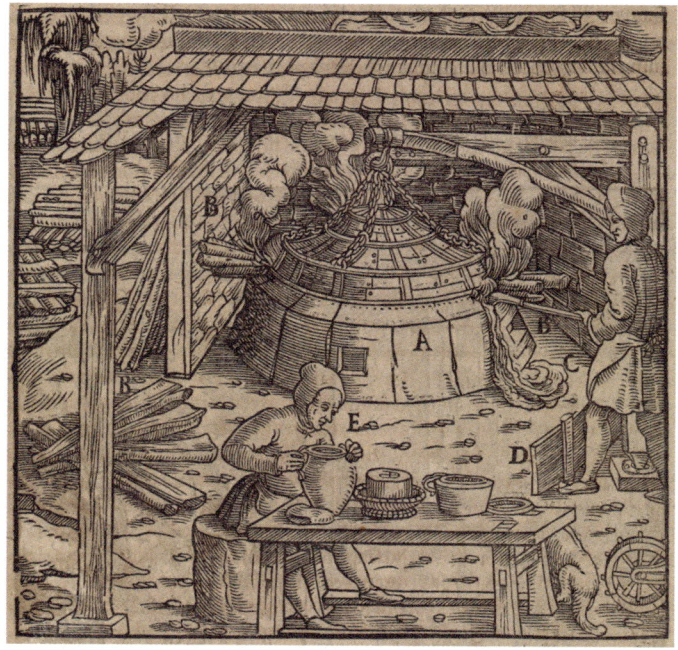

RIGHT A copy in Isaac Newton's hand of an earlier manuscript by George Starkey (under his pen name Eirenaeus Philalethes) describing "the Preparation of the [Sophick] Mercury for the [Philosophers'] Stone," first published in 1678. Starkey, a native of the Americas (born in Bermuda), was a key source of alchemical information for both Newton and his colleague Robert Boyle.

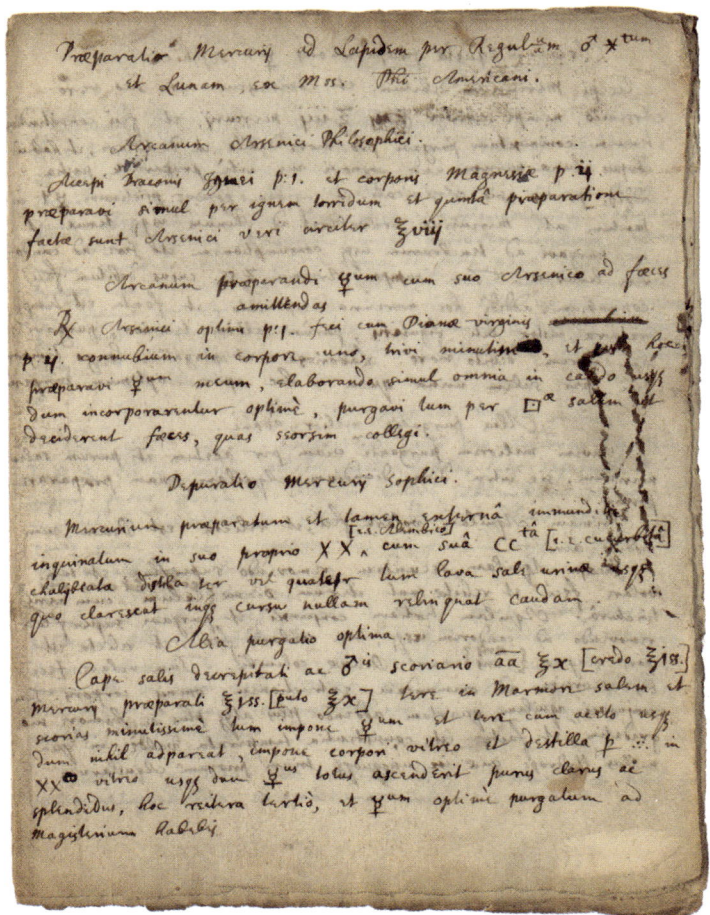

definitions of gold in the early Hellenistic texts of alchemy make no reference to testing by cupellation.

What's more, it was widely supposed that transmutation could happen in stages: a silvery metal that was given a reddish or yellowish appearance might seem to be *approaching* the state of gold. As Joseph Needham says, "it was not thought necessary that all the properties of the yellow metal should be identical with those of natural gold so long as at least one of them was"—heaviness, say, or ductility, but especially color.

It's fair to suppose that in these demonstrations of gold-making there was sometimes intentional trickery on the part of alchemists and credulity on the part of audiences. But that's surely not the whole story: whether alchemy seemed successful or not depended in part on what all parties thought they were doing, and what they were seeing.

PROFILE

Albertus Magnus

(c. 1200–1280)

His common sobriquet Magnus, meaning "the Great," reveals the status awarded to the thirteenth-century German Dominican friar Albert of Cologne. He was considered one of the most learned individuals of his age, and today is regarded as one of the greatest of the medieval philosophers.

In his lifetime, Albert was also known as *Doctor universalis*, testifying to a range of interests that included theology, logic, astronomy, botany and zoology, justice, law, and alchemy. In 1931 the Catholic Church declared him a saint. (An earlier, ultimately unsuccessful attempt in the fifteenth century by the Dominicans to have him canonized was motivated in part by a wish to rescue his reputation from suggestions that he was an impious magician.)

Neither the exact place nor date of his birth is known, although he is thought to have been from Bavaria. Albert studied at Padua and Bologna, and for a time he taught theology at the University of Paris, where Thomas Aquinas was one of his students. In 1248 he established the Dominican Studium Generale in Cologne in 1248, a hub of learning that later became Germany's first university. Despite becoming Bishop of Regensburg for a short period from 1259, Albert spent most of his life in Cologne, where he died around the age of eighty.

Albert had a comprehensive and authoritative knowledge of Aristotle's works, as well as the commentaries on them by the Islamic scholars such as Ibn Sīnā (Avicenna) and Averroes. He was an innovative thinker in the natural sciences, evident in his treatise on minerals *De mineralibus* (c. 1260), which was informed by his visits to mines and discussions with miners. Here he repeats the Jābirian theory that metals were composed of mercury and sulfur, saying that "Sulfur is like the father, and Mercury like the mother"—suggesting that sulfur is like the male seed and mercury the menstrual blood, the mixing of which was thought to engender embryos in living creatures. Albert was clearly familiar with alchemical processes conducted in the laboratory, which he invokes to explain transformations that happen in natural ores in the earth. Thus he accepts in principle that the alchemist recapitulates what happens in nature—although he expresses skepticism about whether alchemists really can achieve transmutation, saying that those who claim to have done so are "deceivers."

It would be surprising if, given the breadth of his learning, Albert had not taken an interest in alchemy. But we don't know just how deeply invested in the topic he really was. Inevitably for such a renowned scholar, he had many alchemical texts attributed to him that he surely did not write. One such is *Libellus de alchimia* (Little book of alchemy). Here the author

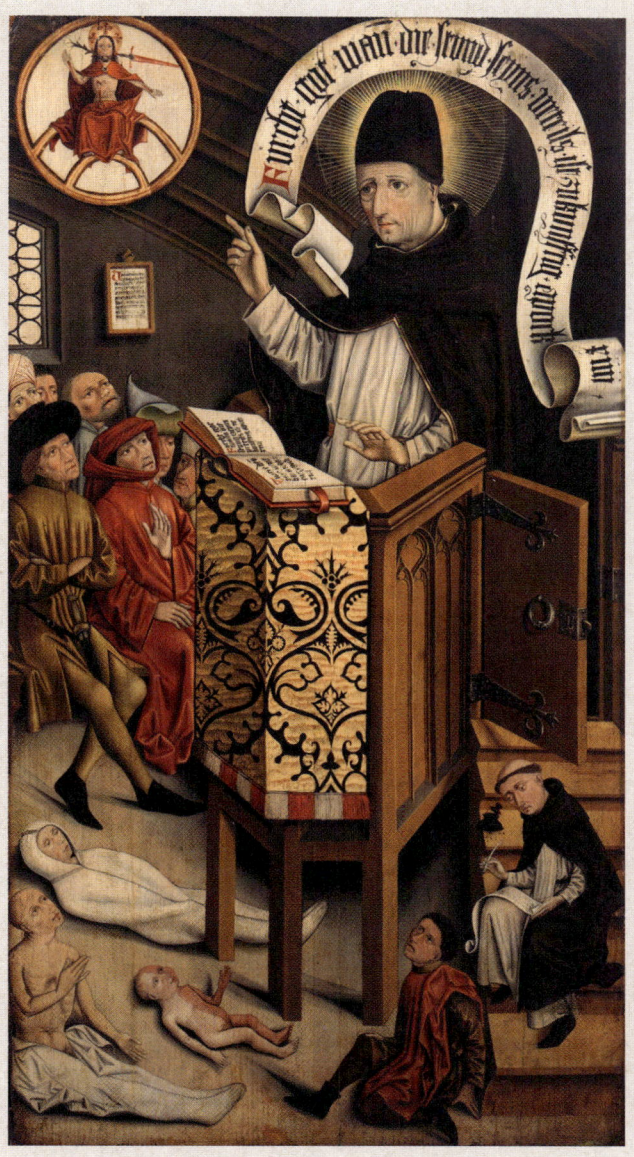

RIGHT Albertus Magnus, depicted in the *Sermon of Saint Albertus Magnus* by Friedrich Walther, c. 1430–95. St Albert was a 13th-century theologian and philosopher of Cologne to whom were attributed several medieval works on alchemy. The inscription on Albert's scroll reads: "Fear God . . . for the hour of his judgment is come."

asserts that metals made by alchemy are the equivalent of the natural versions, barring a few exceptions: alchemical gold, for example, cannot be used as a medicine in the way natural gold can. The author says that he has inspected some alchemical gold that could be melted and reformed six or seven times before finally transforming into a worthless residue. This Pseudo-Albert advises that the alchemist should be discreet and silent about his work, should live in isolation, have diligence and patience—and be rich enough to fund his own studies, without the unreliable patronage of princes and nobles.

A legend says that Albert knew the secret of the philosopher's stone, which he passed on to Thomas Aquinas before his death—an unlikely tale, not least because Aquinas died, to Albert's great dismay, before his mentor did.

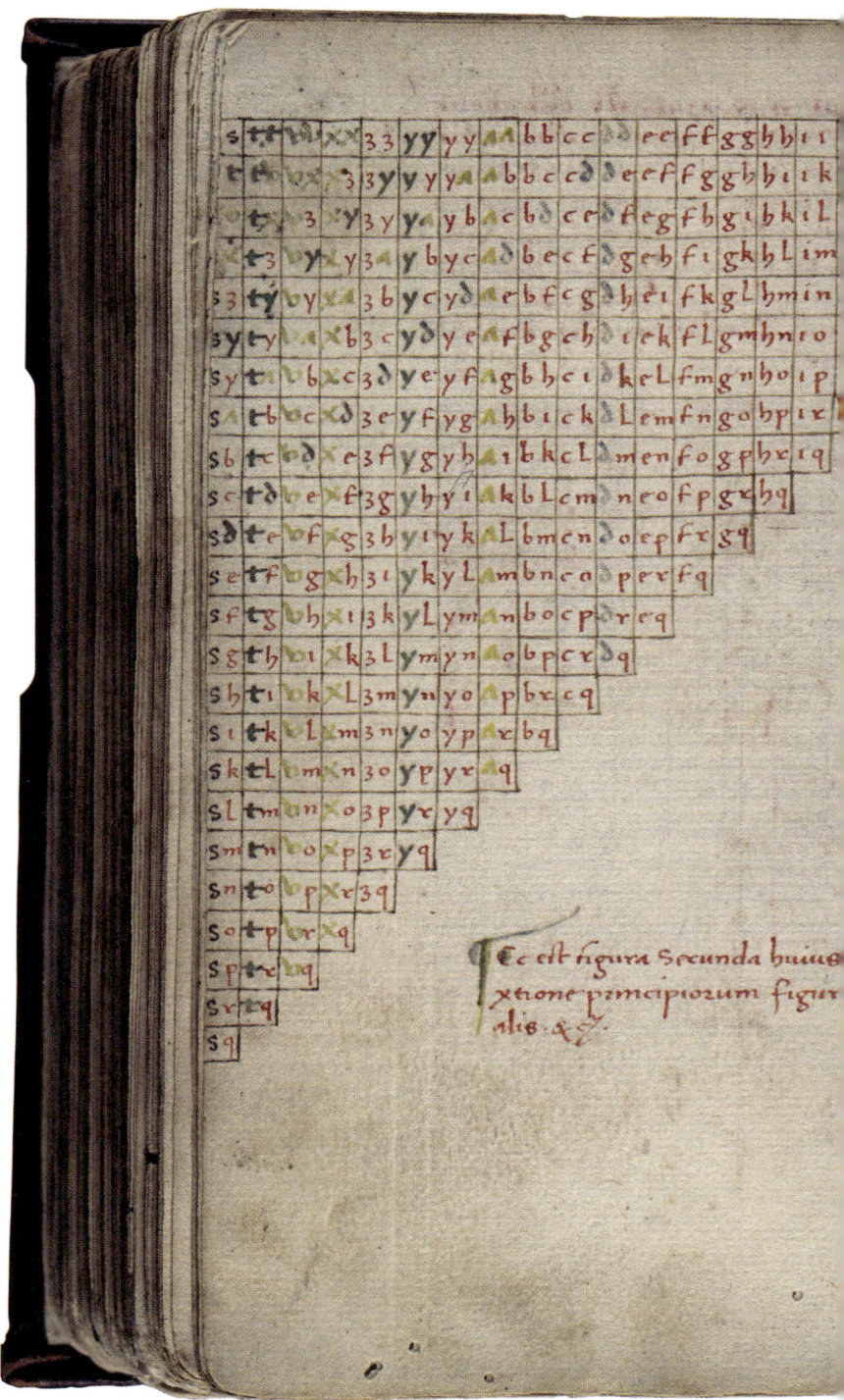

RIGHT Alchemical tables from *Liber de secretis naturae seu de quinta essentia* (Book of the secrets of nature or of the quintessence; 1498), attributed to the Catalan philosopher Ramon Lull. Lull was not the real author—in fact, he seemed to hold a rather negative view of alchemy—and many of the alchemical works attributed to him are now deemed to be by an author designated as Pseudo-Lull (although there were probably several Pseudo-Lulls). This particular work appeared as notions of quintessences (volatile extracts of substances) became widespread due to John of Rupescissa, and the text bears a clear debt to John's work.

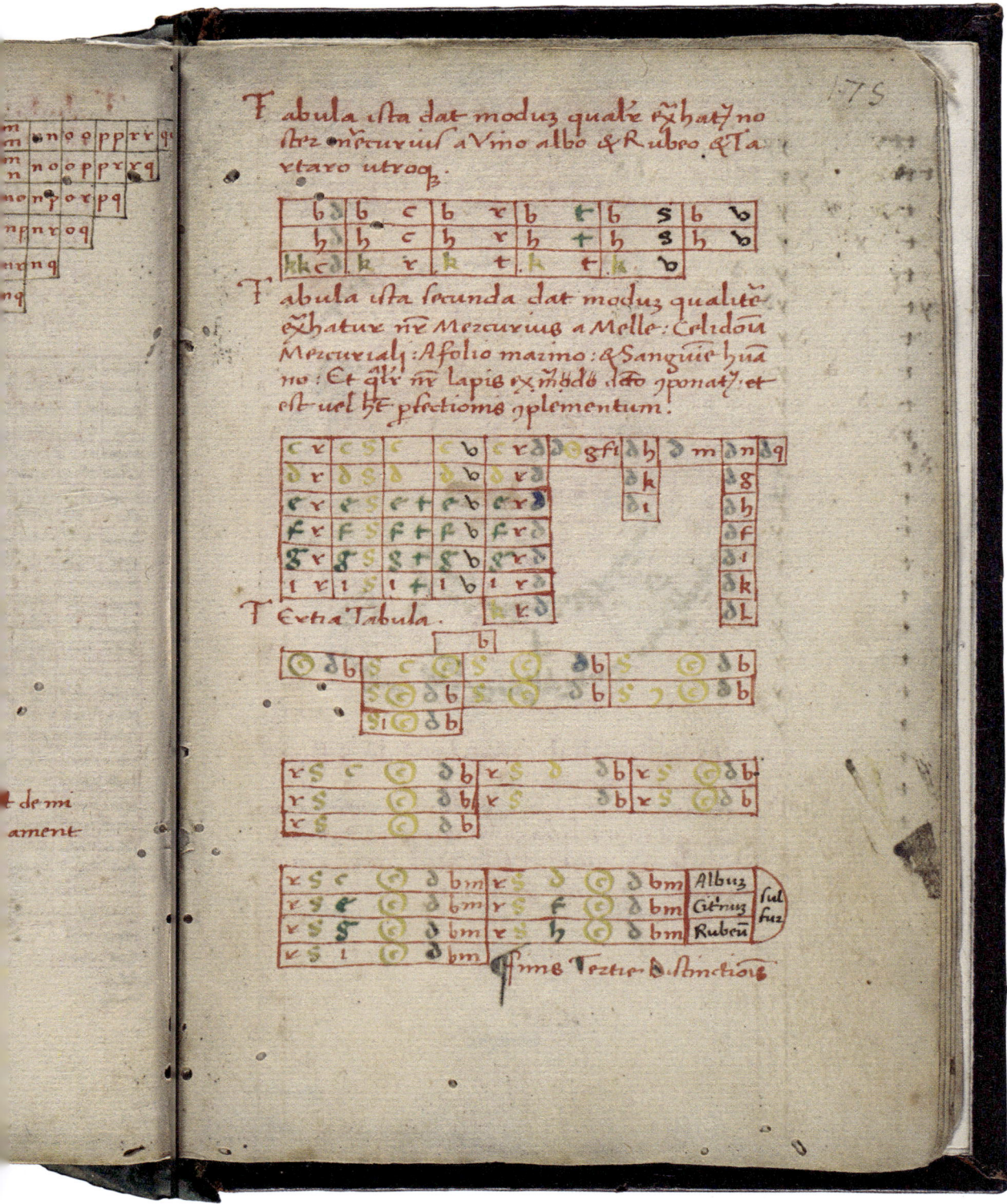

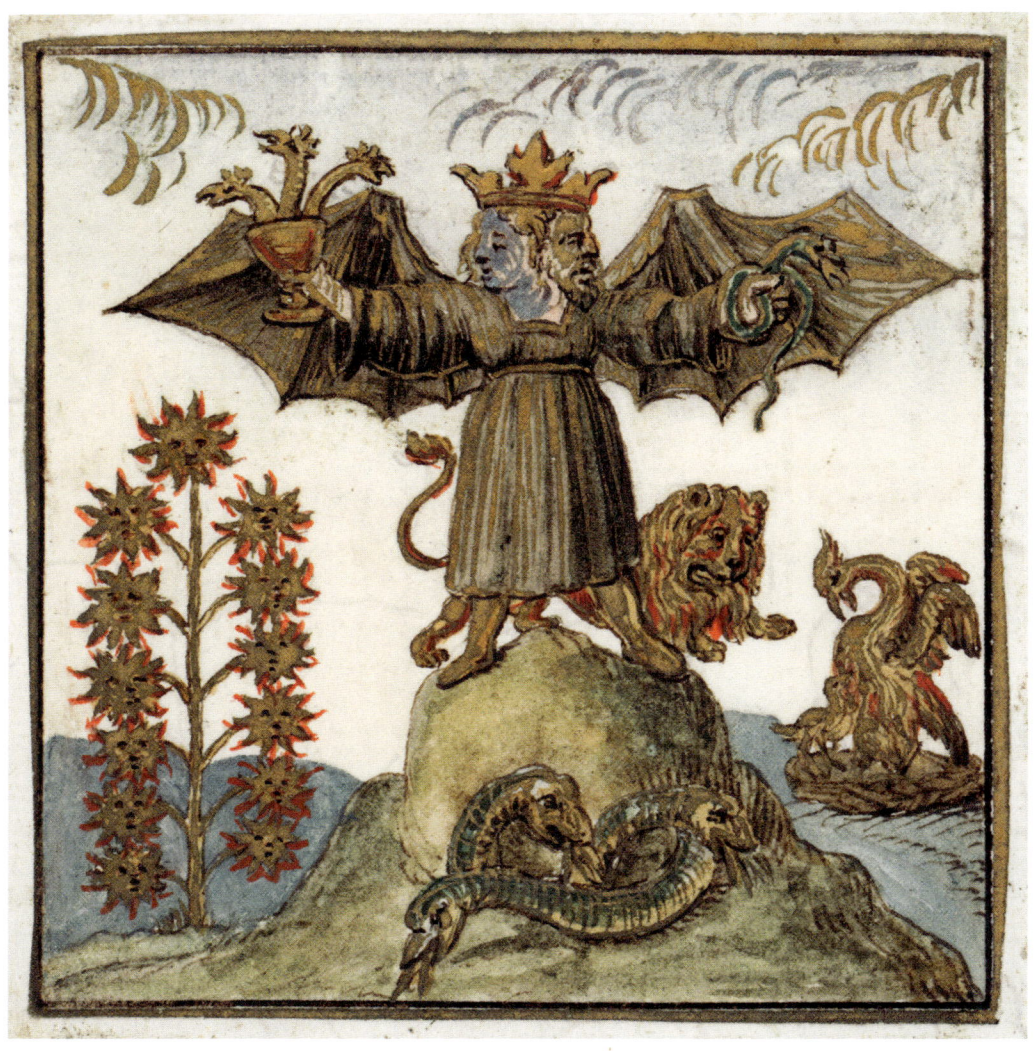

ABOVE "The Demonstration of Perfection": a riot of alchemical imagery, featuring a Lion, Sun Tree, Pelican, and Three-headed Serpent, from an 18th-century manuscript copy of the *Rosarium philosophorum* by the author now denoted Pseudo-Arnald. The text of the *Rosarium* was probably composed in the 14th century, but the allegorical images that accompany it in editions from the 16th century were added subsequently. The first complete edition with images dates from 1550, when it was published in Frankfurt.

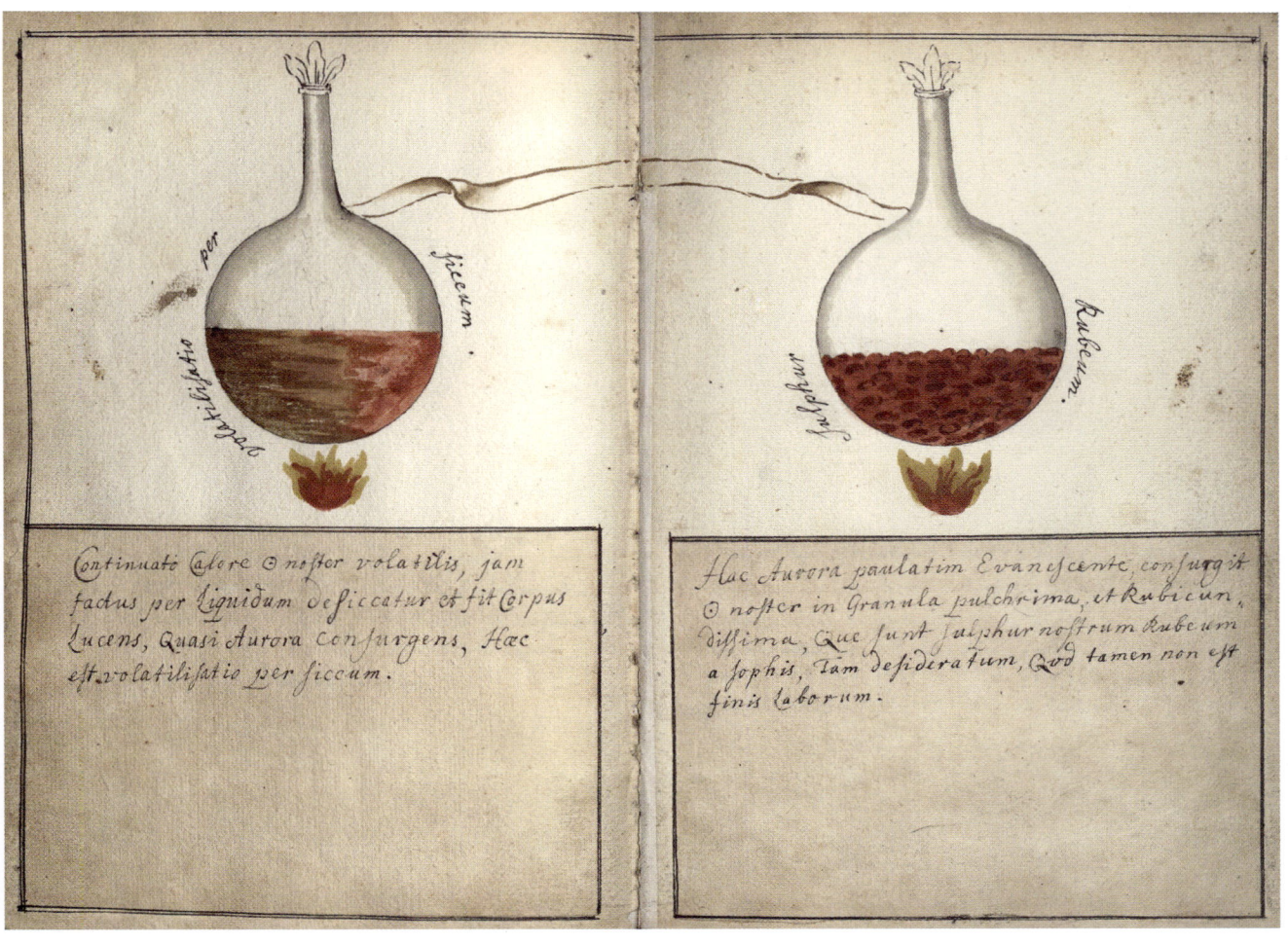

ABOVE One flask holds a winged cherub, symbolizing a volatile spirit, while in the other "the omnipotent king or our stone arises"—in other words, the red substance of the philosopher's stone. This watercolor illustration is from Rabbi Simeon ben Cantara's manuscript *Cabala mineralis* (c. 1675–1700), which reviews alchemical interpretations of the Book of Genesis and is an important contribution to the relationship between alchemy and the Jewish mystical tradition of the Kabbalah. The author was supposedly a medieval alchemist, but is now considered to be pseudonymous.

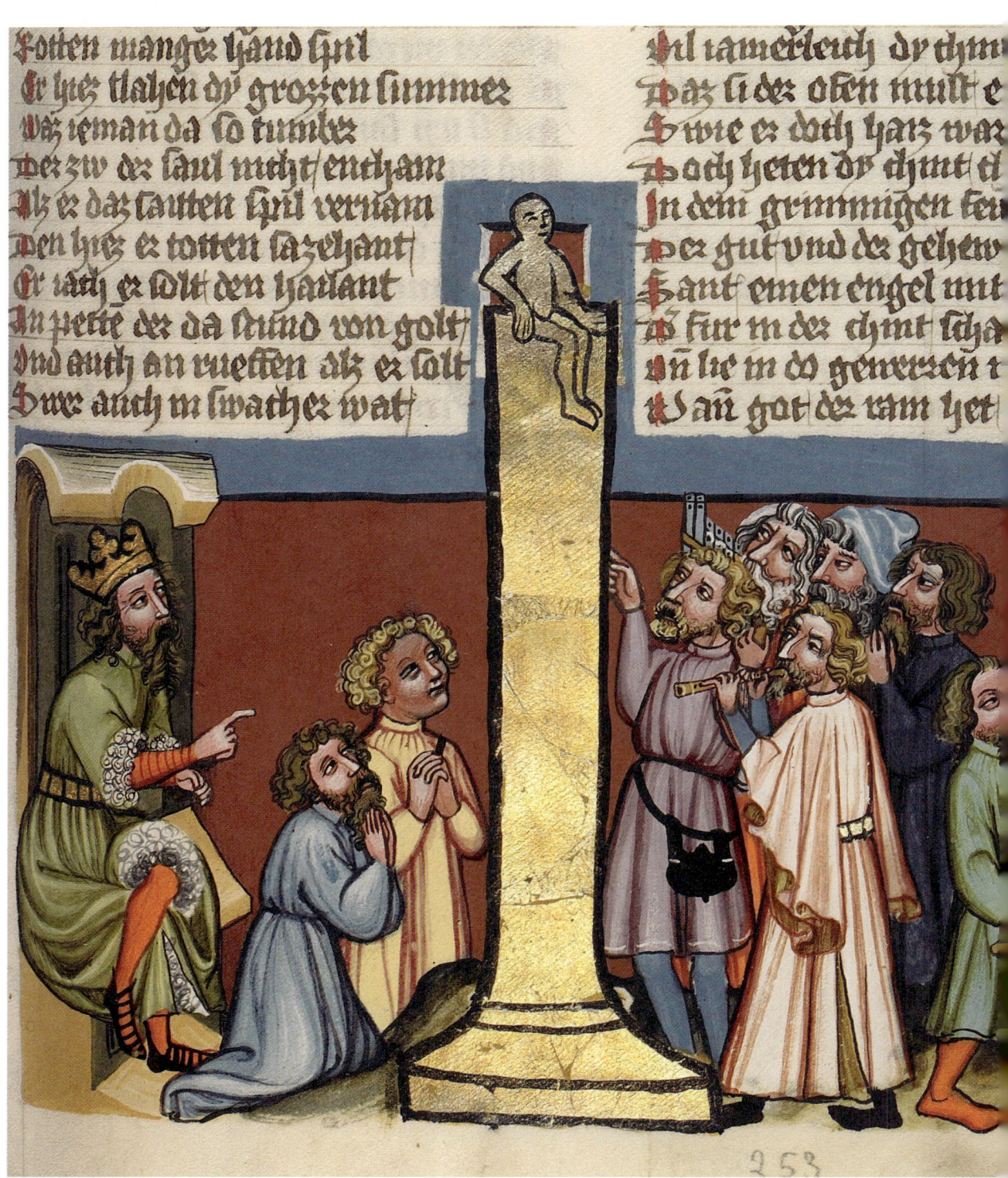

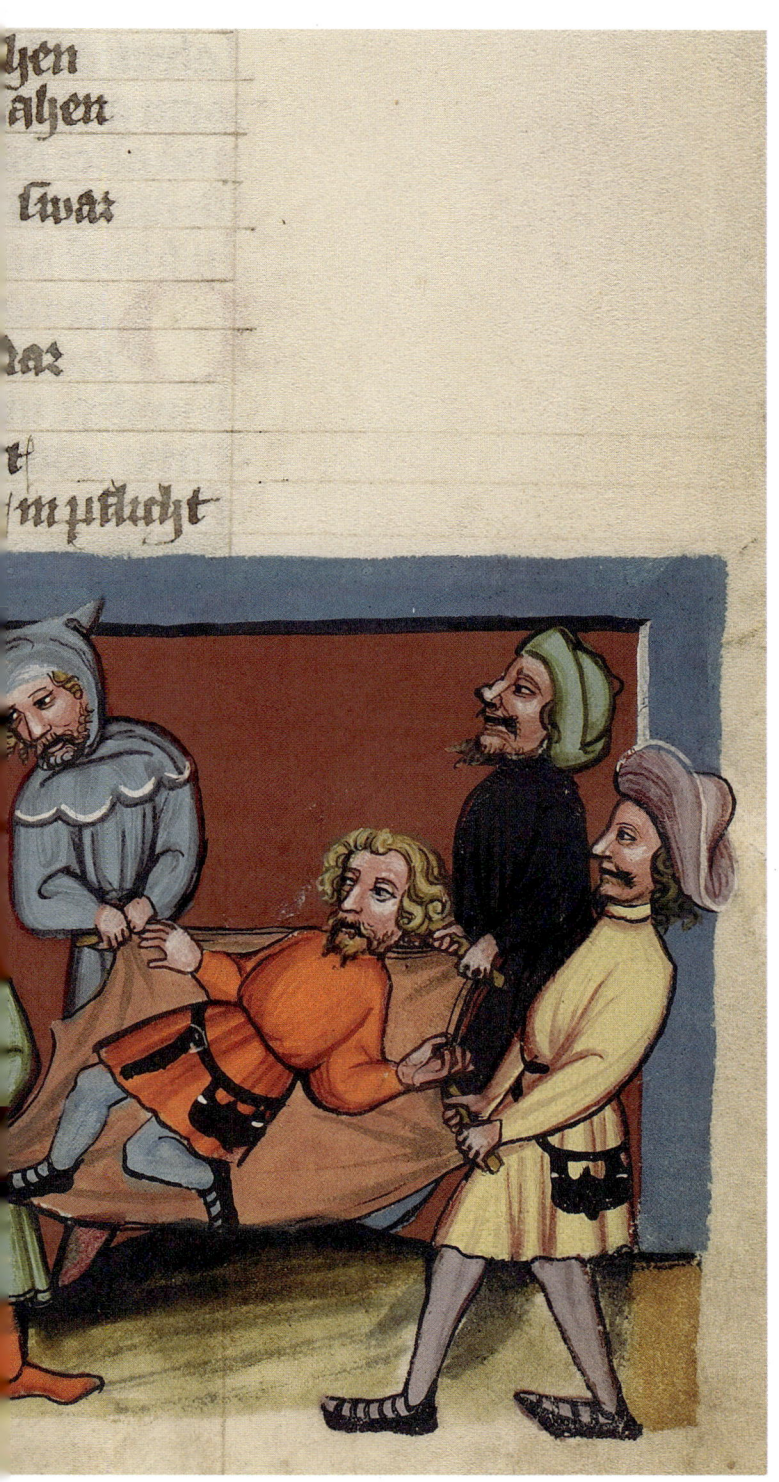

LEFT Making gold by alchemy was frowned upon by many as a dubious occupation, as was the worship of golden idols in the Church. Here the adoration of a golden idol is shown (depicted using real gold leaf), from the illustrated *Weltchronik* (World chronicle), made in Regensburg, Bavaria, c. 1400–1410. A distrust of gold itself was a trope in antiquity: the Greek poet Phocylides was believed to have said "Gold and silver are injurious to mortals; gold is the source of crime, the plague of life, and the ruin of all things," adding ruefully "Would that thou were not such an attractive scourge!" But alchemists were reviled not so much because they made this alluring stuff but because it was suspected either that they did not—that they were deceitful—or that alchemical gold was of inferior quality. Georgius Agricola, the 16th-century expert on mining and metalworking, denounced some alchemists as frauds, but regarding transmutation itself admitted: "Whether they can do these things or not I cannot decide."

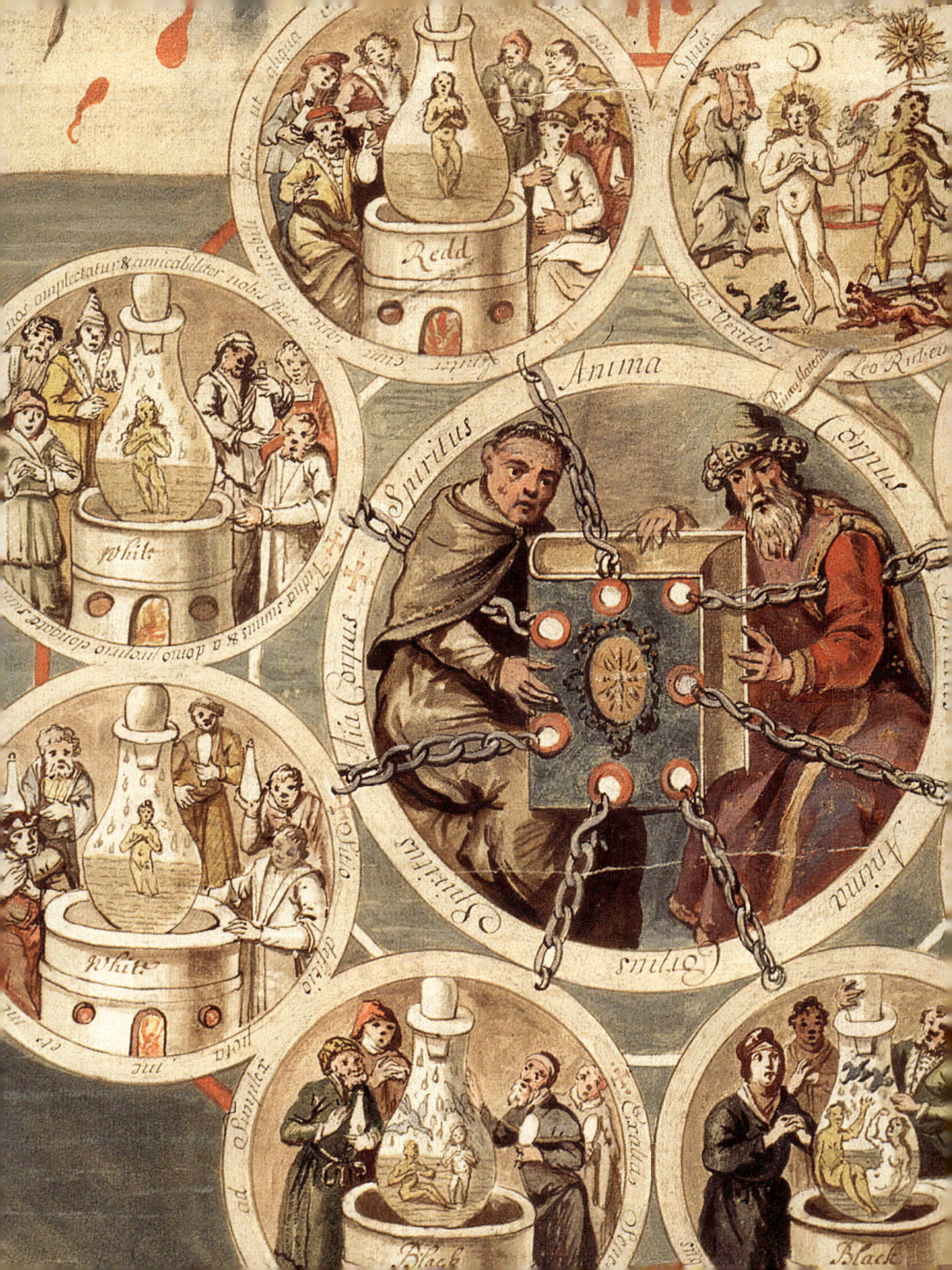

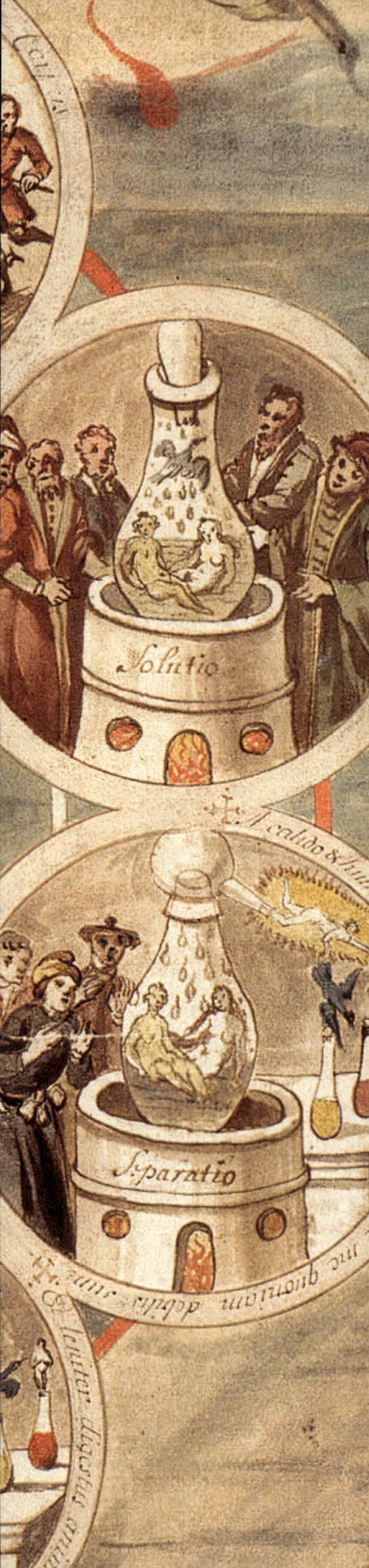

CHAPTER FOUR

Books of Secrets

THE USES OF ALCHEMY

LEFT Alchemists revealing secrets from the *Book of Seven Seals* in a detail from the Ripley Scroll, c. 1600. The Ripley scrolls were richly illustrated compendia of alchemical knowledge attributed to the 15th-century English alchemist George Ripley. Here the imagery hints at the notion that such knowledge was in some sense hidden, revealed, and understood only by true adepts of the art.

CHAPTER FOUR

Alchemists' reputation for concealment led to a popular caricature of the alchemist as a deluded fool trying to make gold in secrecy. Historians of science have now begun to restore recognition of the practical side of alchemy in everyday life: for making useful substances such as dyes and pigments, soaps, and medicines.

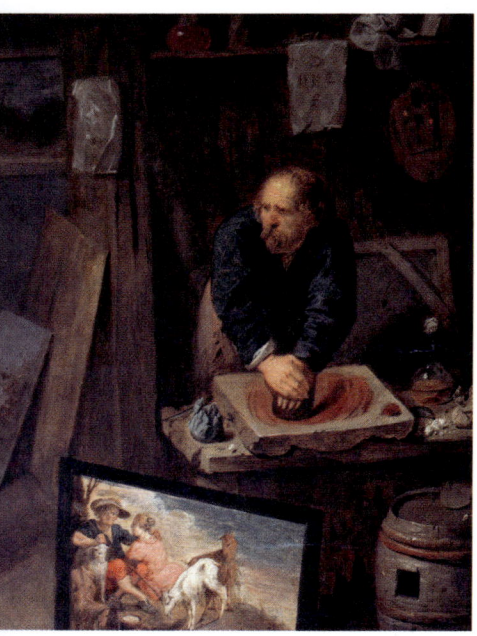

ABOVE This detail of David Ryckaert III's *Painter's Workshop*, 1638, shows a studio assistant grinding a vermilion-colored powder, probably purchased from an "alchemical" supplier and typically sold in pigs' bladders or pouches. The Italian artist Cennino Cennini in the 14th century advised artists making such a purchase to buy the red pigment unground, so that it could not be adulterated with brick dust.

In his *Libro dell'arte* (The craftsman's handbook) the fourteenth-century Italian artist Cennino Cennini offered advice about procuring and using pigments for painting, describing how to make many of them from the raw mineral ingredients. One of the finest pigments was red vermilion, which Cennino simply says is "made by alchemy, prepared in a retort." He adds that if the reader wants to make it themselves, they can get a recipe from the friars. The casual reference to alchemy is a reminder that, however strange and exotic that art might seem today, in the Middle Ages the word could simply refer to what today we would consider to be chemical manufacture.

Cennino's manual exemplifies the "how to" books that became popular during the Renaissance, especially after the advent of the printing press. Many were compendiums of recipes for making domestic substances like soap, glue, dyes, and herbal cures, copied from older sources and perhaps mixed with almanac-style information about the stars and weather. These texts were often sold as "books of secrets"—but that was largely a marketing ploy to boost their allure. They offer a glimpse of how alchemy, far from being a mysterious pursuit conducted solely in secluded and smoky chambers, was a regular part of daily life and the source of materials needed by artisans, industries, and ordinary citizens.

The tradition of secrets

The kind of natural history described by Aristotle and other Greek writers demanded nothing more than careful observation and cataloging of phenomena that might be understood using reason and logic. In contrast, the Hermetic tradition that arose in the Hellenistic culture of the early Christian era asserted that some understanding was made available only to a privileged few and arrived by divine revelation (*gnosis*).

This notion was inherited and consolidated by Islamic scholars, in particular the mystical Ismaili sect, a secret fraternity who believed the Koran contained esoteric knowledge encoded through numerology and symbolism. A strong influence of Ismaili beliefs is evident in the Jābirian alchemical corpus, and the notion of secret knowledge is plain in the most famous work by al-Rāzī: *Sirr al-Asrār* (Secret book of secrets.) Despite its title, the book offered rather straightforward accounts of medical and alchemical substances and how they might be prepared and used.

Al-Rāzī's book is all too easily confused with an Arabic treatise of the same name, purportedly a translation of a Greek original and presenting itself as a letter from Aristotle to his student Alexander the Great. This other *Secret book of Secrets* covers a wide range of topics from politics and ethics to astrology, medicine, and alchemy—a digest of pretty much everything of interest to the medieval intellectual. A thirteenth-century Latin translation,

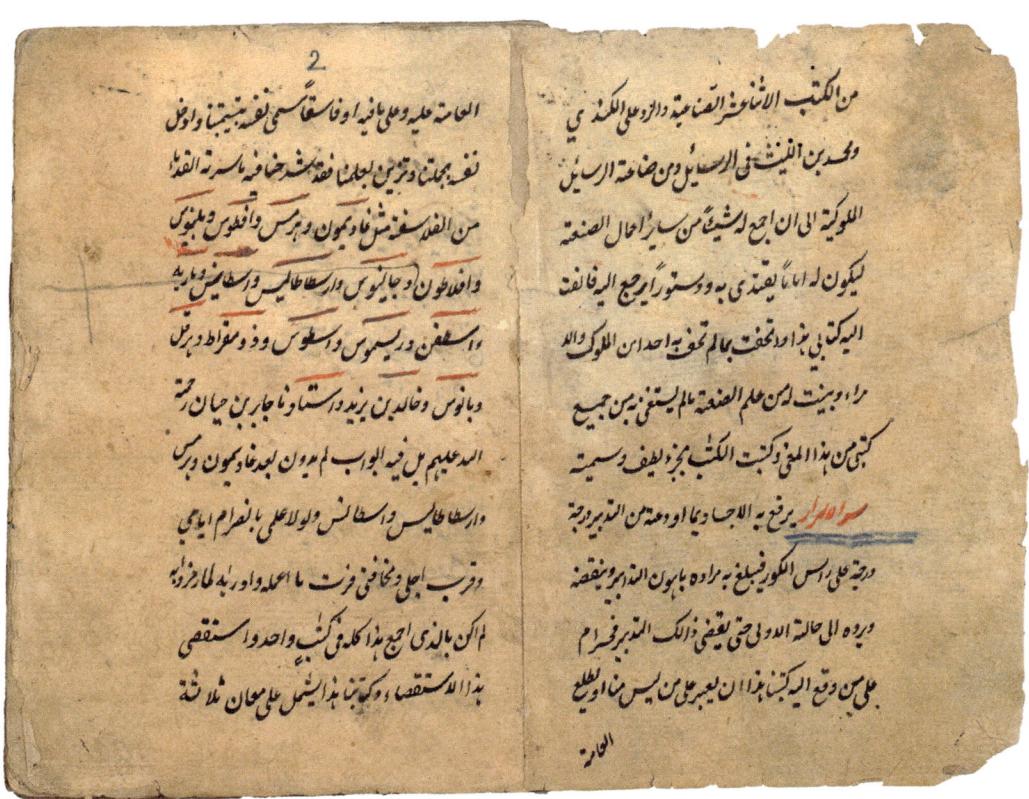

BELOW An annotated 18th-century edition of Muh.ammad al-Rāzī's Sirr al-Asrār (Secret book of secrets). This work by the famed Arabic physician exemplifies the tradition in which alchemical lore was presented as an unveiling of hidden "secrets" of nature.

> "Secret" here tended to connote knowledge that was not truly hidden or forbidden but, rather, confined to experts.

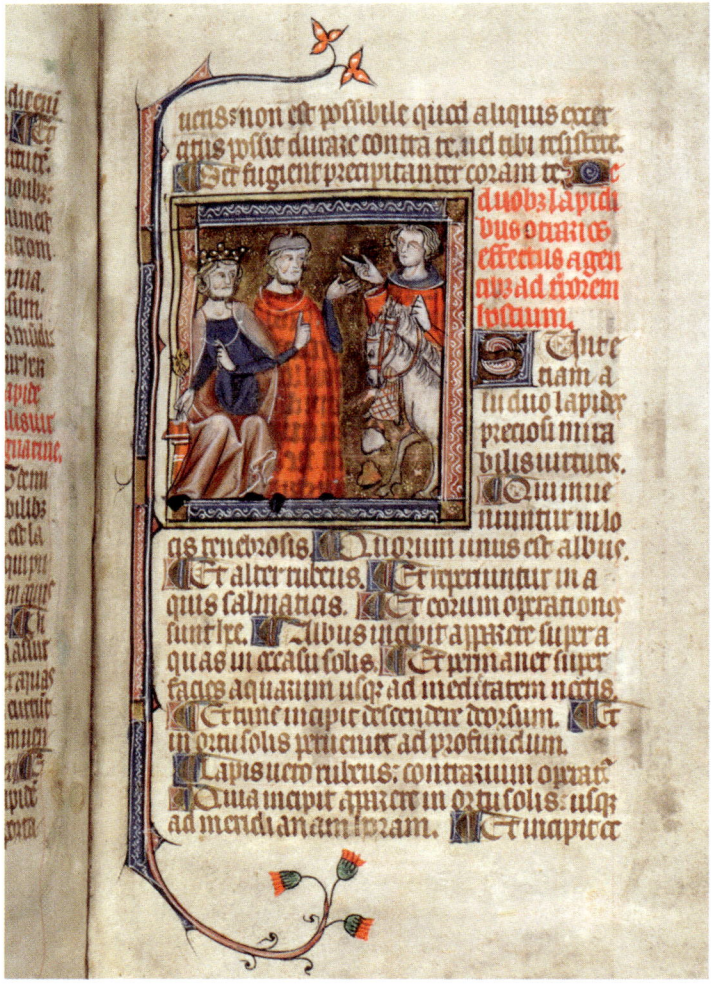

ABOVE RIGHT "Two stones of marvelous virtue" are brought to the king by his servant in this illustration from Pseudo-Aristotle's *Secretum secretorum*, 1326–27, translated by Philip of Tripoli. This work, apparently of Arabic origin (Aristotle himself never wrote on alchemy, which did not exist as such in his time), has been called "the most popular book in the Middle Ages." As well as alchemy, it contained discussions of astrology, medicine, numerology, and other "magical" arts.

Secretum secretorum, is believed to have been one of the most widely read books of the Middle Ages.

Evidently, then, the book was hardly a secret! But the narrative of privileged knowledge was very appealing to the newly emerging class of university-educated medieval scholars, who could bask in the status it conferred on those learned enough to receive it. "Secret" here tended to connote knowledge that was not truly hidden or forbidden but, rather, confined to experts. Some of these medieval books of secrets were essentially handbooks providing "tricks of the trade." But there was another sense in which knowledge could be "secret": it might reveal the mysteries of nature herself, which involve principles or forces that are counterintuitive or invisible to the eye—literally, occult.

90 CHAPTER FOUR

The rhetoric of secrets took on a special complexion in alchemy. Alchemists often went to great lengths to obscure and encrypt their instructions and insights so that they were accessible only to an elite and remained hidden from the uninitiated masses, who could not be trusted to make wise use of this powerful knowledge. Alchemical works deploy symbolic codes for materials—Red King, Green Lion, Black Wolf, vitriol of Mars—as well as for equipment and processes. Once we understand these codes, we can often translate the flowery language of alchemical recipes and procedures into the terms of modern chemistry and figure out what alchemists were actually doing.

Recipe books

The most popular books of secrets of the Middle Ages include *Compositiones ad tingenda* (Recipes for coloring; also called *Compositiones variae*), most probably compiled in the Italian city of Lucca in the late eighth or early ninth century; *De coloribus et artibus Romanorum* (The colors and arts of the Romans), written by the tenth-century Italian monk Heraclius; and *Mappae clavicula* (Little key to

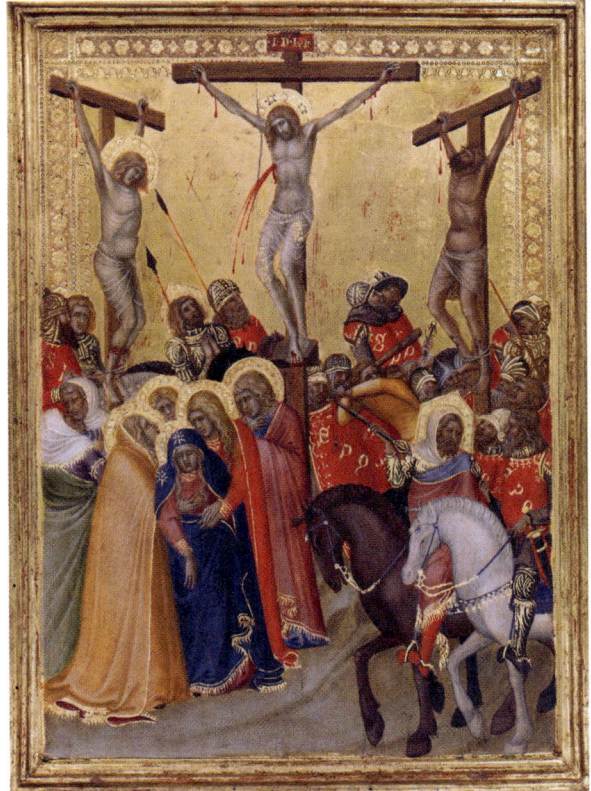

RIGHT Pietro Lorenzetti's Crucifixion of the 1340s connects with earlier sacred art in its lavish use of embellished goldwork (chrysogony) and costly pigments such as lapis lazuli (blue) and vermilion (red).

BOOKS OF SECRETS: THE USES OF ALCHEMY

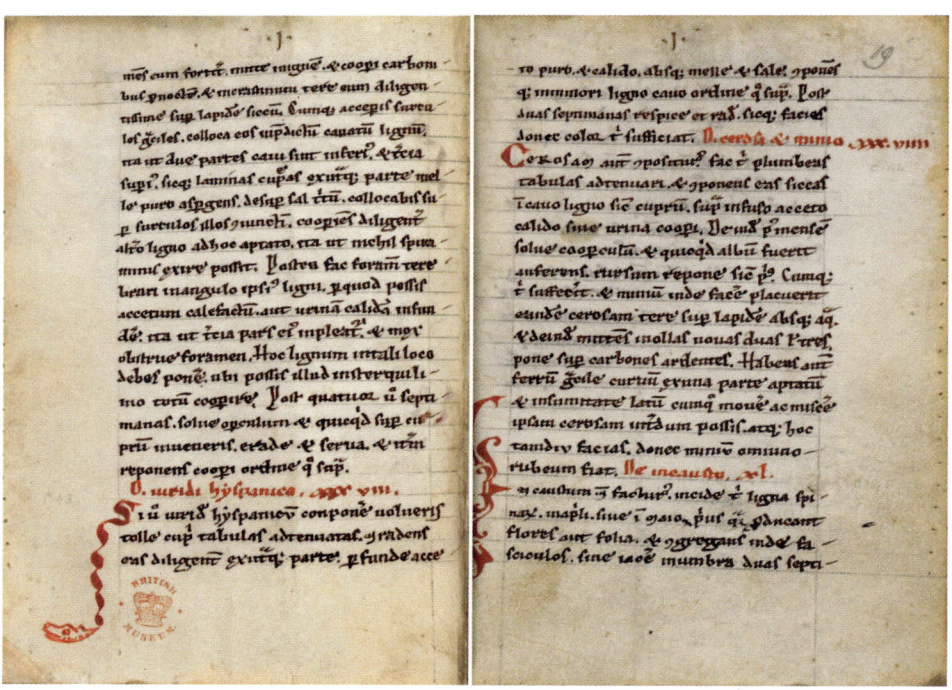

ABOVE Recipes for green, white, and red pigments from an edition of Theophilus's *De diversis artibus* (On diverse arts), probably produced in northwest Germany in the late 12th or early 13th century.

painting), which might originally have been compiled in Alexandria around the seventh century. As the titles suggest, these recipe lists included many for making pigments and dyes, along with tips on glassmaking, metallurgy, and other manual arts. They look like books of practical alchemy.

All the same, they weren't really the workshop manuals they might appear to be. It's not clear that the recipes were widely used, not least because many of them are faulty or useless. It seems that they came more from a literary than a craft tradition: though they masqueraded as how-to guides, they drew on the esoteric notion of hidden knowledge. Thus the distinctions between the practical and the philosophical were not always clear.

One alchemical recipe in particular seemed to promise both the prosaic and the profound. As we saw earlier, Jābir ibn Ḥayyān popularized the idea that all metals are composed of the two philosophical "principles" sulfur and mercury. But, as Cennino explained, combining these real, physical materials produces the valuable red pigment vermilion. The first clear account of this operation appears in the *Compositiones ad tingenda*, and probably the most detailed description in the Middle Ages is given by the German Benedictine monk Theophilus in his book *De diversis artibus*

(On diverse arts), written around 1122. Put the two ingredients in a sealed pot and place them in the blazing coals of a furnace, he says, whereupon "you hear a crashing noise inside as the mercury unites with the blazing sulfur." When it is cool, take out the blackish lump and grind it well, and you will have a brilliant red powder.

Theophilus's book truly does read like a craftsman's manual: he explains how to make colored glass and assemble it into windows, and how to paint doors, cast bells, and build a workshop. All the same, there are signs of the influence of alchemical theory on the work: for example, the proportions of sulfur and mercury recommended by Theophilus for making vermilion don't really work out—there is too much sulfur. Possibly this is because the quantities are derived from Jābirian alchemical theory rather than from practical experience.

What's more, among all this largely sound advice Theophilus suddenly offers what looks like a total flight of fantasy: a recipe for "Spanish gold," probably derived from Islamic sources in Moorish Spain. Take some copper, he says, along with human blood, vinegar,

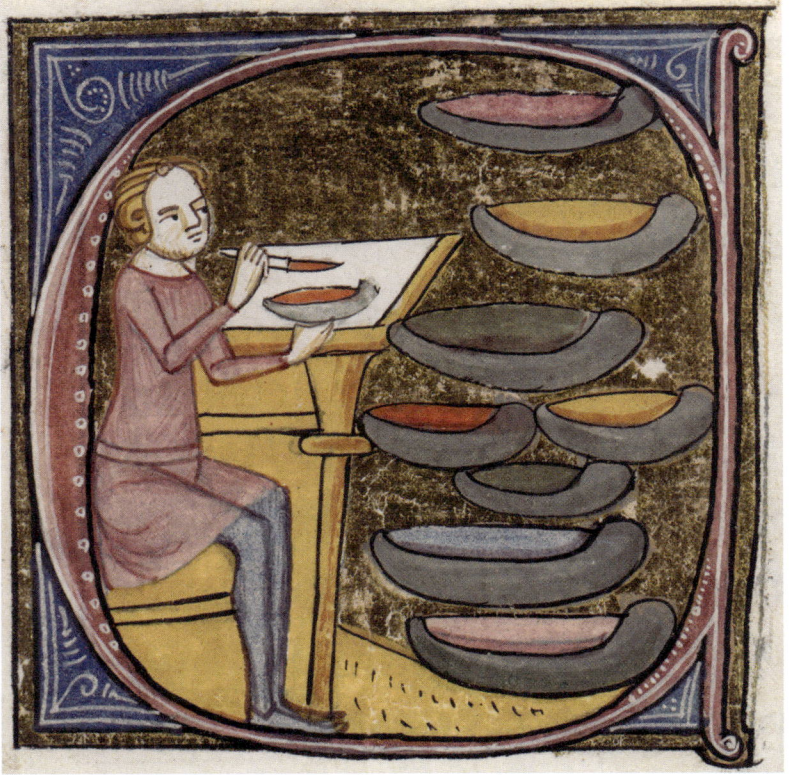

RIGHT An artist mixing colors in the entry on "Color" in James le Palmer's *Omne bonum* (Every good thing; c. 1360–75). Recipes for pigments are likely to have been handed down through texts such as the 9th-century *Compositiones variae*, also known as *Compositiones ad tingenda* (Recipes for coloring).

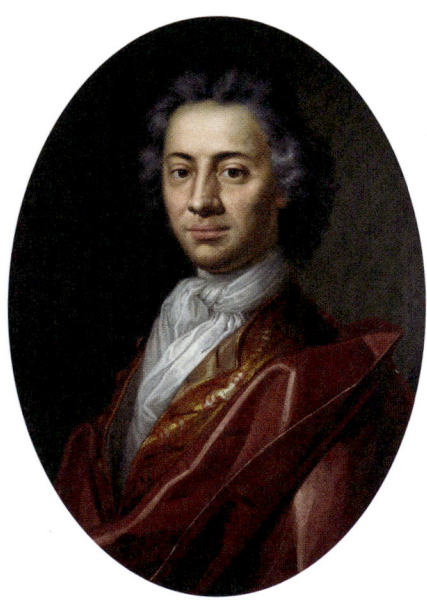

and... basilisk powder! You might suppose it wasn't easy to get hold of the desiccated remains of this fabulous beast, but thoughtfully Theophilus supplies a recipe for that too: if toads fed on bread are allowed to hatch hens' eggs, the chicks will grow serpent's tails after seven days. Needless to say, Theophilus clearly did not check all his recipes before expounding them. Perhaps more to the point, he seems to be taking at face value some of the allegorical language in his alchemical sources.

Given the significance attributed to color changes in the so-called Great Work of making the philosopher's stone, it is not surprising that alchemical experimentation yielded various materials that were valuable to artists. Lead, a common starting ingredient of transmutation, has several strongly colored compounds that found uses as pigments. Lead corroded by vinegar fumes acquires a white coating—lead carbonate, which was used since ancient times as the principal white pigment for painters. The yellow and orange

ABOVE A portrait by an unknown artist of the German theologian, physician, and alchemist Johann Conrad Dippel, 1704. The popular blue pigment that became known as Prussian blue was discovered in Dippel's laboratory around the time that this portrait was painted.

RIGHT Pieter van der Werff's *Entombment of Christ,* 1709, is the earliest painting known to incorporate the pigment Prussian blue, developed in Dippel's alchemical laboratory by Johann Jacob Diesbach.

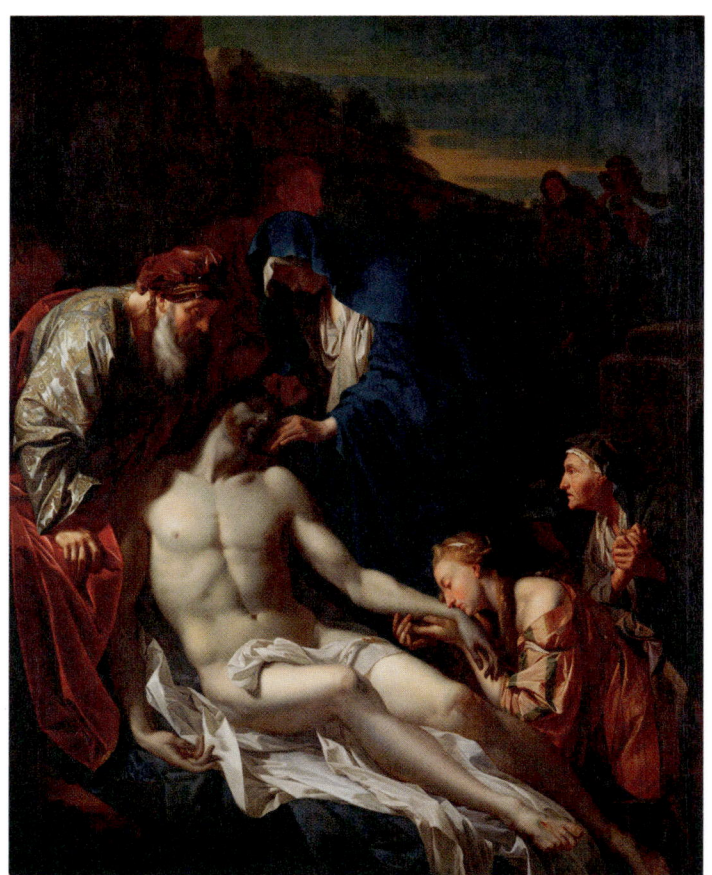

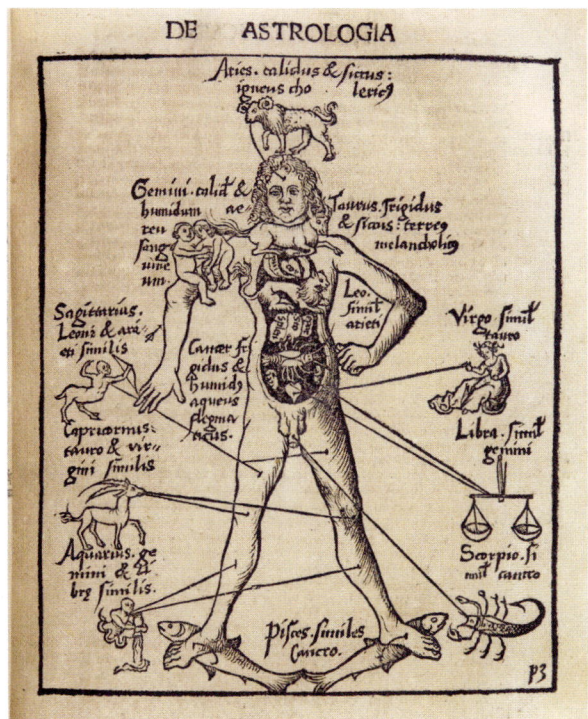

RIGHT The zodiac and the four humors were believed to govern human health, as illustrated in this woodcut of "Astrological Man," from Gregor Reisch's *Margarita philosophica* (Philosophical pearl), published in Freiburg in 1503. Reisch was a Carthusian monk, and his book was intended as a kind of textbook of the seven liberal arts that constituted the standard university curriculum, as well as including discussions of natural and moral philosophy.

pigments known as orpiment and realgar are both compounds of arsenic and sulfur. Regarding orpiment, Cennino warns: "Beware soiling your mouth with it." The name *orpiment* comes from the Latin *auripigmentum*, meaning "color of gold," and in ancient times some suspected that this material, which can be found as a natural mineral, does indeed contain gold. Pliny says that the emperor Caligula extracted the precious metal this way.

The alchemical tradition continued to spawn valuable coloring ingredients even during its swansong. One was an intense blue pigment that became known as Prussian blue, discovered in the early eighteenth century by the German chemist Johann Jacob Diesbach, a manufacturer of dyes and pigments working in the Berlin laboratory of the alchemist Johann Konrad Dippel.

"Don't make gold, make medicines"

Among the books of secrets of the Middle Ages was a genre sometimes known as *antidotaria* or *receptaria*, which were handbooks for making medicines. These texts spoke to a need for cures that did not rely on the learned doctors who studied medicine at the universities.

The formal training of a medieval physician was heavily theoretical, based on a doctrine that dated back to Hippocrates, the most revered Greek authority in the fifth and fourth centuries B.C. The Hippocratic system, supported by the influential Greek doctor Galen in the second century A.D., stipulated that health is governed by four bodily fluids called humors: blood, phlegm, black bile, and yellow bile. Good health demands that these be kept in proper balance, while illness arises when their proportions are awry. One of the most common treatments for all manner of ills was blood-letting: making an incision to drain some blood from the patient to restore humoral balance. The method was potentially worse than useless, enervating the patient and creating the potential for infection.

Procuring the services of a qualified doctor could be expensive, and beyond the means of most people. Doctors might prescribe medicines that could be bought from the druggist or apothecary. Some medicines were minerals, others were extracts of plants or animals. Many were of little medical value; some were toxic. Among the frightful potions a patient might be expected to swallow was a popular cure-all, dating back to antiquity, called theriac: a concoction of many ingredients that allegedly included viper flesh, mixed with honey. From the sickly syrups of theriac comes the word *treacle*.

It was no wonder, then, that there was a big market for books describing how to make your own medicines. While the techniques of alchemy had long supplied oils and other distilled "essences," it was not until the Renaissance that medicine began to rival chrysopoeia as the focus of Western alchemy. That transformation was due in large part to the Swiss alchemist and physician Paracelsus (see page 102). "Don't make gold, make medicines," he wrote.

The demand was never more urgent. The bubonic plague (Black Death) that arrived in Europe in the mid-fourteenth century left populations in constant dread: its symptoms were truly horrible and often ended in death. A further scourge at the end of the fifteenth century was syphilis, probably borne by troops returning from the New World, which left flesh rotting to the bone. The humoral theory of Hippocrates and Galen seemed of no help in the face of such maladies. "The patient asks for cure and not for theory," Paracelsus wrote.

He extended the Jābirian sulfur-mercury theory to include a third "principle": salt, a generalized category that encompassed a variety of chemical compounds, as it does in chemistry today. This system of *tria prima* embraced all substances, including plants and animals, rather than metals alone.

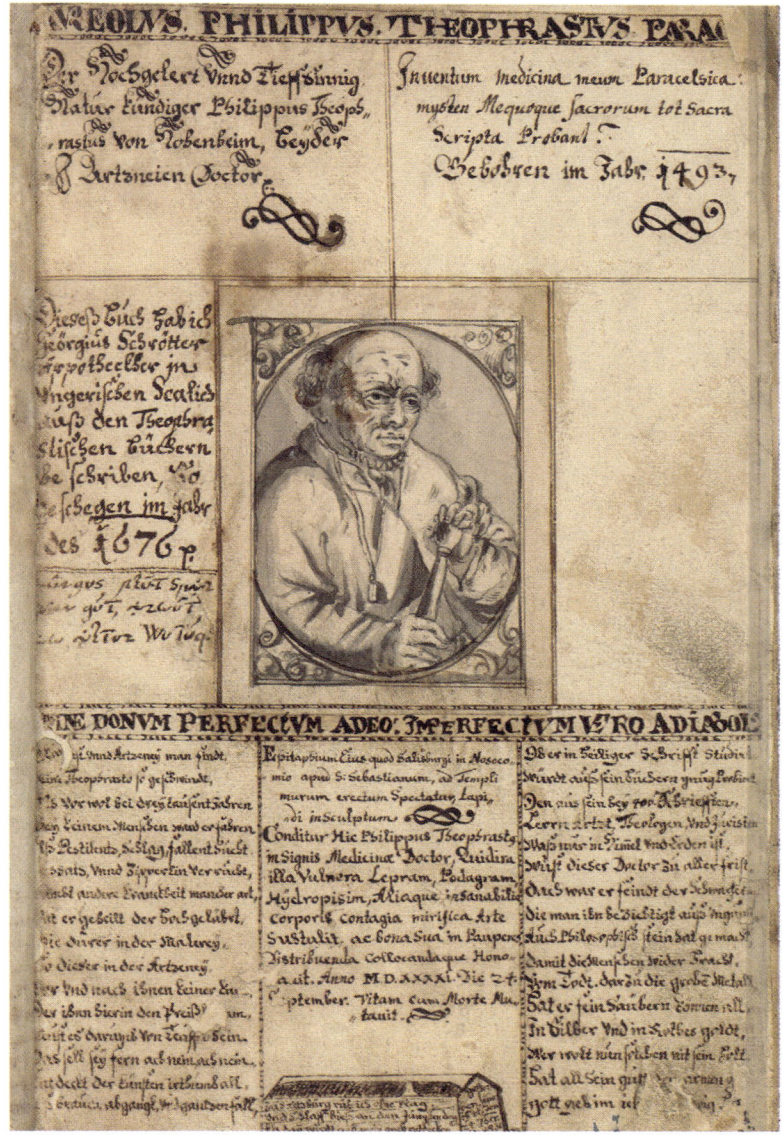

RIGHT Annotated page with a drawing of Philippus Aureolus Theophrastus Bombastus von Hohenheim, known as Paracelsus, from a collection of his writings compiled by the pharmacist Georgius Schrotter in 1676. Few of Paracelsus's writings were published during his lifetime, but his reputation rose posthumously owing to the publication of various compendia of his works.

After Paracelsus died, his works were collected and published by several dedicated editors in the late sixteenth century, creating a boom in Paracelsian (al)chemical medicine; this became known as iatrochemistry, from the Greek *iatros*, meaning "physician." Supporters of the "old" (humoral) and "new" medicine battled it out for decades. Despite considering Paracelsus to be a liar and in league with the devil, the Saxon physician Andreas Libavius was a staunch supporter of iatrochemistry and wrote an influential textbook on pharmacy and metallurgy called *Alchymia* (1597);

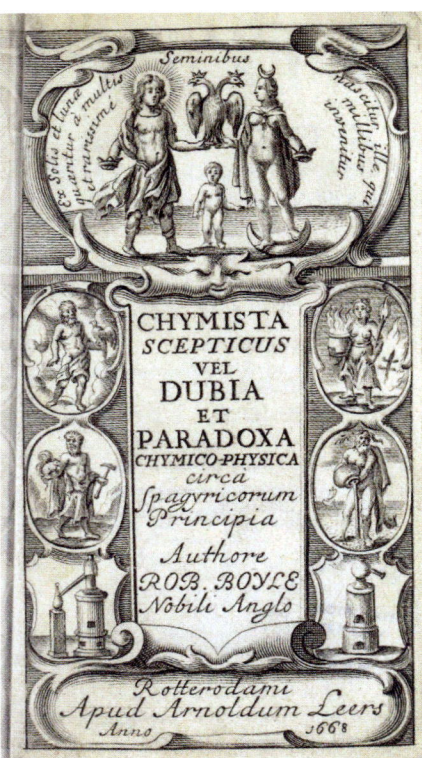

ABOVE The title page of *The Sceptical Chymist* (1661) by the Anglo-Irish natural philosopher Robert Boyle features alchemical motifs such as the double-headed eagle, which symbolizes transformation, and various pieces of apparatus.

OPPOSITE The title page of *Alchymia* by the Saxon physician Andreas Libavius (Frankfurt, 1606) proclaims its historical antecedents with portraits of Aristotle, Galen, and others. Inside, woodcuts provide practical illustrations of alchemical techniques.

the title demonstrates that alchemy could, even at that stage, be seen not as incorrigible mysticism but as a practical, down-to-earth chemical art.

Secrets and commerce

The invention of the printing press in the mid-fifteenth century meant that books could be produced and acquired relatively cheaply, and it became common for them to be written in the vernacular rather than Latin so that they could be read by the general public. An Italian book called *Secreti*, first published in 1555, was a huge hit, still in print more than a hundred editions and a century and a half later. The author was advertised as one Alessio Piedmontese, who deployed the old sales pitch that the book contained knowledge once deemed too powerful for widespread disclosure. In fact, the contents were mostly mundane and hackneyed recipes for medicine, perfumes, soaps, and cosmetics, as well as cooking recipes and old accounts of metallurgy. *Secreti* really belongs to the popular genre of arts manuals known as *Kunstbüchlein* (manuals of the arts). Some historians suspect that the book's real author was a courtier and poet named Girolamo Ruscelli, who claimed to have founded an Academy of Secrets in Naples. Ruscelli's academy might have been little more than a figment of his imagination, but such exclusive, proto-scientific "secret societies" flourished in the late sixteenth and seventeenth centuries. One of the most famous of these was the Accademia dei Lincei, of which Galileo became a member in 1611.

Alchemy was only one strand of this ongoing tradition of secrets, but it retained an aura of knowledge too dangerous for the rabble. The Anglo-Irish natural philosopher Robert Boyle, in his 1661 book *The Sceptical Chymist*, criticized the obscure language of alchemists, but nonetheless he considered transmutation of metals to be possible and shared the view that such alchemical secrets should be hidden from the common people.

Boyle's own alchemical knowledge seems to owe a debt to the American George Starkey (see page 212), the son of a Scottish minister who was born in Bermuda. When he moved to London in 1650, Starkey adopted the secretive practices of alchemy, publishing his tracts under the grand pseudonym Eirenaeus Philalethes. He considered his knowledge to be divinely sanctioned, and used it to sustain a business of selling oils and perfumes. In Starkey we can see how the old tradition of gnosis and secrets of nature blends with the early modern entrepreneur's drive to conceal trade secrets.

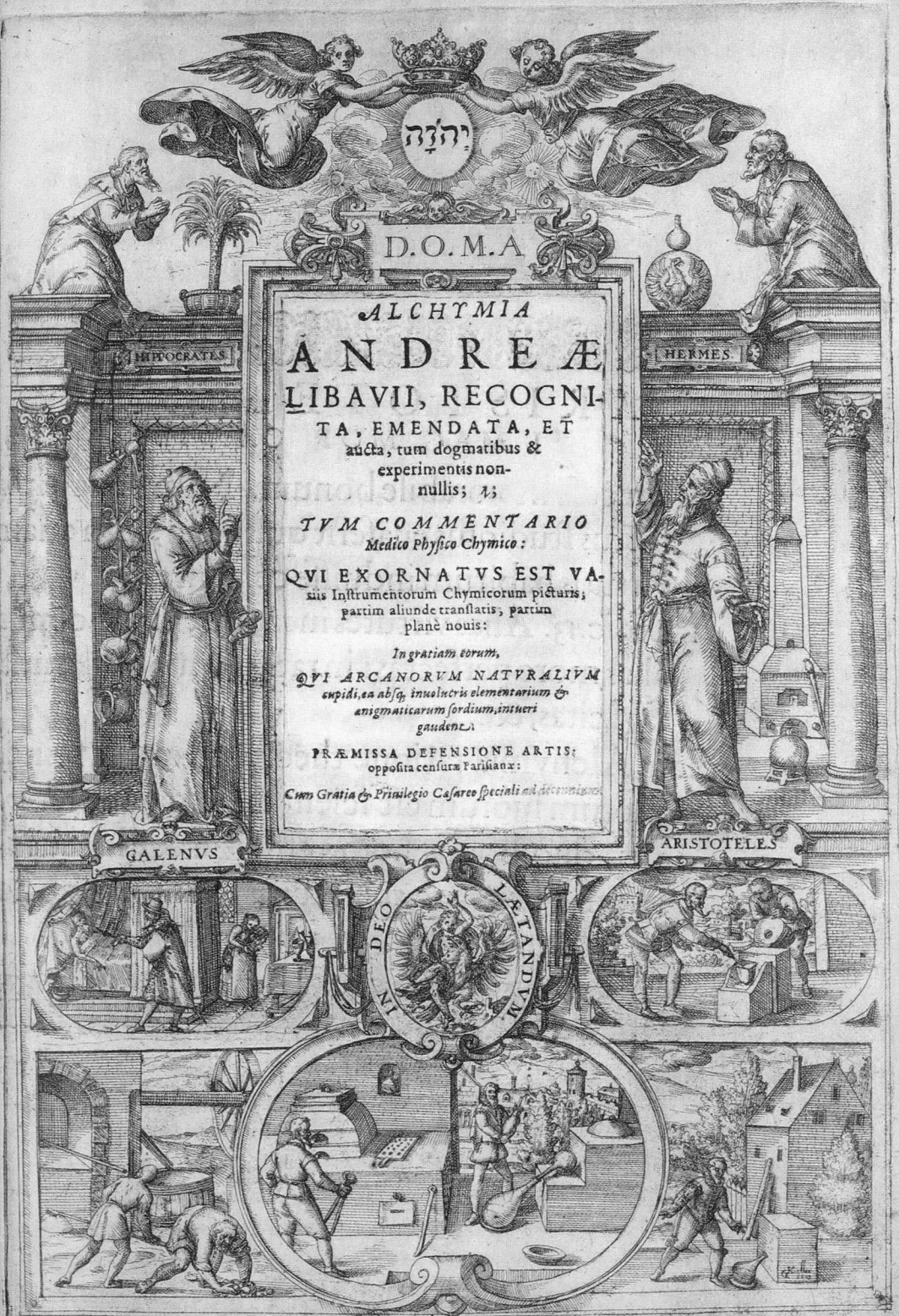

FRANCOFVRTI,

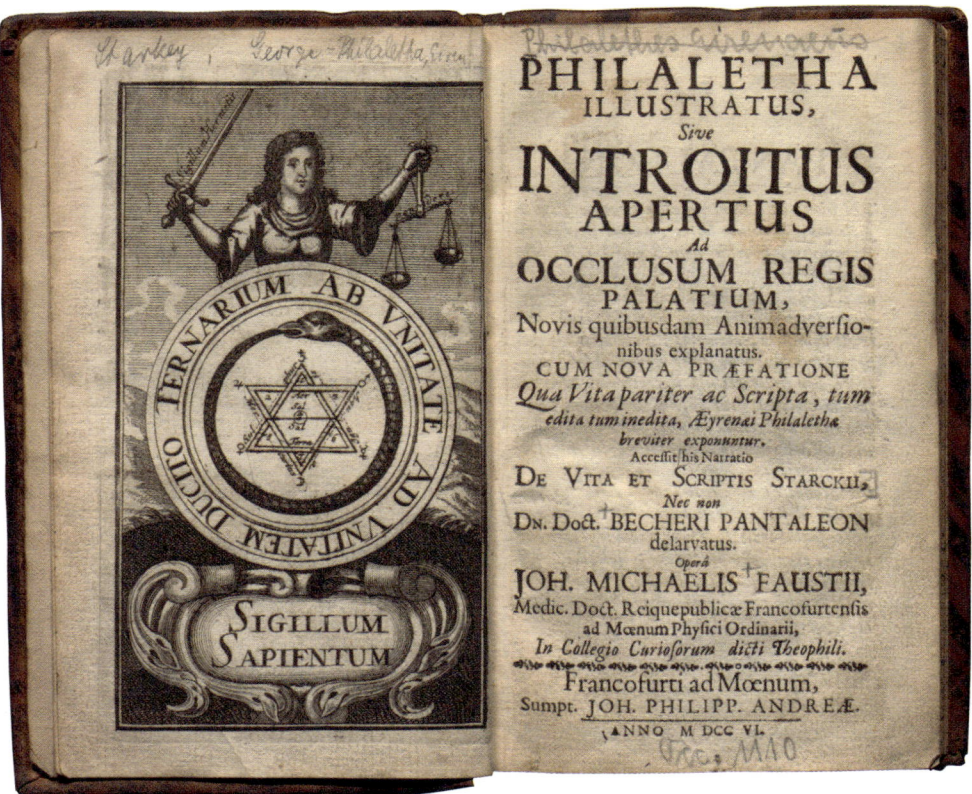

ABOVE George Starkey's *Philaletha illustratus*, published under the pseudonym Eirenaeus Philalethes (Frankfurt, 1706). Starkey was considered to be a master of the alchemical art when he came to England in 1650, where he became associated with the circle of progressive scholars around the Prussian exile Samuel Hartlib.

White gold

One of the most valuable trade secrets of the early eighteenth century emerged from alchemical investigations. In 1702 King of Poland and Elector of Saxony Augustus II established an institution called the Royal Contubernium that he charged with the task of making the philosopher's stone. Over time this college adopted the more modest and achievable goal of producing substances that would contribute to the economic well-being of Saxony, among which was porcelain—"white gold," as some called it.

Porcelain was desired in Europe ever since imports from China began in the fourteenth century. No one in the West knew how to make a product as fine and hard as that found in the East—until the work of the German mathematician and inventor Ehrenfried von Tschirnhaus. In the 1690s Tschirnhaus experimented with high-temperature firing of ceramics containing a high proportion of kaolinite clay, a key ingredient of porcelain, using lenses and new types of kilns. He found that adding alabaster, limestone, or chalk to the mixture could yield a hard, white, slightly translucent material much like Chinese porcelain. But to progress he needed more funds and a proper laboratory; these became available when Augustus II put Tschirnhaus in charge of the Royal Contubernium.

In 1704 Tschirnhaus acquired an assistant, an alchemist named Johann Friedrich Böttger of Berlin. Hearing of Böttger's claim to have transmuted silver into gold, Augustus II had the man kidnapped and brought to Saxony, where he was ordered to repeat the transmutation. Böttger had no success and kept trying to flee from his captors until eventually he was handed over to Tschirnhaus for "protective custody." Tschirnhaus argued that Böttger's services might be better employed in assisting with the synthesis of porcelain.

The recipe might finally have been discovered by accident, when Böttger found during an alchemical experiment that his crucible itself—a "Hessian" crucible made from a clay with a high kaolinite content (see page 119)—had been transformed by the intense heat into a white ceramic. Which of the two men truly deserves the credit for cracking the secret of porcelain remains disputed. Tschirnhaus died suddenly and mysteriously in 1708 after allegedly succeeding in making a true porcelain cup (subsequently, the story goes, stolen by a laboratory servant) using a recipe he would not share with Böttger; but in 1709 Böttger wrote to the king saying it was he alone who had deduced the formula. The case has all the elements of a murder mystery, infused with the secretive spirit of alchemical tradition.

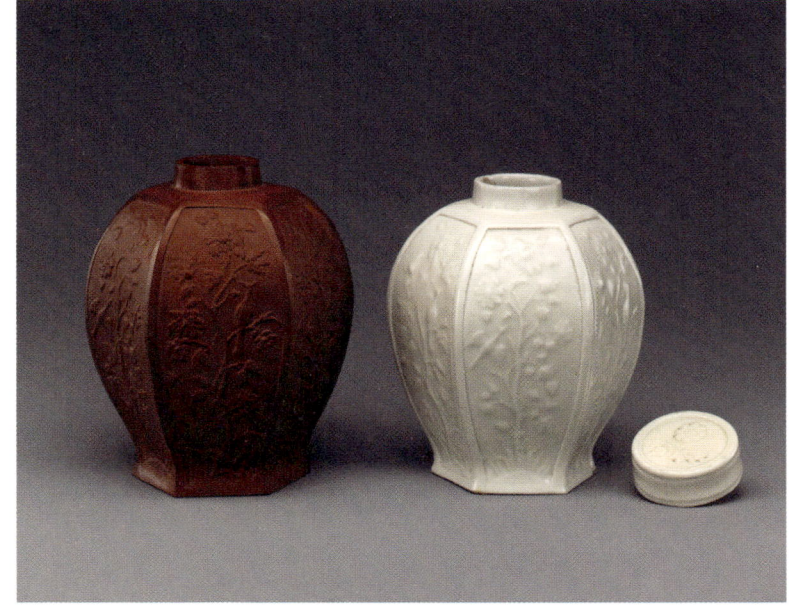

RIGHT Tea caddies of the earliest Meissen red stoneware and the new white "Böttger porcelain," developed by Johann Friedrich Böttger and Ehrenfried Walther von Tschirnhaus, c. 1710–13. Porcelain, a commodity of great commercial value, had previously been imported from China, as the method of making it was unknown in the West. Its discovery in Meissen was the result of alchemical experiments in an institution created by the ruler of Saxony, Augustus II. The celebrated porcelain is still produced today.

BOOKS OF SECRETS: THE USES OF ALCHEMY

PROFILE

Paracelsus

(1493–1541)

A key question for the modern historian of alchemy is: What can alchemy tell us about the evolution of ideas, and of efforts to make sense of the world? The Swiss physician and alchemist Paracelsus supplies one of the best reminders that we should not expect the answers to be simple, or linear, or to divide neatly into "right" and "wrong"—or indeed to be easily separable from their historical context.

During his wild and tempestuous life—and in the decades after it ended—Paracelsus accrued stories worthy of a legendary wizard. He was said to ride a magical horse that bore him all over Europe, to consort with demons, and to carry an elixir of life in the pommel of his broadsword. Legend has it that he sought esoteric lore in Alexandria and among Russian shamans, and that he made an artificial man called an homunculus. Yet some now credit him as the father of toxicology and even of the modern notion of chemotherapy: preparing chemical cures for specific diseases. He personifies all the strife and turmoil of his times, when terrible wars, popular uprisings, and religious rancor wracked Europe while explorers journeyed across the seas to new lands. Even now, it is hard for historians to know quite what to do with Paracelsus. Once dismissed as a braggart, windbag, and charlatan, he is today awarded a significant place in the history of science. His voluminous writings present us with a mixture of poetry, fantasy, and invective, marked with occasional startling and prescient insights.

Paracelsus was not his real name but rather his self-styled Latinate appellation in the Renaissance manner. Some claim it means "beyond Celsus," surpassing the famous Roman philosopher of that name, though there is debate about this. He was christened Philippus Aureolus Theophrastus Bombastus von Hohenheim, born into a family of Swabian nobles fallen on hard times. His father Wilhelm, the town physician in the Swiss village of Einsiedeln, named his son after the ancient Greek scholar Theophrastus, a disciple of Aristotle and author of an influential treatise on minerals.

When political conflict forced Wilhelm to move to the mining town of Villach in Austria in 1502, Paracelsus learnt the alchemical art of metal-making as well as picking up a rough-and-ready knowledge of Latin and the basic scholastic syllabus in the local monastery schools. Around 1507 he set out for the universities of Germany

OPPOSITE The Swiss philosopher, alchemist, and physician known as Paracelsus, engraved by Theodor de Bry, c. 1598. Paracelsus was one of the most influential figures in Renaissance alchemy and medicine, who advised that alchemy might be used to prepare new drugs targeted at specific ailments.

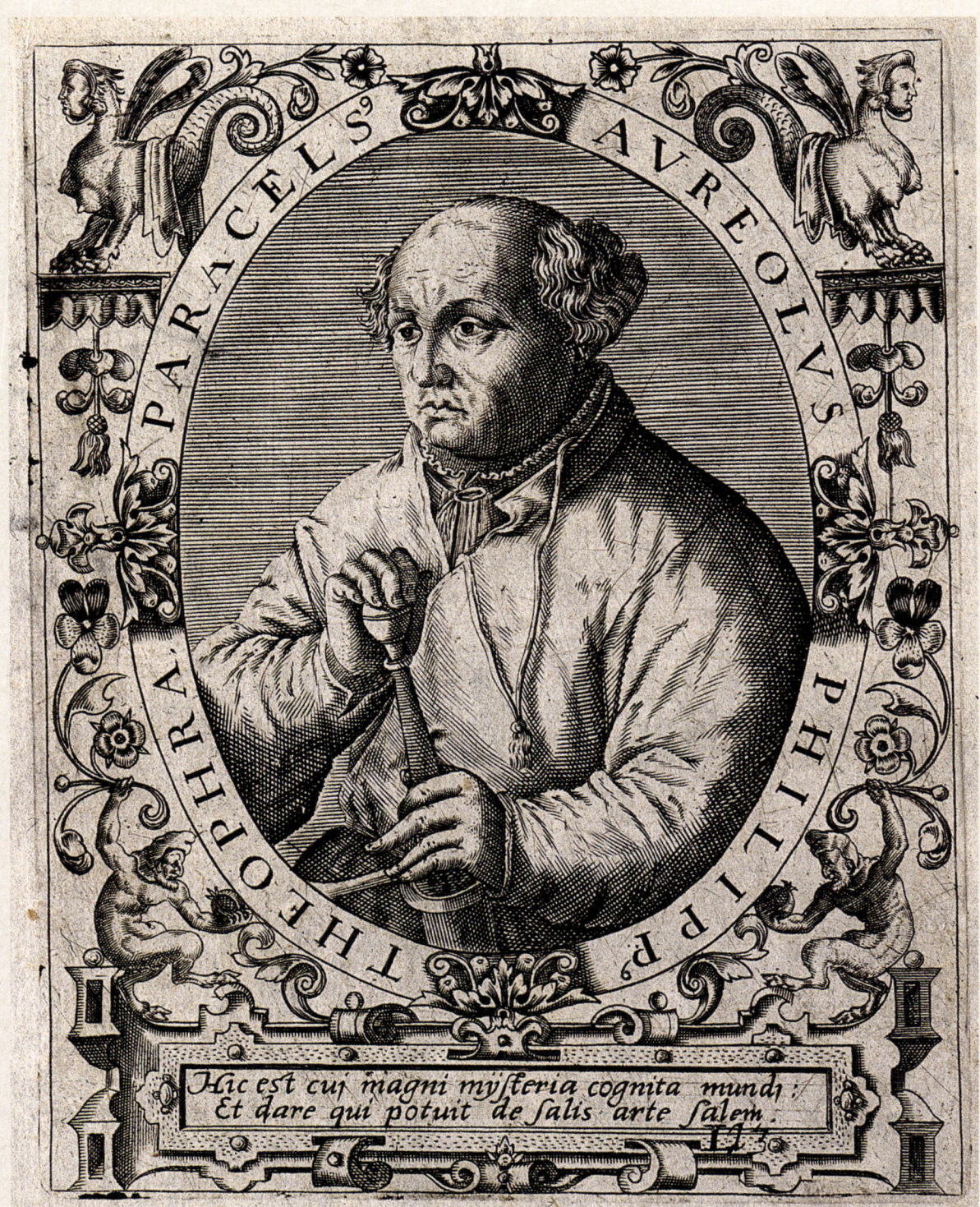

PROFILE

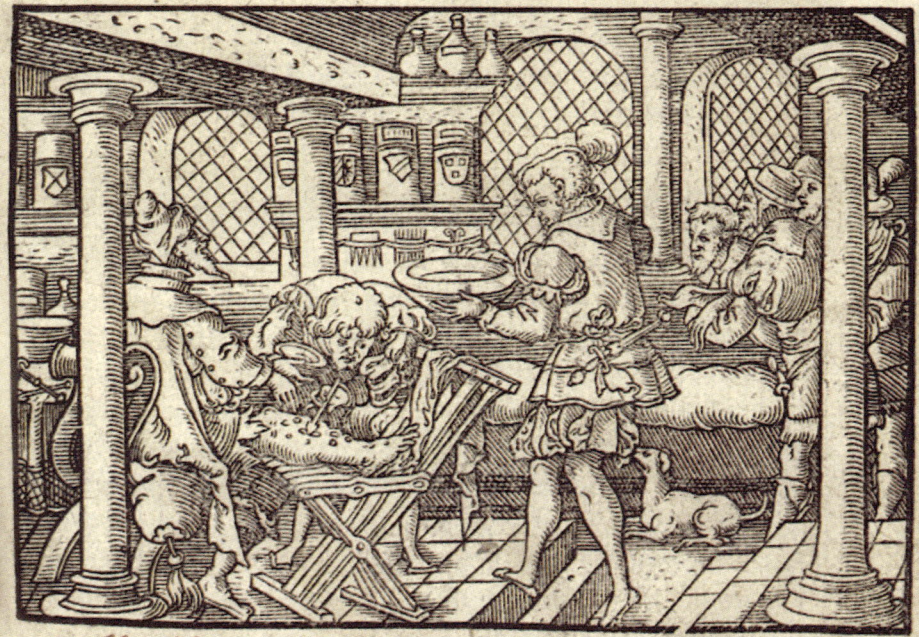

ABOVE Advice on treating sores, such as those caused by syphilis, from *Der grossen Wundartzney* (The great surgery), a work by the Swiss alchemist and physician Paracelsus (depicted in the center of the page), published in Frankfurt in 1562.

> True medicine was to be found in the world at large—among "barbers, bathkeepers, learned physicians, women, and magicians who pursue the art of healing."

to study medicine, only to find the rigid training, rooted in the old theories of Hippocrates and Galen, not at all to his liking.

Health was then still generally considered to be governed by the balance of four bodily "fluids" called humors. The task of the physician was to restore this balance, often by the practice of blood-letting. Those who could afford a doctor (the fees were high) were often prescribed medicines that could be bought at apothecaries. Paracelsus rejected the idea that healing should be learned from ancient books, and he despised the lofty airs and fine clothes of well-paid physicians, whom he derided as "high asses." He considered that true medicine was to be found in the world at large—among "barbers, bathkeepers, learned physicians, women, and magicians who pursue the art of healing." He went, he said, "to alchemists, to monasteries, to noble and common folk, to the experts and the simple."

Having completed his medical training at the University of Ferrara (though it remains unclear whether he ever formally graduated as a doctor), he traveled all over Europe and beyond. He seems to have served as an army surgeon in Italy, then made his way through France and Spain, allegedly to Britain, the northern German states, and Scandinavia, and perhaps to Russia, Constantinople, Cairo, and Greece, although the evidence for some of these journeys is flimsy at best. His fortunes rose and fell, not least because his irascible nature and his outspoken contempt for medical orthodoxy landed him repeatedly in trouble. In 1527 he was appointed town physician in Basel, only to so antagonize the university faculty that he had to flee for his life under cover of darkness.

Throughout his wanderings he wrote ceaselessly, explaining his ideas about medicine, alchemy, and the secrets of nature. His manuscripts were a strange mixture of antiestablishment polemic, practical recipes for medicines, and speculative theories on the cosmos, astrology, supernatural beings, and mental diseases. He struggled to get much of his work printed during his lifetime, but after his death (from uncertain causes) in Salzburg in 1541, his reputation soared. In the 1560s and 1570s, several publishers enthusiastic about his ideas collected and reproduced his writings, stimulating a wide interest in "Paracelsian" medicine. For later scholars, the task became that of sifting what seemed useful and effective in these colorful tracts from what was deemed superstitious, bombastic, and obscure.

phlozn / quadmodu cm epohsit sal alkali / a ca
lice vina sul tine clauellato / vel sal tartari
ab ipso tartaro calcinato / per humoes intue
nientens donec nichil remanat acuitatis. Ω Sal
mu̅m / qd est tinctura / exphitur a calce metallo
ru̅ et putrefactoem iuxta totu̅ qpositum expuet
suam nam / et induct alienam / de tali sale d̅t
Senioz / primo sit vinu̅ / postea sal / et de illo
sale / p diuisos opacoes pitur phloru̅ / Sa
lia no non sp̅a / ponut p spirib; / sal p acie
acuitatis a corosiuis / corosiuis sicut e salarmo
qd est nobilius / ecp totum e spirale / post h̅ sal
gu̅me / et sal vrine / Nam Galazno de scdm
electus in h̅ ope nouu̅ / testante d̅r̅o in li. do
preptoru̅ vbi d̅r̅ cp alimzadir tibi solu̅ desca-
uiat piam ipsi corpa soluit et liquefac̅ et
spialia reddit. Alium phorum posserint
ee totum aerexum in eo / et vocauint ipm di-
uisis nomib; / ipsi t̅n̅e sic excelleys but in lib
Sur phoz scribientur / et quidam vocauit ipm
a loco / ut termini spu̅tu̅ lune / quidam a co-
lore / ut saguis / vrina thami / epinosa an̅a

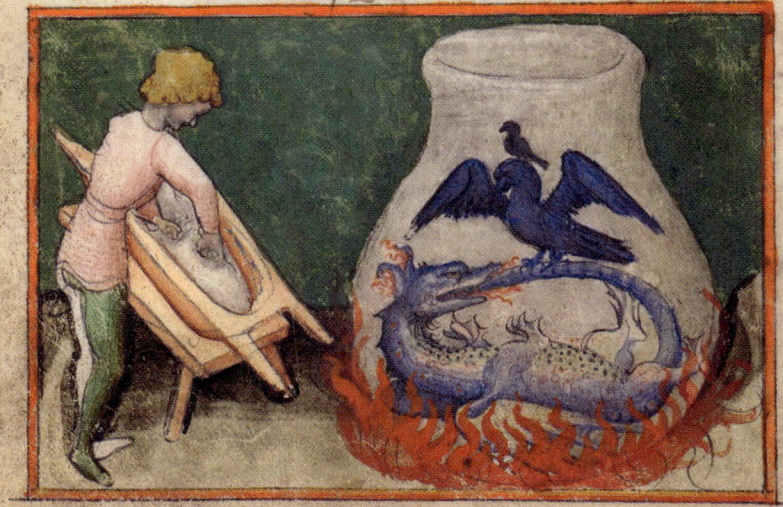

28

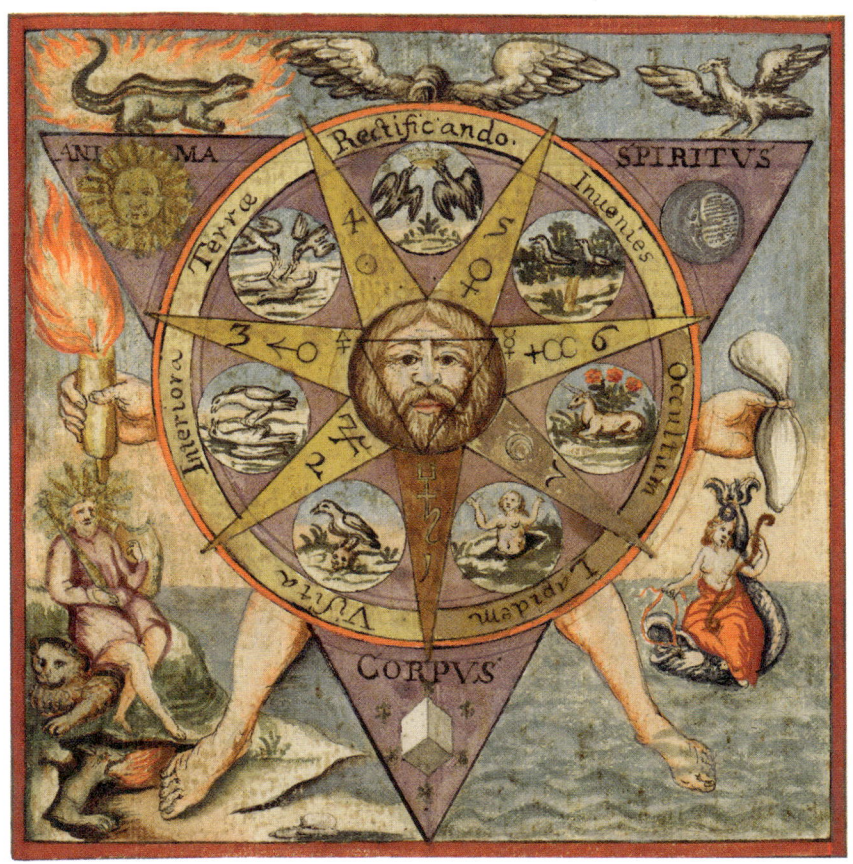

OPPOSITE The 15th-century alchemical text *Aurora consurgens* (sometimes translated as "Rising dawn") contained particularly fine illuminations. Although the work was attributed to the influential 13th-century Dominican theologian Thomas Aquinas, it is unlikely that he was its true author—there is no good evidence of his interest in alchemy.

ABOVE "Vitriol," from Georg Hellmerich's *Musaeum hermeticum* (Hermetic museum; 1692), a colored reproduction of a woodcut in *Azoth, siue aureliae occultae philosophorum* (Azoth of the philosophers; Frankfurt, 1613) by the pseudonymous German alchemist Basil Valentine. *Azoth* was one of the hazily defined "power words" of alchemy, referring to a substance lodged somewhere between a spiritual concept and a tangible chemical agent of transformation. The word is derived from the Arabic *al-za'buq*, which can refer to mercury, and it features in the works of Jābir. An Azoth could be an agent of transmutation, a universal solvent (mercury itself was known to be capable of dissolving gold to form an amalgam), or a kind of "cure-all" elixir. By starting with the letters *A* and *Z*, Azoth symbolized a sense of totality—from alpha to omega—alluding to the completion of the Great Work of alchemy, the creation of gold.

BOOKS OF SECRETS: THE USES OF ALCHEMY

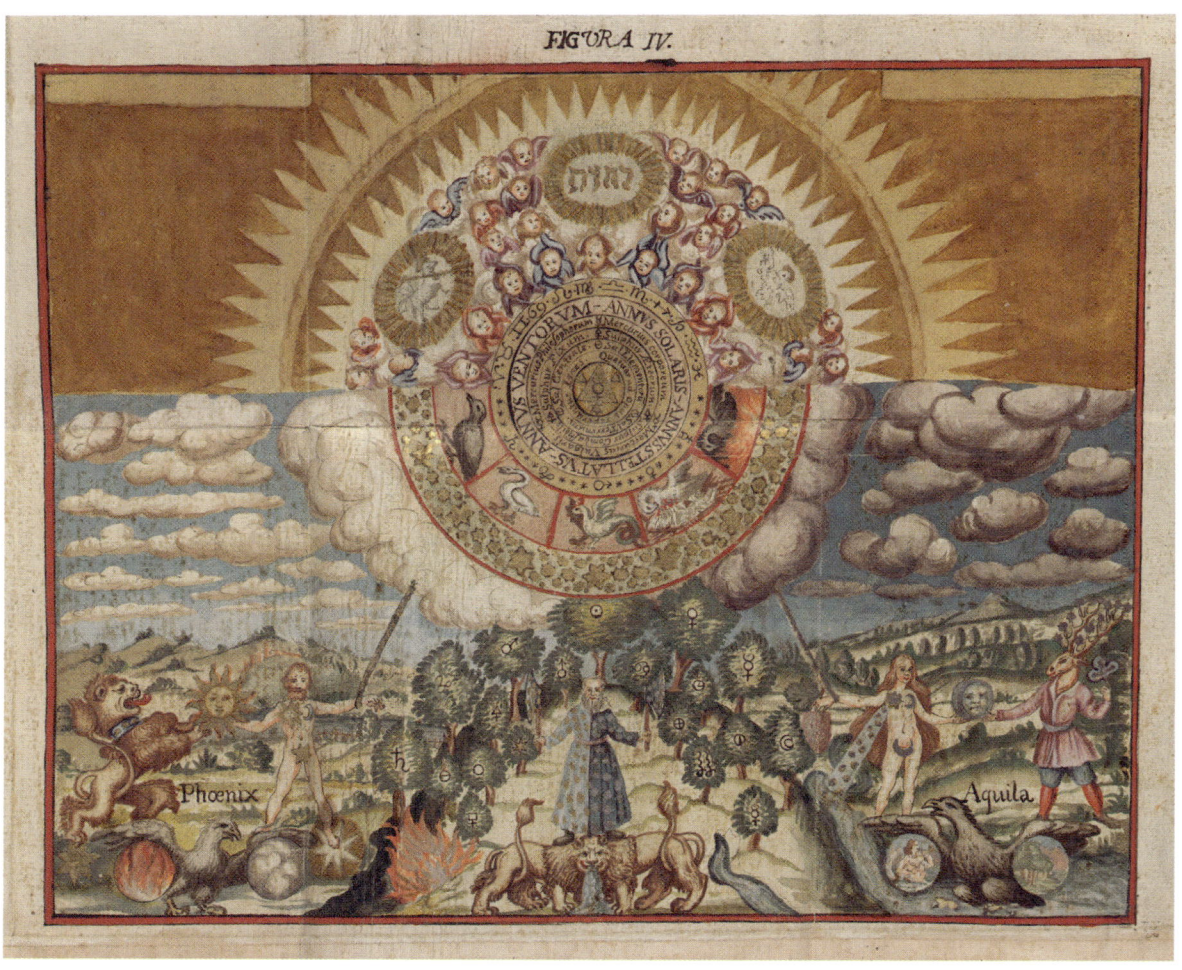

ABOVE "Theosophy and Alchemy," from Georg Hellmerich's *Musaeum hermeticum* (Hermetic museum; 1692). The book is a rather haphazard collection of alchemical texts, including works by Basil Valentine, George Ripley, and Michael Maier, as well as a transcript of the *Emerald Tablet*.

OPPOSITE Alchemists revealing secrets from the *Book of Seven Seals*, in a detail from the Ripley Scroll, c. 1600. Few copies of the Ripley Scroll, depicting the works of the 15th-century alchemist and canon George Ripley, still exist; they are believed to have belonged to aristocrats and nobles. The scrolls contain many standard alchemical symbols, but their meaning is sometimes obscure. The psychiatrist Carl Jung was one of those who speculated about their meaning.

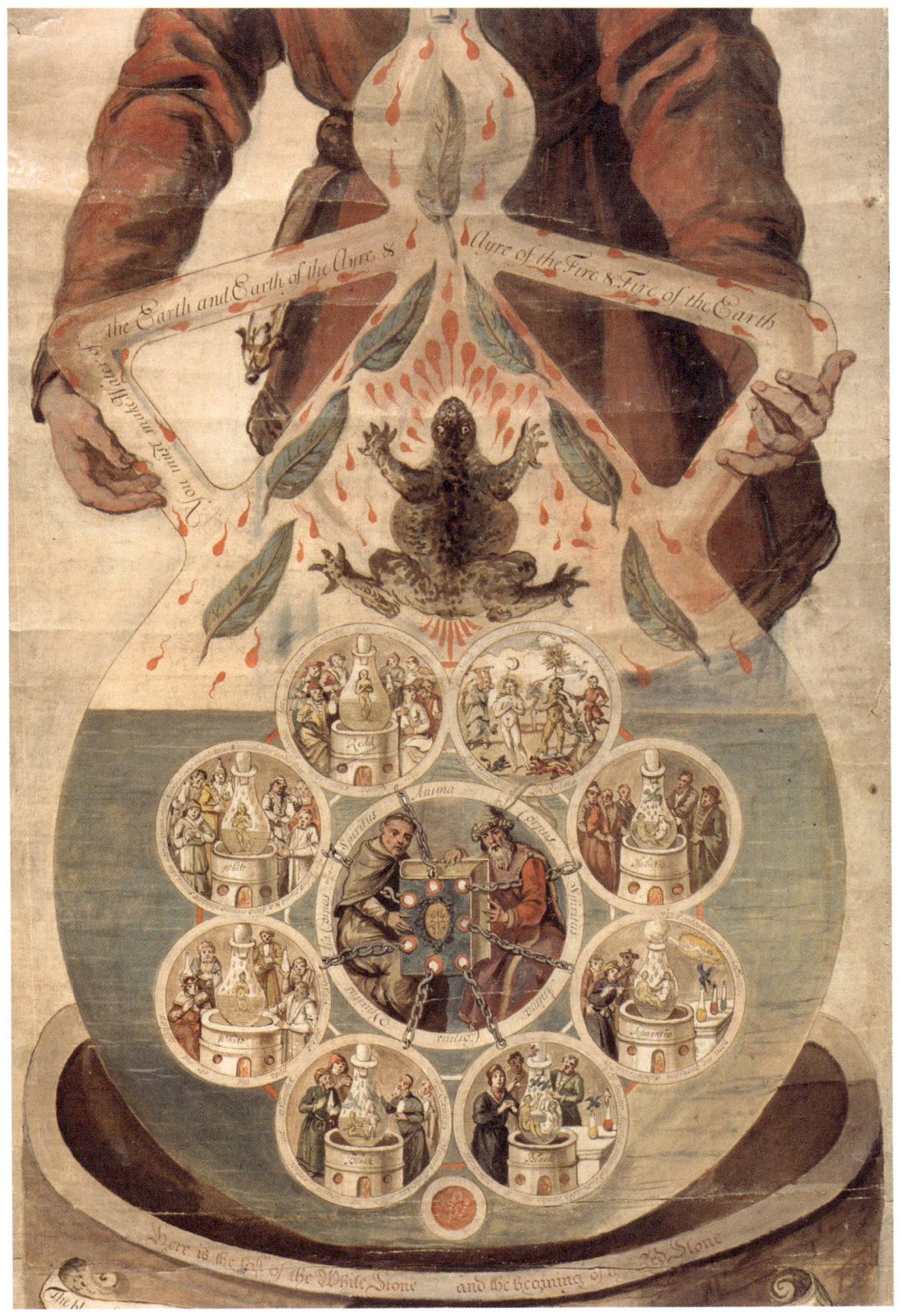

BOOKS OF SECRETS: THE USES OF ALCHEMY

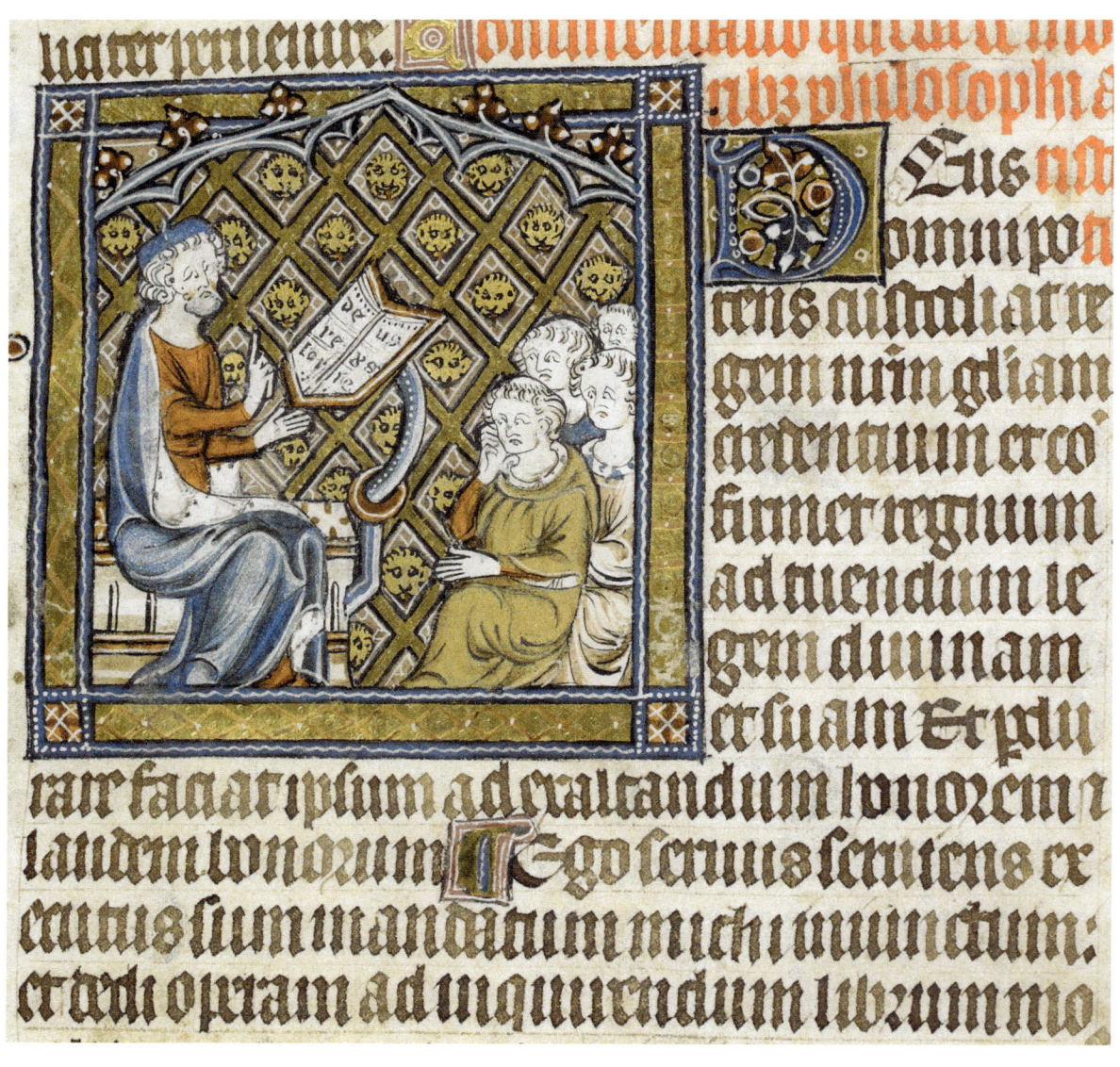

ABOVE Aristotle teaching students, who perhaps include Alexander the Great, in an illustration from Pseudo-Aristotle's *Secretum secretorum*. The book, translated by Philip of Tripoli from an Arabic manuscript probably of the 10th century, purports to be in the form of a letter from Aristotle to Alexander. As well as material on alchemy, it discusses magic, astrology, and other esoteric topics. Philip of Tripoli, an Italian monk, made the first complete translation into Latin in the early 13th century.

OPPOSITE The mythical basilisk—a serpent-type creature hatched from the egg of a snake or toad by a cockerel. Mixed with copper, vinegar, and human blood, its powder was listed as an ingredient in Theophilus's "Spanish Gold," an unusual digression into the fanciful for Theophilus, whose handbook *De diversis artibus* (On diverse arts) generally has a practical flavor far removed from the cryptic language commonly associated with alchemical texts. This colored engraving is from Friedrich Justin Bertuch's *Bilderbuch für Kinder* (Picture book for children; Weimar, 1806).

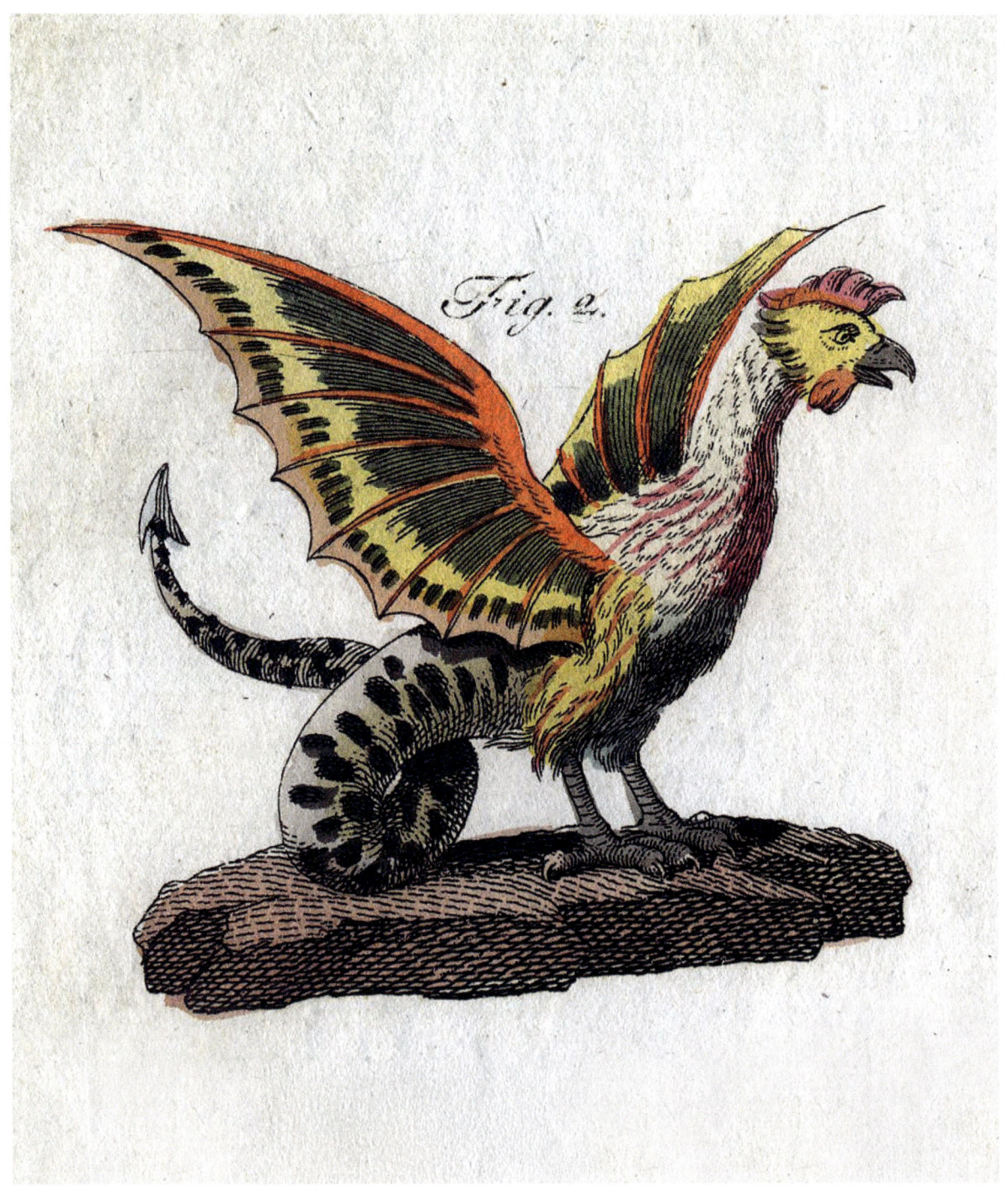

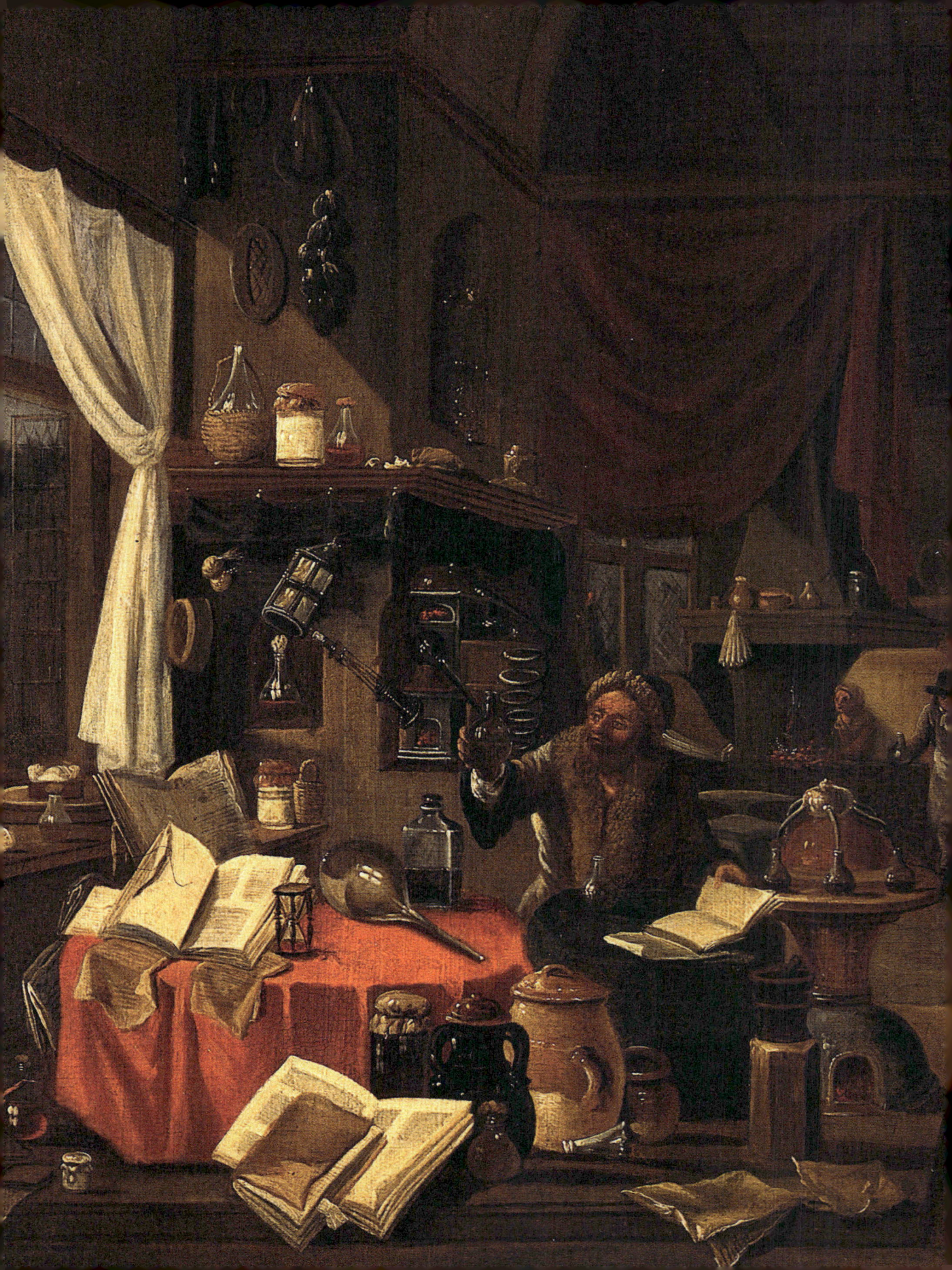

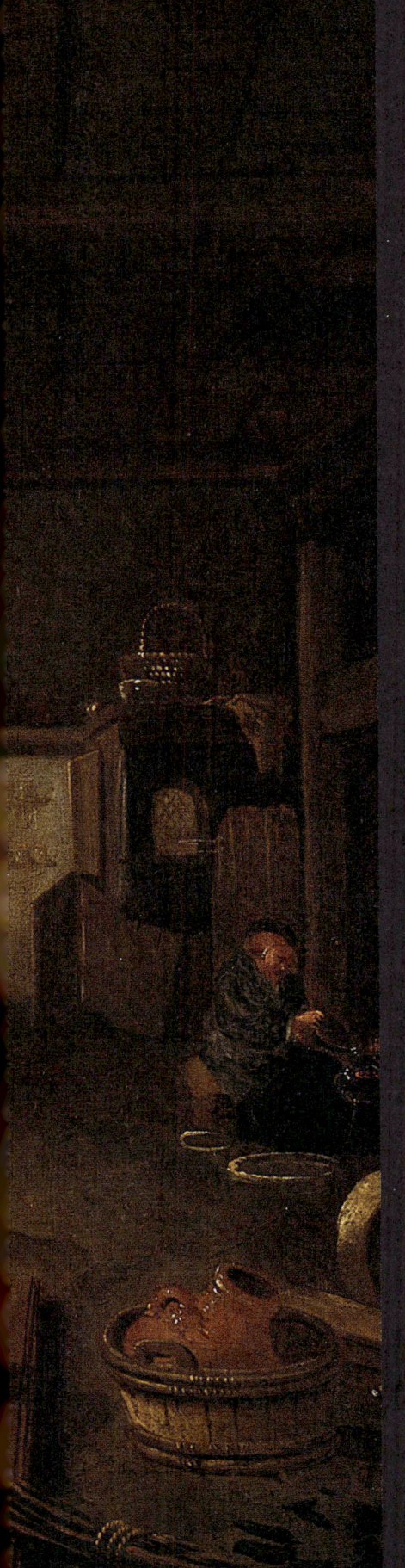

CHAPTER FIVE

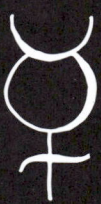

Puffers
THE ALCHEMICAL LABORATORY

LEFT An alchemist's laboratory, with all its tools and equipment, in a detail from a painting by an unknown artist, probably Dutch, c. 16th century. Alchemical laboratories were a popular topic for Dutch painters of this period, and although they cannot be considered accurate depictions of such workplaces (their intent is often satirical), they offer some indication of the kinds of apparatus that alchemists and their successors, the chymists, might use.

CHAPTER FIVE

Alchemy can be considered a form of early experimental science, demanding specialized apparatus that testified to the underlying philosophy: to understand nature, it is not enough simply to observe; one must also *transform*. In this respect, alchemy was truly the antecedent to modern chemistry: it is all about change, about making. Alchemy thereby helped to establish the notion of the laboratory as a special environment, stocked with tailor-made equipment, for exploring the world.

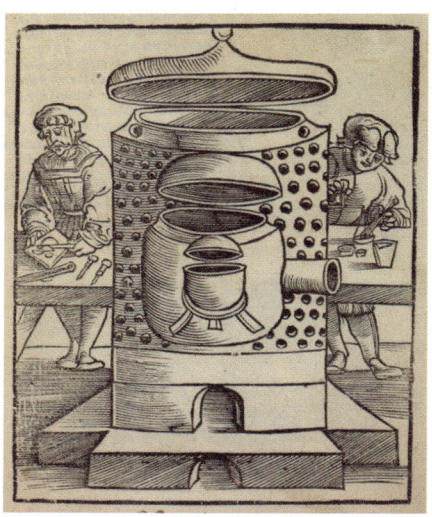

ABOVE An alchemical furnace (athanor or *piger henricus*), from the title page of Pseudo-Geber's *Geberi philosophi ac alchimistae maximi, de alchimia libri tres* (The three books on alchemy by Geber, the great philosopher and alchemist; Strasbourg, 1531). Thea furnace was the alchemist's most important piece of equipment, as heat was the key agent of transformation.

To judge from illustrations in early alchemical texts, some of the equipment of the Hermetic art remained unchanged for centuries. The key agent of transformation was heat, and so the centerpiece of the alchemical workspace was the furnace.

This was typically a tall brick-built structure called an athanor or *piger henricus* (lazy Henry), fueled by coal or charcoal; the word derives from the Arabic for furnace, *al-tannur*. Some treatises, such as that of the medieval Pseudo-Geber, give careful instructions, including dimensions, for building these furnaces.

The athanor could generate strong, steady heat. But alchemists needed to be able to vary and control it: from gentle warmth for distilling volatile oils and essences, to the intense temperatures needed to smelt ores and melt metals. "Nothing may let [hinder] more your desires, than ignorance of Heates of your Fiers", wrote the fifteenth-century alchemist Thomas Norton. One way to produce sustained low heat was to bury a vessel within rotting animal dung. A common alternative was the water bath or *bain-marie*, a term still preserved today in cooking. This technique was attributed by Zosimos of Panopolis to a mysterious alchemist known only as Maria Judea (the Jewess)—hence the name. Zosimos evidently holds Maria in high esteem but sadly supplies no further details about her.

For intense, steady heating, the alchemist might use an ash or sand bath, surrounding a vessel with hot ash or sand inside a furnace. To create still higher temperatures, the sand might be replaced by iron filings. But alchemists had no temperature scale,

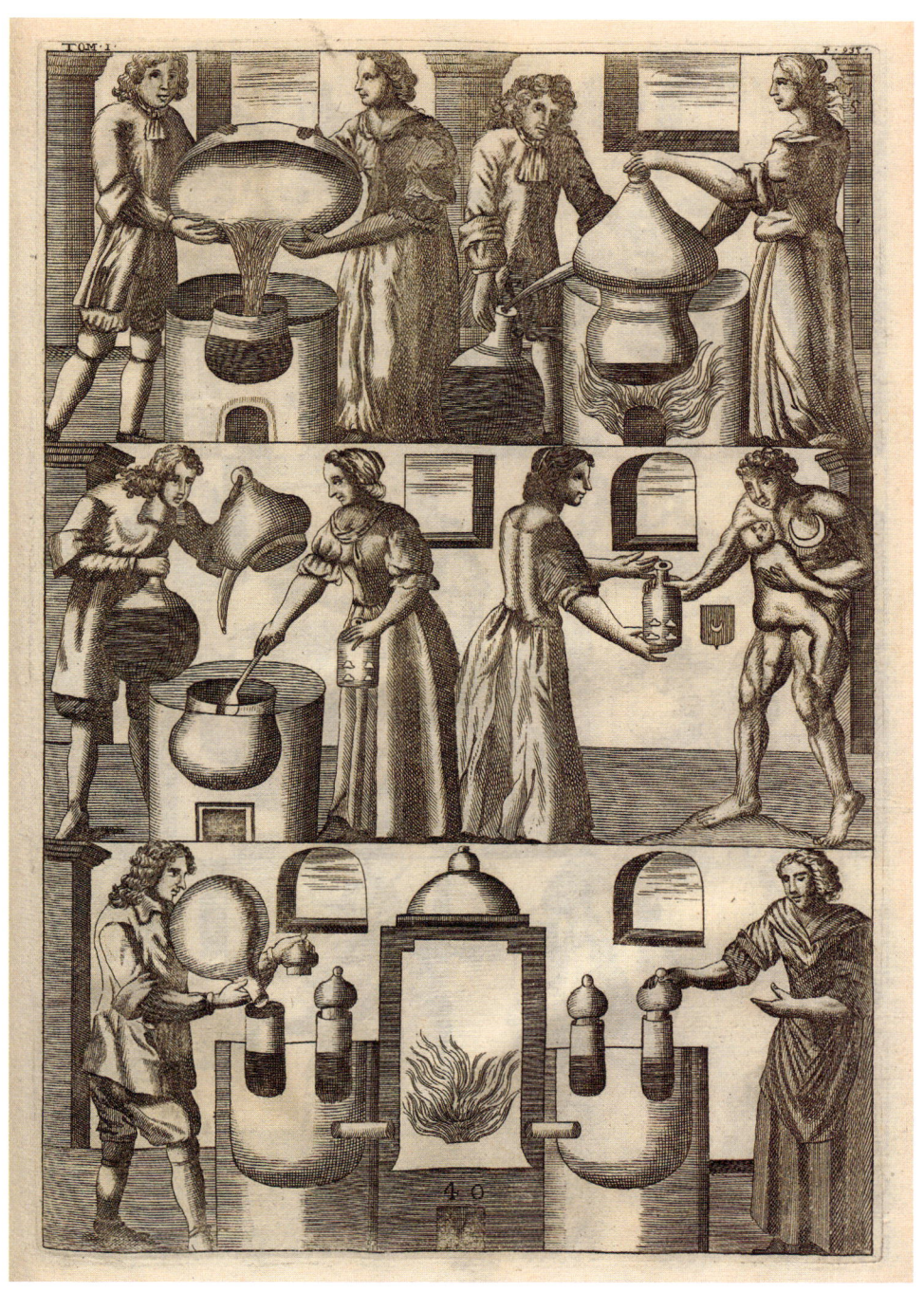

ABOVE A *bain-marie* in use. The invention of the water-bath is attributed to the alchemist known as Maria the Jewess, mentioned by Zosimos of Panopolis. The technique provided a means of heating a substance gently, and is still known by that name and used in cooking today.

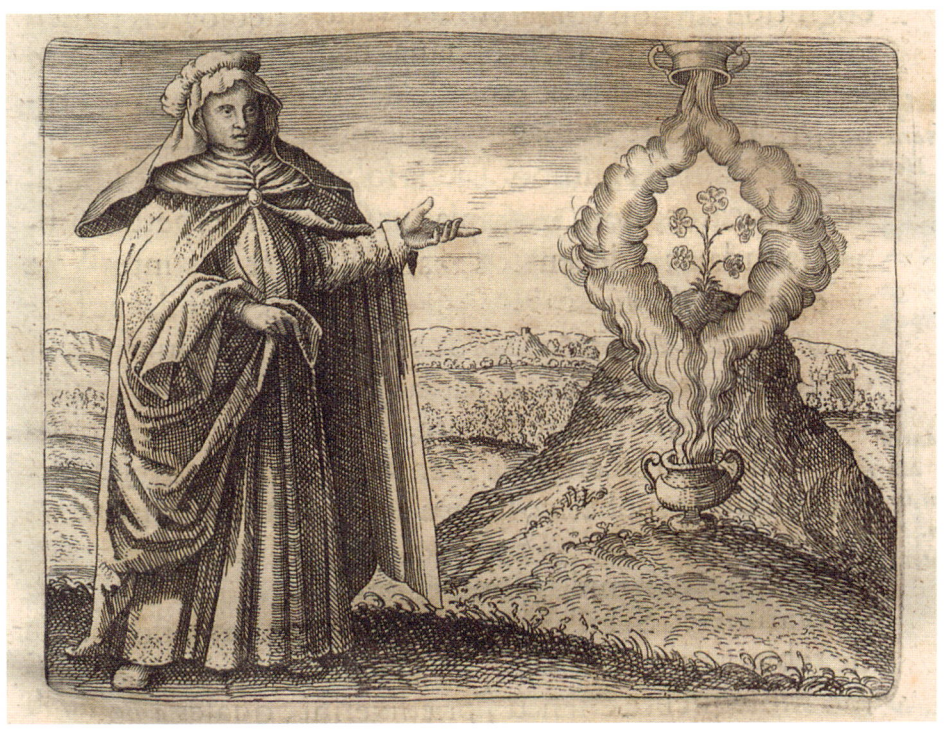

ABOVE A copper engraving of Maria Judea (Maria the Jewess), a legendary alchemist of antiquity about whom nothing is known beyond the mention of her name by Zosimos, from Michael Maier's *Symbola aureae mensae* (Symbols of the golden table; Frankfurt, 1617). The image shows twin trails of fumes from an earthly vessel, perhaps representing Philosophical Sulfur and Mercury, reuniting in an inverted "celestial vessel" in the sky. In the accompanying text, Maria asserts that "Art," meaning alchemy, can accomplish things that Nature cannot.

nor any way to quantify degrees of heat beyond the color changes that occurred as objects or materials were heated, becoming red- and white-hot. Nor did they have much fine control over heat until the seventeenth century, when the German chemist Johann Glauber introduced a furnace chimney that could be used to alter the air flow over the fuel.

The Arabic physician al-Rāzī drew up a systematic list of his equipment and materials, giving us some indication of what a well-stocked workplace would contain. He mentions beakers, flasks, phials, basins, lamps, various types of furnaces (including the athanor), water baths, alembics, cucurbits, mortars and pestles, funnels, and more. He also created a taxonomy of materials, dividing them into categories such as stones, metals, spirits, and salts. They include ores like pyrite (iron), malachite (copper), galena (lead), cinnabar (mercury), and stibnite (antimony), as well as minerals such as gypsum, alum, natron, and borax, and materials made by chemical transformation such as lime, potash, white and red lead, iron and copper oxides, and perhaps mineral acids. Needless to say, to stock and maintain such an array of materials and methods, you needed either to be rich or to have a wealthy patron.

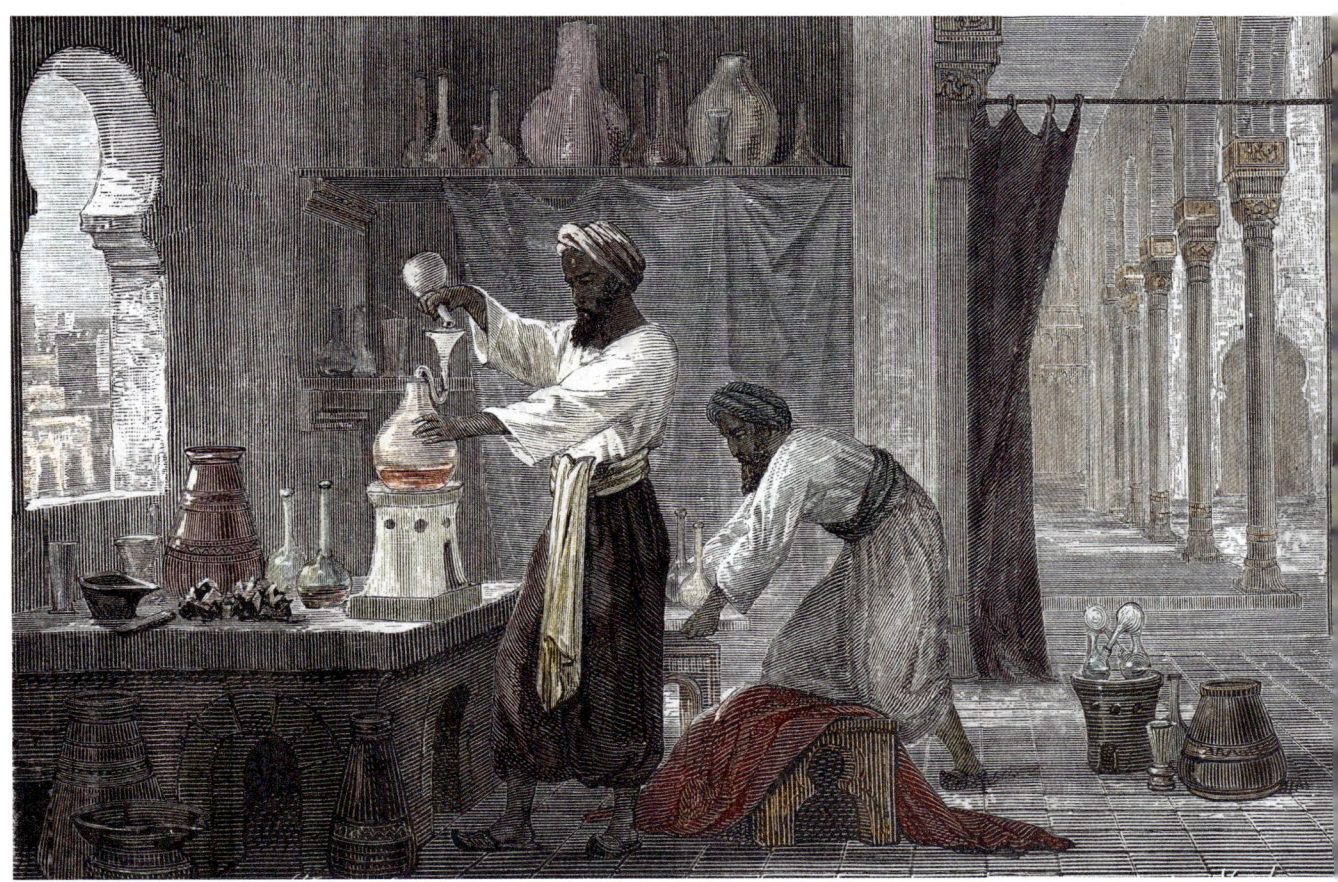

ABOVE A representation of the 10th-century alchemist and polymath al-Rāzī at work in his well-equipped Baghdad laboratory, from Louis Figuier's *Vies des savants illustres du moyen age* (Lives of illustrious scholars of the Middle Ages; Paris, 1867). Al-Rāzī was one of the most important alchemical and medical practitioners during the "golden age" of Islamic learning. He was revered in the West during the Middle Ages under the Latinized name Rhazes.

Devices

What are these oddly named instruments? The alembic has become the iconic piece of kit for the alchemist: it is basically a still for distillation. The liquid sat in a round-bottomed vessel with a narrow neck, shaped like a gourd and named after the Latin word for those plants: cucurbit. On top of the neck was a bulbous cap with a long, downward-sloping beak-like tube, called the ambix (the alembic itself gets its name from the Arabic form of that word, *al-anbīq*). The ambix is inserted into the neck of another flask called the phial or phiole, where the condensed liquid collects. Zosimos asserts that Maria the Jewess invented the alembic too; it was certainly in general use by his time, and the Arabic alchemists deployed it for distilling essences such as oil of roses. Alembics were generally made from glass: the technique of blowing molten glass through tubes to make such domes and flasks was seemingly devised in the Middle East around the first century B.C.

PUFFERS: THE ALCHEMY LABORATORY   117

Distillation was also carried out in a vessel called a pelican (so named because of its supposed resemblance to a pelican bending its neck to peck at its breast and nourish its young with its own blood), in which dual tubes from the neck fed back down into the main body, allowing any distillate that condensed in the neck to return to the liquid. At face value this seems to defeat the object of distillation—but the process can separate very volatile substances, which escape from the open neck, from less volatile ones, which become concentrated in the original liquid. It can also allow chemical reactions to proceed in the flask by continual input of heat without boiling the vessel dry. This process was called cohobation by alchemists; modern chemists know it as reflux distillation.

Another piece of equipment named from the Arabic is the aludel: a series of funnel-shaped, open-bottomed ceramic or glass pots that fitted one atop the other to make a column with a tapering neck.

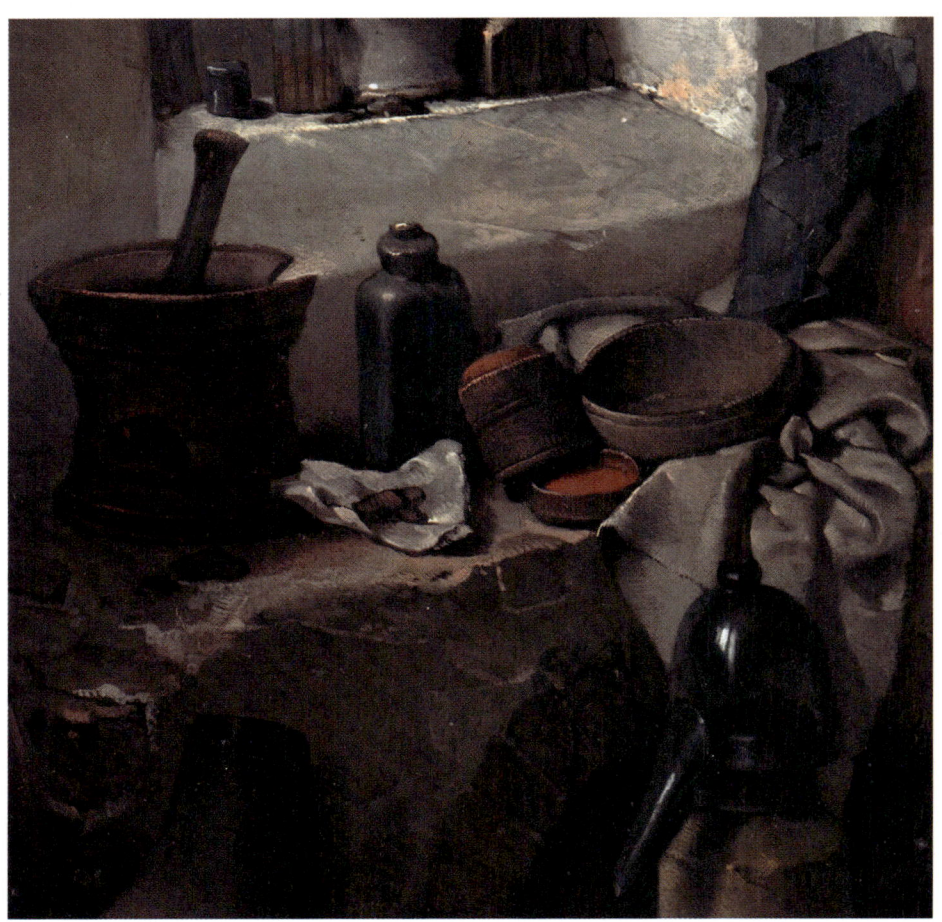

BELOW Detail from Cornelis Bega's painting *The Alchemist*, 1663. Among the alchemist's accoutrements on and around the furnace are a wooden box containing the vestiges of a red powder (red sulfur or perhaps the philosopher's stone), a mortar and pestle for grinding minerals, and a pelican vessel.

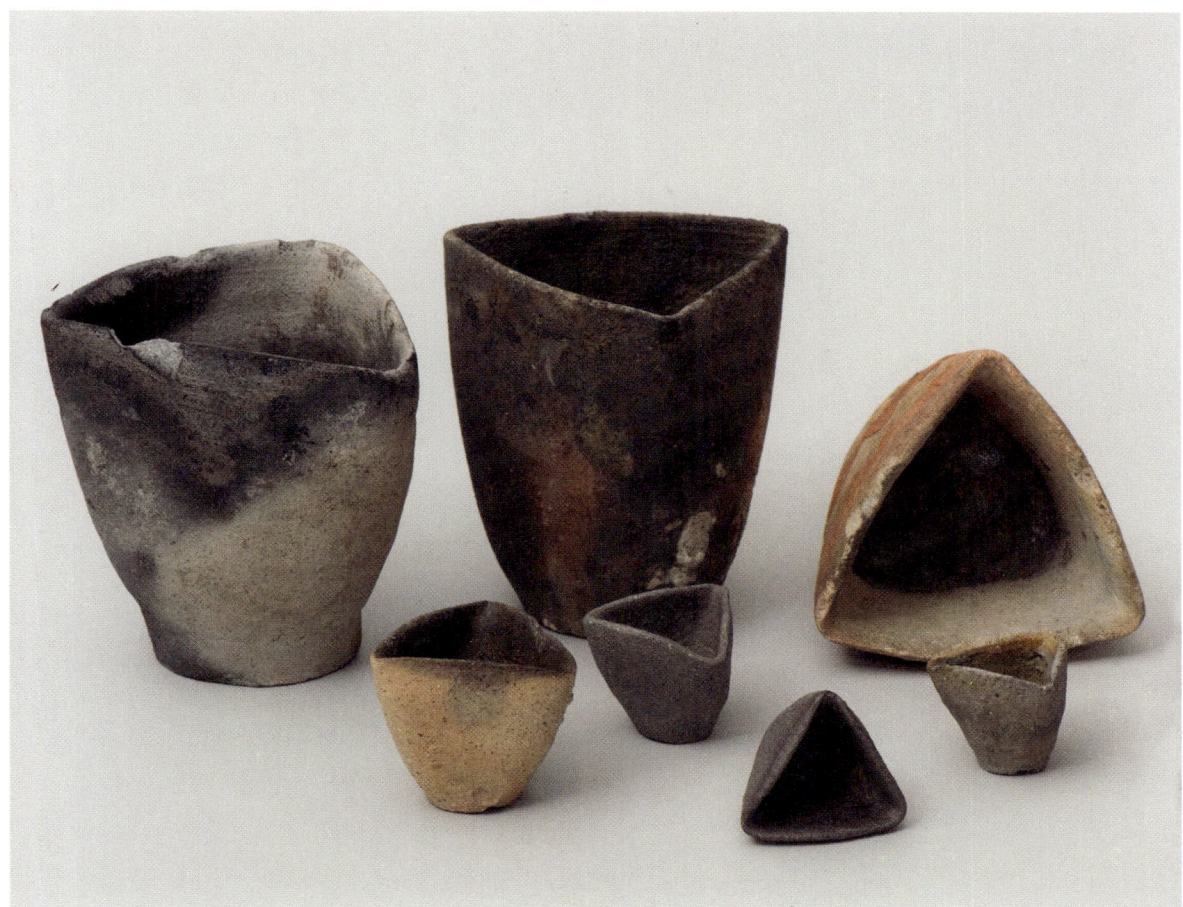

ABOVE Small triangular crucibles from a 16th-century laboratory in Oberstockstall, Austria. The crucibles, shipped worldwide, were often pinched into triangular shapes to make pouring spouts. The large stash of discarded equipment found under a damaged floor in Castle Oberstockstall in Kirchberg offers one of the best glimpses into the apparatus and methods of alchemists. Fragments of around 800 artifacts were found, including crucibles, alembics, flasks, retorts, aludels, and more. They are thought to have come from the laboratory of an alchemist named Urban of Trennbach, who was active in the late 16th century and followed the teachings of Paracelsus.

The lowest pot did have a bottom, and in this would be placed a solid substance to be sublimed, meaning that it was heated in the furnace until it turned directly into a vapor. The vapor would rise up and then condense into a purified solid at the top. You might think of it as a kind of "distillation for solids," used, for example, to separate the components of arsenic minerals such as orpiment, or to extract mercury or sulfur.

The alchemist's most important item of equipment was arguably the crucible: a simple cup-like container that could withstand intense heating. (The name comes from the Latin *crucibulum*, "little place of torment.") Crucibles would be used for mixing and calcining (heating in air) ingredients, and for evaporating, melting, and smelting metals. They came in various sizes, and from the Middle Ages they typically had a triangular cross-section to give them pouring spouts.

PUFFERS: THE ALCHEMY LABORATORY

Crucibles were made from baked clay. Documentary records and archeological evidence suggests that nearly all crucibles used by alchemists from the Mediterranean to Scandinavia, Britain, and even the young American colonies, came from just a few manufacturers in the German states and in Bohemia and Upper Austria. Indeed, most written documents refer to just one provenance: the villages of Almerode and Epterode in the state of Hesse. The method of their manufacture was kept secret for hundreds of years: the English natural philosopher Robert Plot, the first professor of chemistry at Oxford University, wrote in the mid-seventeenth century about the "mystery of the Hessian wares."

Hessian crucibles had a rough, sandy texture and an orange color. They were fired at very high temperatures from kaolin clay mixed with ordinary quartz sand, which creates a tough aluminosilicate mineral called mullite. The other common type of crucible was made in Bavaria from a clay naturally rich in graphite flakes. These flakes acted a little like fibrous reinforcement when the clay was fired, making it very tough and giving it a dark, smooth surface. In the late eighteenth century, crucible manufacturers deduced the secret of the Bavarian method and began mixing graphite intentionally with clays.

Spirit and essence

Techniques such as distillation and sublimation played a central role in alchemy. The principle of distillation is simple: the liquid is heated until it (or some volatile component within it) evaporates, and then the vapor is cooled to condense it back to liquid. It is an ancient art: a primitive kind of distillation seems to have been practiced in the late second millennium B.C. in Mesopotamia to extract oils from plants for making perfumes. These extracts were often described as "spirits": the supernatural or religious connotations were not incidental, for the substances (invisible until condensed) were considered a kind of purified essence of the raw material. The term, along with the notion of an "oil" as any kind of distilled extract, persists in popular names for some chemical substances today:

Hessian crucibles had a rough, sandy texture and an orange color. They were fired at very high temperatures from kaolin clay mixed with ordinary quartz sand.

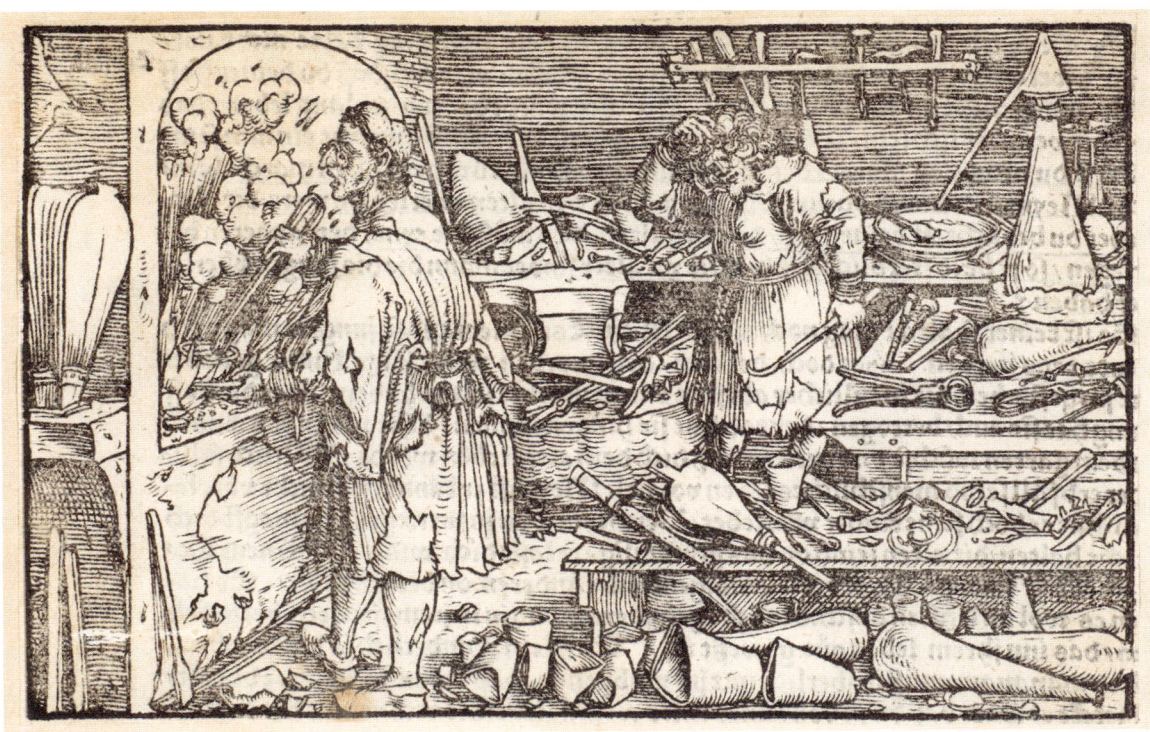

ABOVE "The art of alchemy is deceptive to many": a hectic alchemical laboratory, with metal smelting taking place. Note the crucibles stacked on the floor. Engraved by Hans Weiditz, from *Francisci Petrarche, des hochweisen fürtrefflichen Poeten und Oratorn, zwei Trostbücher, Von Artznei und Rath beyde im guten und widerwertigen Glück* (Two books of consolation, on medicine and advice in both good and bad fortune, by Francesco Petrarcha, the highly wise and excellent poet and orator; Frankfurt, 1559).

"spirit of salt" is hydrochloric acid, for example, while "oil of vitriol" is sulfuric acid, made by dry distillation of vitriol (sulfate) salts. It was widely believed that pretty much any substance, even minerals and metals, yielded a spirit or oil by distillation.

Today, of course, the "spirits" prepared by distillation are often of the alcoholic variety. Spirit of wine became "alcohol" via a curious etymology. The Arabic word *al-kohl* or *al-kuḥl* was derived from ancient Assyrian for a black cosmetic eye paint (which we still call kohl) made from the mineral stibnite (antimony sulfide). So "alcohol" once referred to that black powder, then to any powdered "essence," and finally to the essence of wine, in medieval Latin *alcool vini*, which is first described in alchemical recipes around the twelfth century. It might also be known as *aqua vitae*, "water of life"— a term preserved, as it were, in the French fruit brandy *eau de vie*.

The use of distillation to make spirits, particularly alcohol, was popularized in the thirteenth century by works attributed to the physician Arnald of Villanova, and then by the fourteenth-century Franciscan John of Rupescissa (see page 130). For Arnald (or rather, the writer using his name), the spirit of wine was a kind of condensed sunlight accumulated by grapes. He claimed that distillation of human blood could also yield a potent essence with medicinal properties. John of Rupescissa's book *De consideratione quintae essentiae* (On the consideration of quintessences) extolled the medical virtues of the spirits and elixirs that could be made by distillation, including a red liquid extracted from antimony sulfide that he deemed a powerful alchemical agent. John considered that the essences extracted from matter by distillation were akin to the incorruptible fifth essence or quintessence (aether) of Aristotle. The spirit of wine, he observed, could preserve meat from decay and shield the body from illness.

In his *Liber de arte distillandi* (Little book on the art of distillation; 1500, expanded in 1512), the German surgeon Hieronymus Brunschwig of Strasbourg offered a rather more prosaic celebration of distillation, free from theological allegories of purification. Brunschwig's treatise was very much a practical how-to handbook

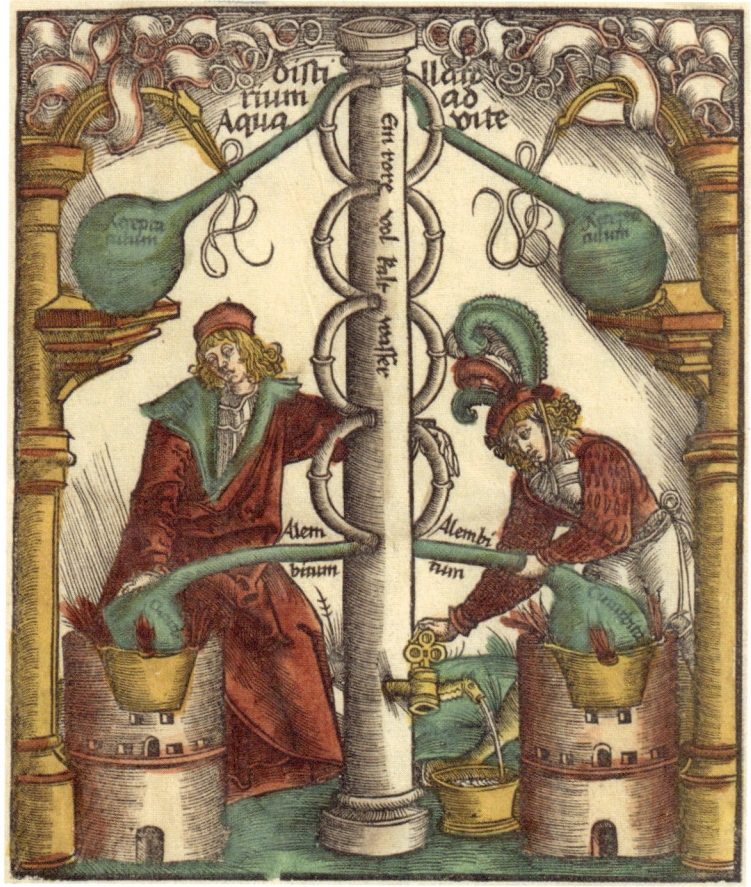

RIGHT Alchemists distilling compounds in Hieronymus Brunschwig's *Liber de arte distillandi* (Little book on the art of distillation; Strasbourg, 1512). Brunschwig's book was a practical handbook on this important chemical technique, with none of the esoteric terminology or symbolism typical of alchemical texts. It explained how plant oils and essences could be prepared using alembics and water baths. Brunschwig was a surgeon of Strasbourg, and his book was influenced by the works of John of Rupescissa.

for the artisan: written in vernacular German and aided by splendid woodcuts, it explained how medicinal oils and essences could be prepared using equipment such as alembics and the *bain-marie*.

An important alchemical process for extracting and purifying silver and gold from alloys with other metals such as copper, lead, and antimony was cupellation. It was done in a shallow dish called a cupel, made of a porous material such as bone ash, and relied on the relative inertness of silver and gold to oxidation (reaction with air). If impure metal was melted in the cupel, the impurities oxidized and the reaction products were absorbed by the dish, leaving behind pure silver or gold.

Invention of the laboratory

What did the alchemist's workplace look like? It's not easy to be sure, for no undisturbed site has ever been found. Pieter Bruegel the Elder's sixteenth-century drawing *The Alchemist* presents a rather unflattering view of a workshop littered with equipment and blighted by fumes. The alchemist sits at his desk reading a manual and giving orders to his hapless assistants, while children add to the chaos by climbing into a cupboard to root around in the pots and flasks. All this was intended as satire, illustrating the ruinous folly of the quest to make gold—it is an image made with moral instruction in mind rather than as an accurate record. The same may be said of the paintings of alchemical laboratories by the seventeenth-century Flemish artist David Teniers the Younger. Teniers was a prolific

RIGHT This unflattering view of an alchemist's workshop by Pieter Bruegel the Elder, titled *The Alchemist*, 1558, was later widely circulated in a popular engraving by his son Pieter Bruegel the Younger. As the alchemist directs futile labors in his chaotic laboratory, his children are seen through the window being led to the poorhouse.

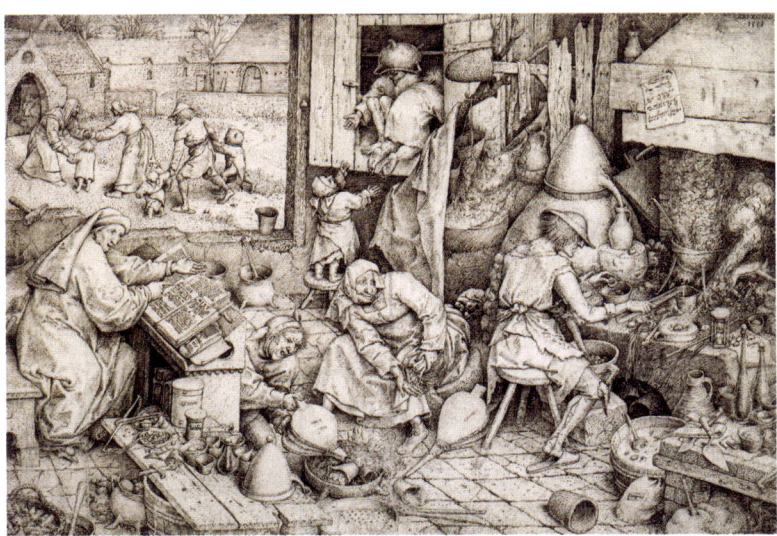

BELOW Lithograph, 1835–40, by Carl Straub of a painting by the 17th-century Flemish artist David Teniers II entitled *Ein Chemiker in seinem Laboratorium* (A chemist in his laboratory). Teniers painted several images of alchemists at work, generally with an air of gentle mockery of this futile pursuit.

painter of rustic life, and his portraits of alchemists are gently mocking, in contrast to Bruegel's biting satire, showing earnest but deluded old men with avuncular beards. Their workspaces are liberally scattered with pots, flasks, and other equipment that allowed Teniers to demonstrate his skills at depicting the gleaming surfaces of glass, pewter, and glazed ceramics. In one such picture painted in 1649, Teniers depicts a white-bearded alchemist using bellows to raise the heat of his furnace, showing why alchemists were often derisively called "puffers" and why their workplaces were notorious for being sooty and unhealthy.

The first known occurrence of the word *laboratory* in English is in a 1592 book by the Elizabethan magus and alchemist John Dee

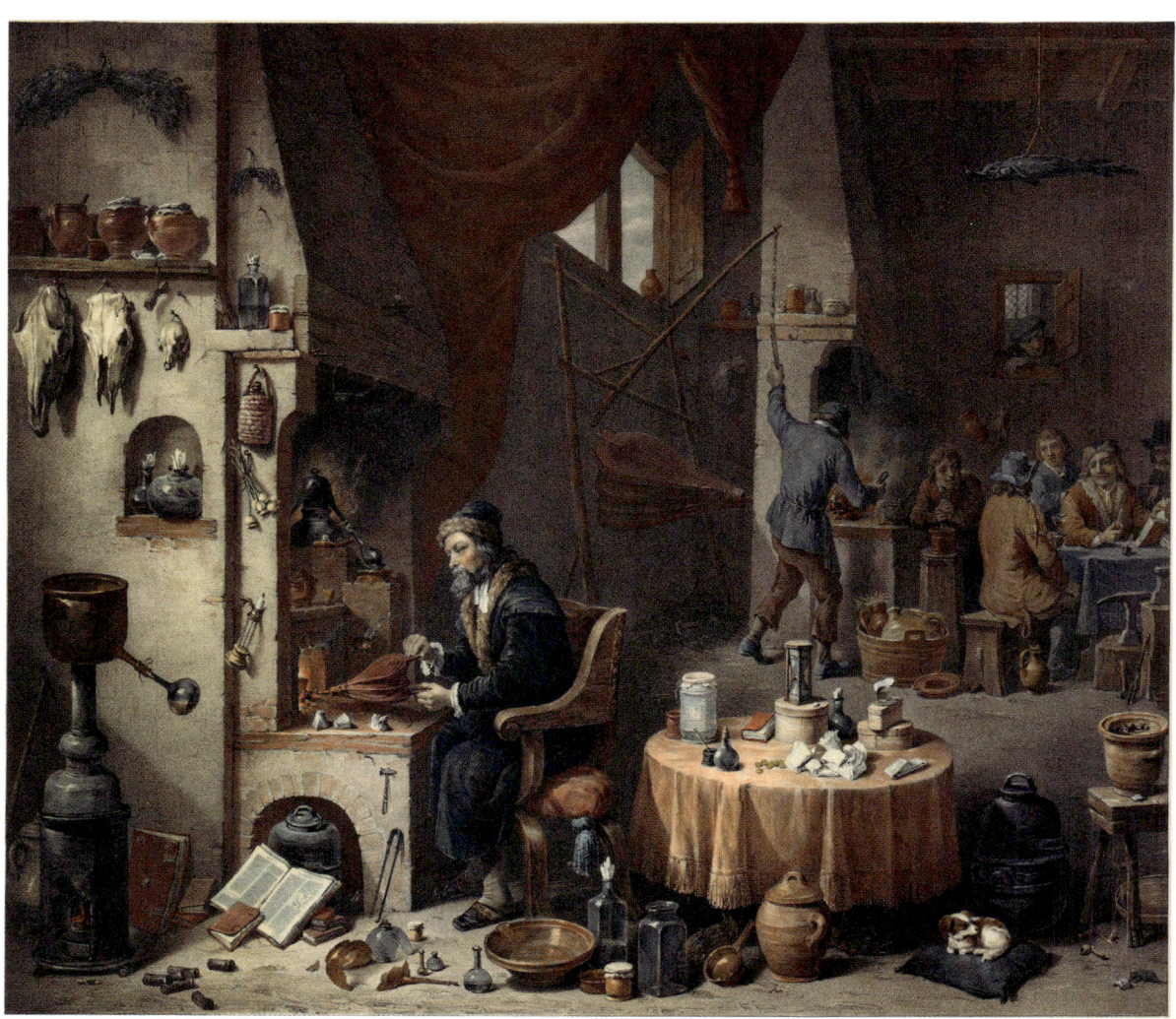

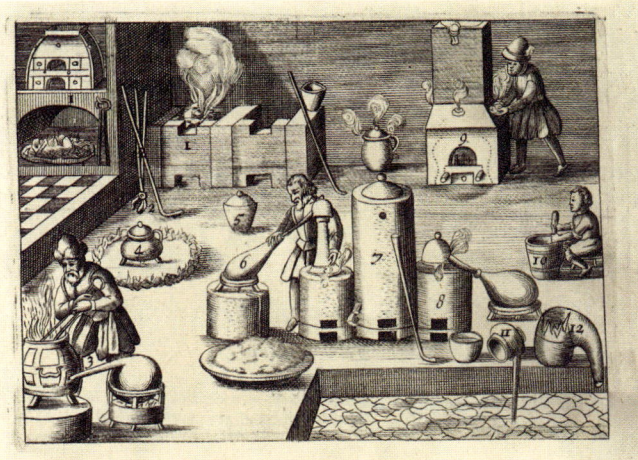

Deciphered,
1. The form of an *Athanor* or great *Furnace*.
2. The *Forceps* or *Tongs* and *Fork*.
3. The *Coppel* or *Test*, with Philosophers *Bellows*.
4. The *Digestive Pot* with its *Cover* and *Fire* about it.
5. A cover'd *Crucible*.
6. The long Bell, or *Matras-Glass* on a *Sand Furnace*.
7. The *Wind Furnace* with a *Blow-pipe*.
8. A *Furnace* with a *Copper head*, and its *Receiver*.
9. A *Furnace* with a *naked* and open *Fire*.
10. The *Pestle* and *Mortar*, with one beating the *Metals*.
11. The *Owl's Head*, or another form of a *Cover* to the Figure 8.
12. A *Retort*.

RIGHT A well-organized workshop illustrated in Lazarus Ercker's treatise on metallurgy, reproduced, with the equipment helpfully labeled, in *Fleta minor*, the English translation by Sir John Pettus (London, 1683).

(see page 173). At that time, it could refer to the working space of a "manufacturing" pharmacist who prepared medicines. Whereas alchemy in the Middle Ages was often conducted in a makeshift environment, pharmacies needed a purpose-built site, with a long bench and shelves for jars and bottles, that might have served as a model for the (al)chemical laboratory. There was probably also some similarity with other professional workplaces for chemical processing: distilleries, soap-boilers' workshops, and, perhaps most of all, metalworking operations. The excellent woodblock illustrations in treatises on mining and metals written by the German scholars Georgius Agricola and Lazarus Ercker in the sixteenth century show us what these might have been like.

They are typically well-organized spaces, with furnaces in neat rows and racks on the wall where retorts and other equipment are stored. Ideally, such places would be well ventilated, since everyone knew that the fumes from metalworking could be choking and toxic.

By the seventeenth century it was clear that alchemists needed a bespoke and orderly place for their work. In 1606 Andreas Libavius published a blueprint for a "chemical house": an entire building dedicated to the Hermetic art. From the outside it looks like a regular house of the kind that a well-to-do German burgher might own, but inside is a suite of specialized rooms: storerooms for materials, including wood, wine, fruit, and vegetables (for extracting oils and alcohol), rooms for weighing out ingredients and for performing operations such as crystallization, as well as a bathroom and toilet. Libavius's dream seems never to have been realized, but perhaps that is just as well given that one facility he did not really take into consideration was proper ventilation.

A more ambitious scheme for an alchemical laboratory was described by the German alchemist Johann Joachim Becher in his 1680 book *Chymisches laboratorium* (Chemical laboratory). Becher (see page 138) was convinced that making gold could be a regular

BELOW Details from Andreas Libavius's *Alchymia* (Frankfurt, 1606), showing Libavius's scheme for a "chemical house"—a purpose-built workplace for a systematic chemical technology. The plan was apparently never executed.

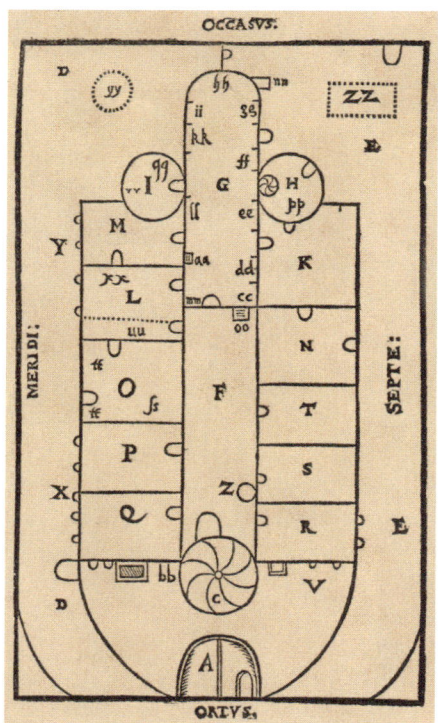
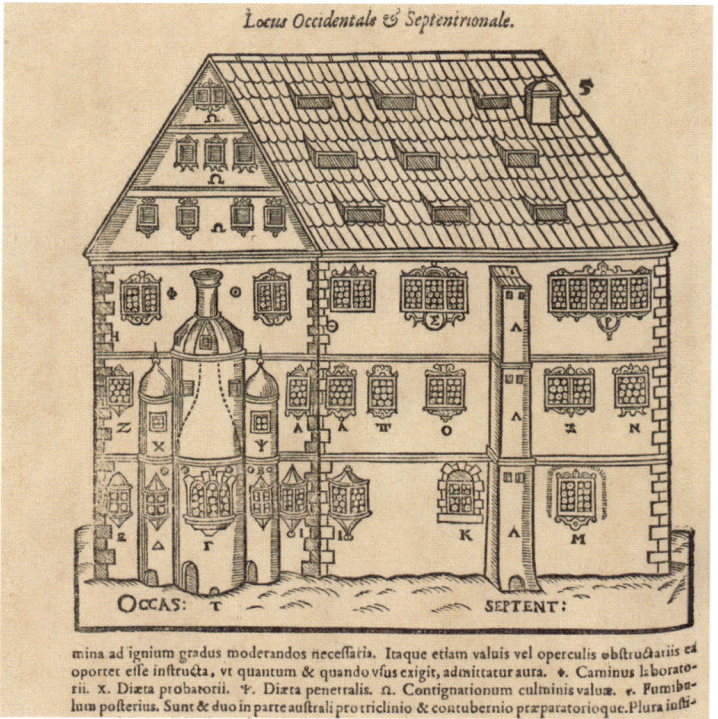

and important contribution to a state's economy: a God-given source of wealth for financing public works that did not depend on taxing the populace. In his vision, the alchemist becomes akin to a civil servant, performing industrious and virtuous labor for the common good. That could never happen in the chaos of the traditional alchemical workplace, which Becher describes as follows:

The used glass vessels have been smashed into bits, so that they cannot be rinsed out. For each operation they take new crucibles and glasses so that they do not need to clean the old ones. Whole things, broken bits, clean, dirty, new, used, prepared materials, raw materials, wooden, clay, and glass utensils are standing all mixed together. The windows, tables and floor are full. And nothing is labelled as to what it is . . . The stink from the furnaces, the soot and dust from the coals, the sand, water, and lime do not help the work so much as aid the confusion.

Bringing order and method to alchemical production did not mean, however, that state-funded alchemical operations should work openly. Becher's plan included measures to ensure that the techniques of gold-making were not leaked: for example, keeping manual workers ignorant of the details of the process (they would preferably be illiterate) so that they weren't tempted to sell their knowledge. Alchemy's clandestine practices were, in this view, like the trade secrets jealously guarded by guilds—indeed, not unlike the intellectual-property protection of companies today. Becher saw alchemy as a nascent industrial process involving hierarchical division of labor, and his laboratory anticipated the age of the factory, with each facet of production allotted its proper place.

Bearer of light

The lone alchemist seeking gold was not yet obsolete in Becher's day. This is nowhere better illustrated than in the story of German alchemist Hennig Brandt, who validates chemist Marcelin

Alchemy's clandestine practices were like the trade secrets jealously guarded by guilds—indeed, not unlike the intellectual-property protection of companies today.

> The light streaming from [Brandt's] flask gives the scene the air of divine revelation, firmly locating alchemy as a pre-scientific mode of understanding the world.

Bertholet's judgment in the nineteenth century that chrysopoeia stimulated all manner of fruitful chemical experimentation.

Brandt lived and worked in Hamburg, where he apprenticed as a glassmaker but dedicated himself to alchemy, financed by the generous dowry of his first wife and the inherited wealth of his second. He personifies the caricature of the alchemist who fritters away his money in doomed pursuit of gold.

Having decided that the crucial ingredient for transmutation might be found in urine, in 1669 Brandt distilled great volumes of the stuff until he isolated a soft residue that caught alight when it came into contact with air. This substance glowed spontaneously and gave off an unpleasant garlicky smell: it was the chemical element that we today call phosphorus, from the Greek meaning "bearer of light."

Although Brandt kept his discovery closely guarded, the news of so marvelous a substance came to the attention of Johann Kunckel, professor of chemistry at the University of Wittenberg, who told fellow chemist Daniel Krafft in Dresden. Kunckel went in search of Brandt, only to discover the alchemist to be chemically untutored and reluctant to divulge further details. Krafft had similar ideas: the somewhat unlikely story is that he tracked down Brandt first and was in the midst of negotiating a price for his secret when Kunckel turned up at the door. In any event, Krafft seems to have departed from Hamburg with most, if not all, of Brandt's stock of phosphorus, while Kunckel set about trying to make it himself from distilled urine. He was eventually successful and began demonstrating the remarkable glowing material to German nobles.

When Krafft displayed phosphorus to the Duke of Saxony, the duke's librarian Gottfried Leibniz (soon to become a celebrated mathematician and philosopher) sent the news to the Royal Society in London. In September 1677, Krafft was invited to London, where he was hosted by Robert Boyle. He exhibited phosphorus before a select group of guests at the house of Boyle's wealthy sister Lady Ranelagh, using flashy showmanship honed in the courts of Europe. He cast pieces of glowing phosphorus like tiny stars over the expensive carpets and dipped his finger into the liquid to trace out the word *Domini* in what Boyle called "a mixture of strangeness, beauty and frightfulness."

Krafft was cagey about the secret of making phosphorus, but Boyle deduced that urine was the raw ingredient, and employed an assistant to conduct the smelly distillation in his lodgings. In 1680

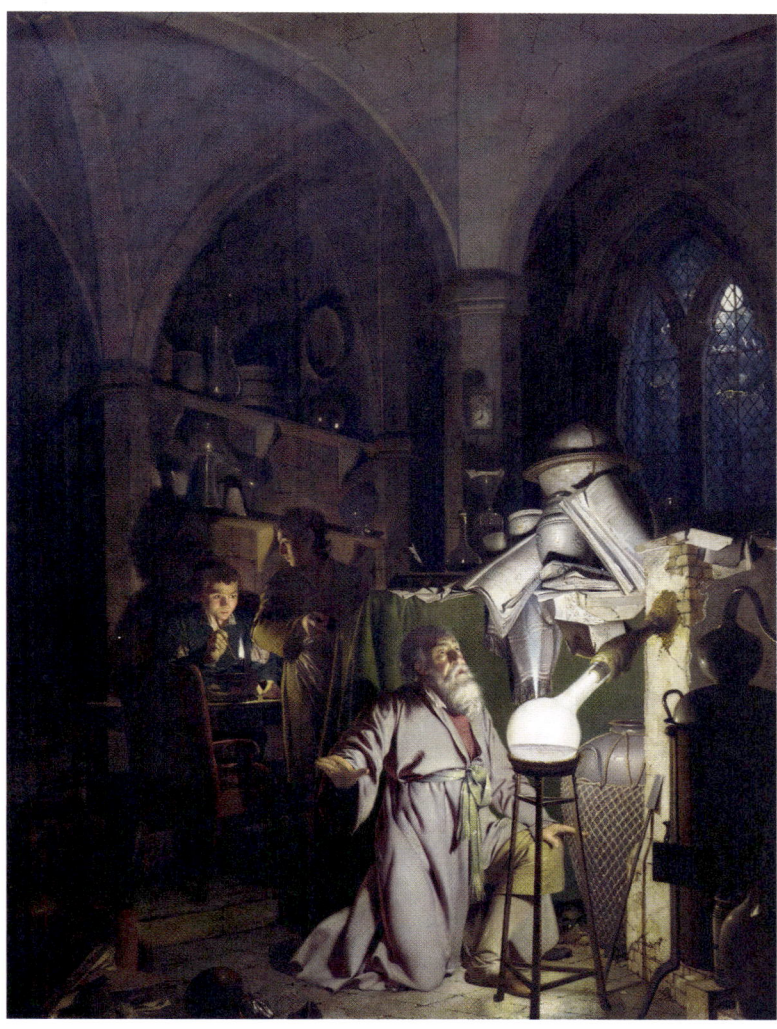

RIGHT Hennig Brandt's discovery of phosphorus is immortalized in Joseph Wright of Derby's 1771 painting *The Alchymist, in Search of the Philosopher's Stone, Discovers Phosphorus, and Prays for the Successful Conclusion of His Operation, as was the Custom of the Ancient Chymical Astrologers,* now in Derby Museum and Art Gallery. Wright depicts this discovery as an act of almost divine revelation.

Boyle deposited a sealed letter with the Royal Society, to be opened only on his death, describing how the substance was made: the secretive impulses of alchemy were still felt.

Brandt's discovery was depicted in a highly romanticized manner by the English painter Joseph Wright of Derby in his 1771 work *The Alchymist, in Search of the Philosopher's Stone, Discovers Phosphorus.* Brandt is shown dressed like a monk, laboring in a vaulted chamber as though in an abbey. The light streaming from his flask gives the scene the air of divine revelation, firmly locating alchemy as a pre-scientific mode of understanding the world. That image was neither fair nor accurate, but it was an indicator of where the reputation of chrysopoeia was heading.

PROFILE

John of Rupescissa

(c. 1310–c. 1366)

Alchemy held a particular attraction for unorthodox religious thinkers over the ages, especially those of a millenarian disposition—that is, believers in impending apocalypse or radical social change. The fourteenth-century Franciscan friar John of Rupescissa exemplifies this stereotype of the alchemist: his denunciations of ecclesiastical hierarchy brought him into frequent conflict with the church, and he spent the last two decades or so of his life in prison.

Rather little is known about John of Rupescissa's life. He was known also as Jean de Roquetaillade, although he seems not to have been a native of that town in southern France; some say he was born in the province of Girona in Catalan Spain. He studied at the University of Toulouse before entering a Franciscan friary at Aurillac. He became associated with a group within the order called the Spirituals, who called for a return to the ideals of poverty advocated by the order's founder, St. Francis of Assisi. The Spirituals believed that the apocalypse foretold in the Book of Revelation, when the Antichrist would appear on Earth, was near at hand, and John prophesized to that effect—for example, in his 1349 tract *Liber secretorum eventuum* (Book of the secrets of things to come).

In his *Liber lucis* (Book of light, c. 1350), John explains that his interest in alchemy was prompted by the hope that it could supply the church with wealth when it "shall be tormented and have all [its] worldly riches despoiled by tyrants" upon the arrival of the Antichrist. This mission fueled his quest for the philosopher's stone, which he believed could be made from a purified form of mercury combined with a "Philosophical Sulfur." Drawing on earlier works spuriously attributed to Arnald of Villanova (see page 71), John gives rather detailed (but somewhat cryptic) practical instructions for making the stone, a transformation that he likens to the sufferings of Christ.

Other of John's alchemical works are concerned with making a life-prolonging elixir and medical remedies. These recipes appear in his book *De consideratione quintae essentiae rerum omnium* (On the consideration of the fifth essence of all things), which describes his belief that a *quinta essentia*, analogous to the imperishable Quintessence from which Aristotle supposed the heavens to be made, could be extracted—generally by distillation—from all manner of substances. Such purified essences could, he said, restore vigor to the aging body.

John noted that the potent liquid or spirit (*aqua ardens* or *aqua vitae*) that could be distilled from wine had the power to preserve meat from decay, and he claimed to have extracted a similar miraculous essence from

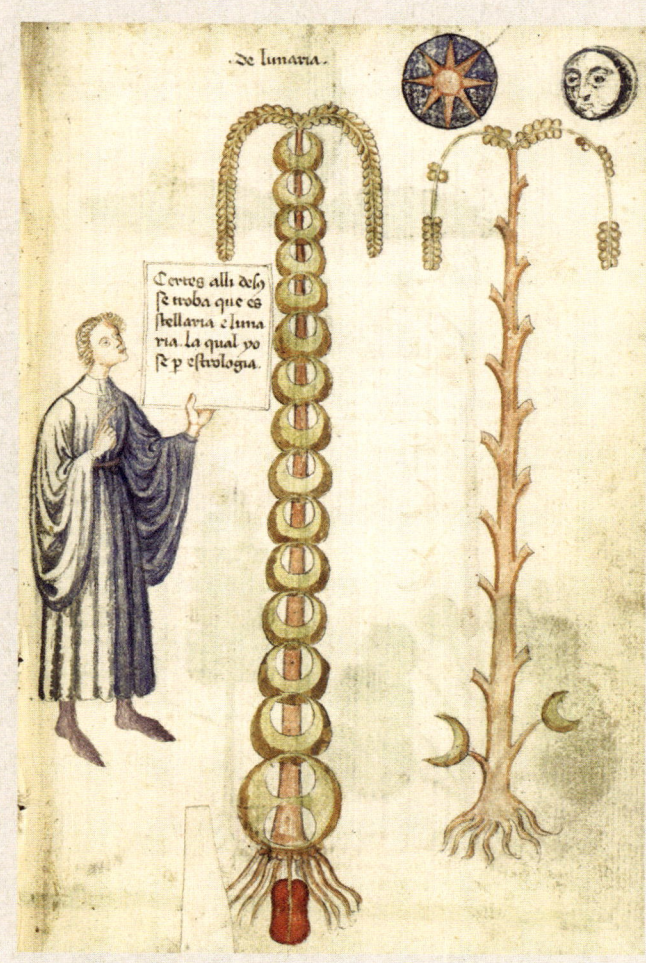

RIGHT John of Rupescissa, in a woodcut from an album containing scenes from Hartmann Schedel's *Weltchronik* (World chronicle), known as the *Nuremberg Chronicle,* from the workshop of Michel Wolgemut and Wilhelm Pleydenwurff, 1493.

antimony: "a treasure which the whole world cannot equal," and which "surpasses the sweetness of honey." This quintessence of antimony, he says "takes away pain from wounds and heals marvelously. Its virtue is incorruptible, miraculous, and useful beyond measure." He says that the medicine can be prepared also in powdered form, and can be used to treat leprosy, paralysis, consumption, and even possession by demons. Leprosy, John adds, can also be cured by a kind of "water" prepared from ripe strawberries. Evident here are the beginnings of both a focus on alchemical medicine in the West and the notion that remedies come from separating what is pure and virtuous in substances from what is noxious and evil.

While a life in prison does not sound conducive to this kind of discovery, John, it seems, was able to pursue his studies while incarcerated. In one manuscript he attests to the help he received from his jailers in obtaining alcohol (for medical purposes!), and he appears to have had access to books, to have received visits from friends, and even to have been consulted for prognostications by cardinals and other dignitaries. His internment in various church institutions was perhaps less a punishment than a way to keep the unruly scholar out of trouble.

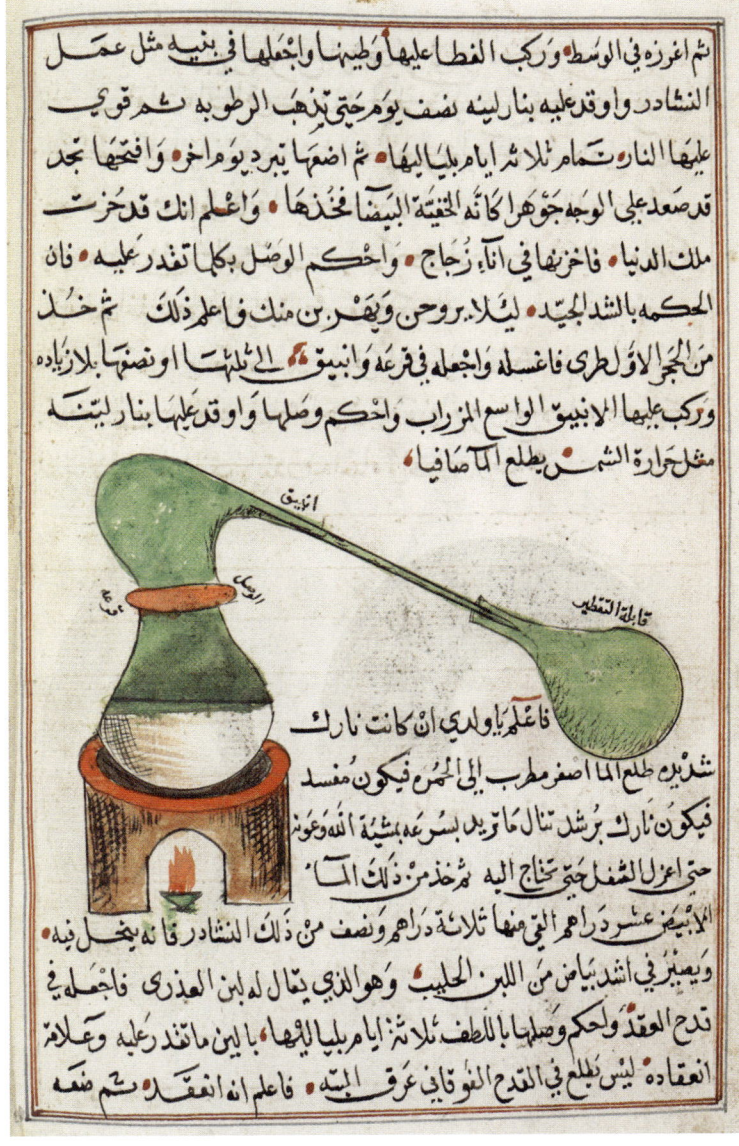

ABOVE Distillation apparatus in an 18th-century copy of *Kitāb al-aqālīm al-sabʿah* (Book of the seven climes), a 13th-century text by Abū al-Qāsim al-ʿIrāqī, known as al-Sīmāwī, meaning "practitioner of white or natural magic." Al-Sīmāwī lived in Egypt, but very little is known about him. His book is focused entirely on alchemical illustrations, generally reproduced from earlier Arabic texts and annotated with al-Sīmāwī's own interpretations of their symbolism.

OPPOSITE Alchemical vessels, including distillation helmets, retorts, and a crucible, in Carlo Lancillotti and Johann Lange's *Der brennende Salamander* (The burning salamander; Lübeck, 1681). These instruments, generally made from glass, are specialized for different operations. The crucibles were made from fired clay, with a triangular cross-section for pouring, and were used to mix and heat ingredients. The border of the page shows alchemical symbols for materials and processes.

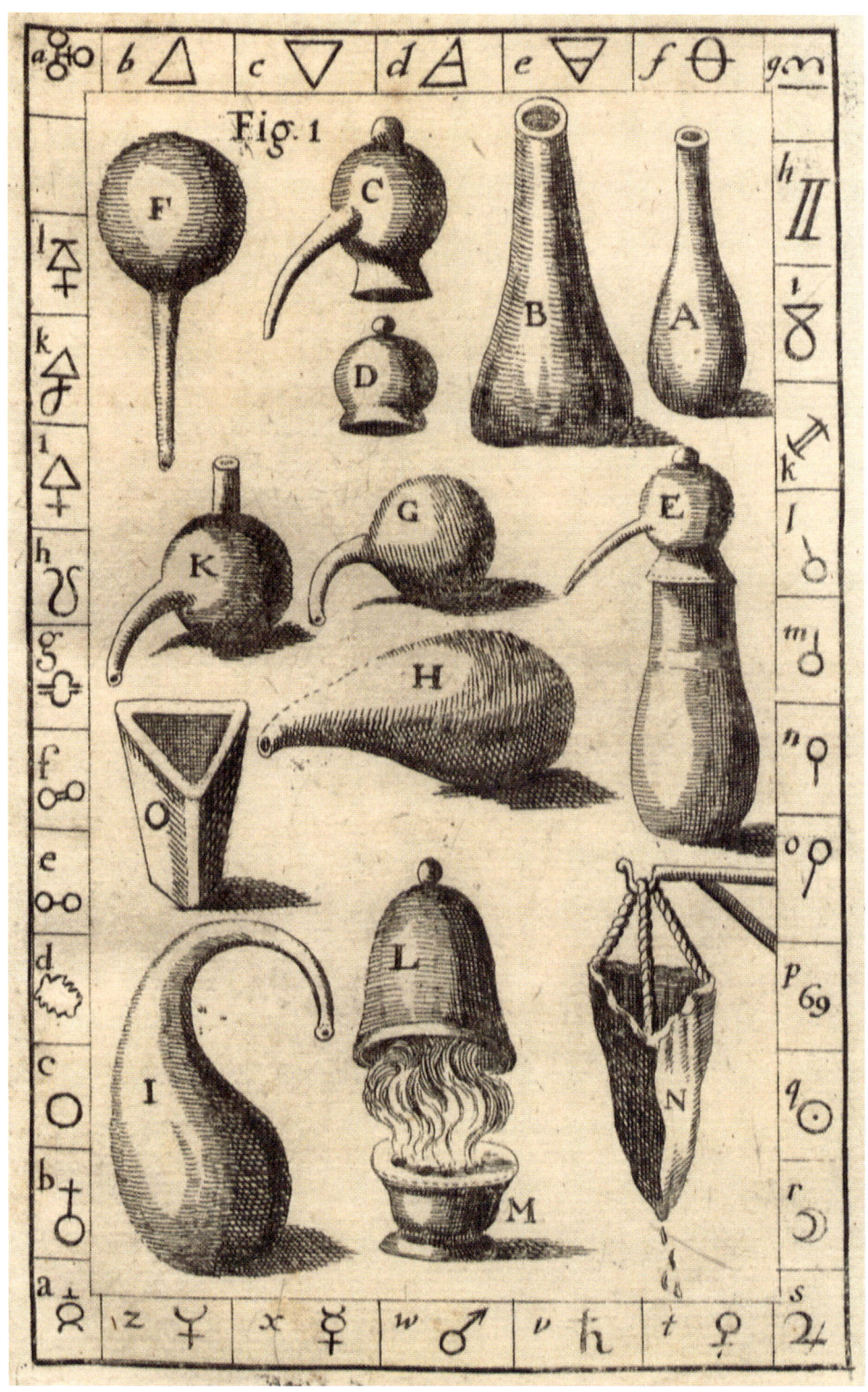

PUFFERS: THE ALCHEMY LABORATORY

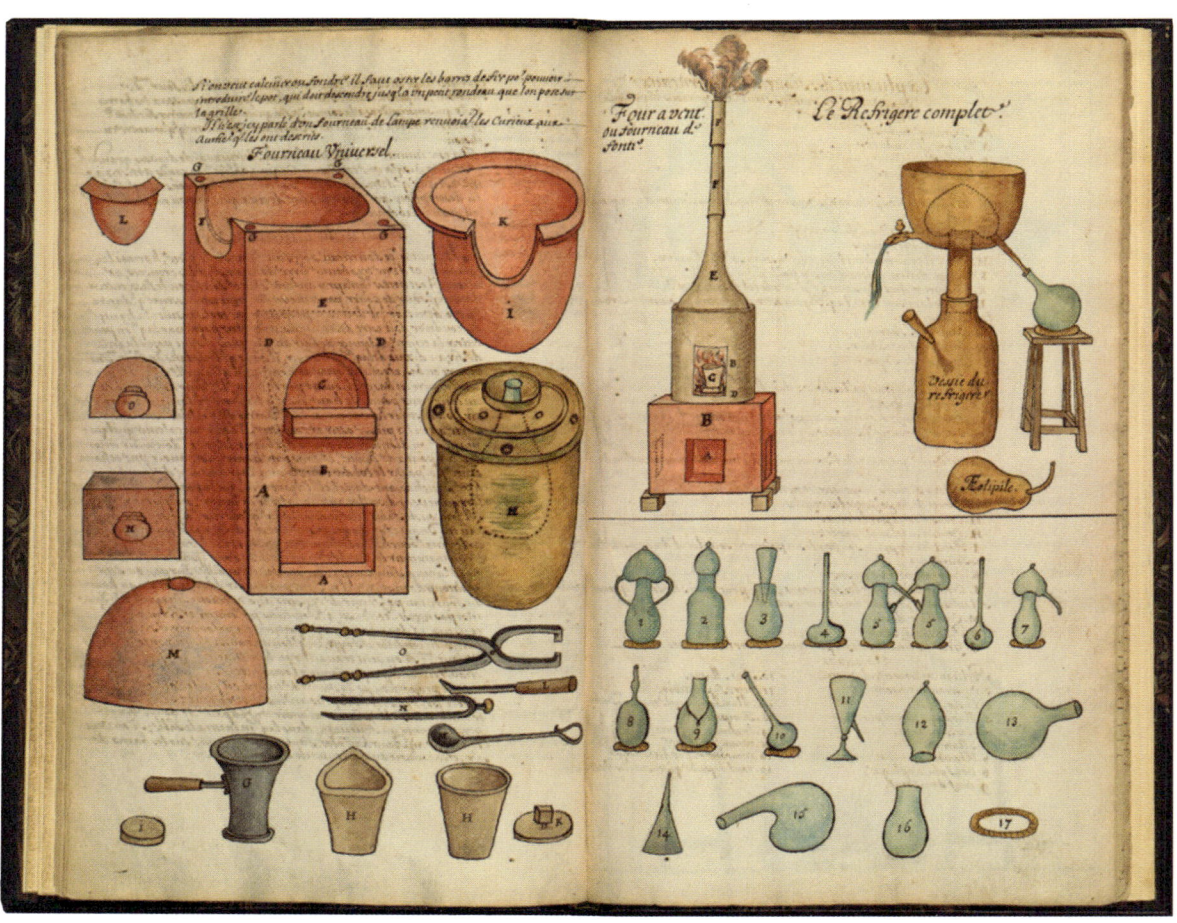

ABOVE Alchemical equipment including furnaces, crucibles, and alembics, from an anonymous French manuscript, *Traité de chymie* (Treatise on chemistry; c. 1700). By the early 18th century, books of this kind were often more in the manner of practical textbooks, with little of the cryptic symbolism and terminology associated with alchemy in previous times. This transitional discipline between alchemy and chemistry is now generally called chymistry.

OPPOSITE Two monks in a well-equipped alchemical workshop, one at the furnace, from a collection of alchemical tracts mainly in German, *Miscellanea Alchemica XXIV* (1543). Such an explicit depiction of monastic alchemy is relatively unusual, although many alchemists during the Middle Ages were clerics or in holy orders—there was no conflict between religious beliefs and an exploration of chemical transformations.

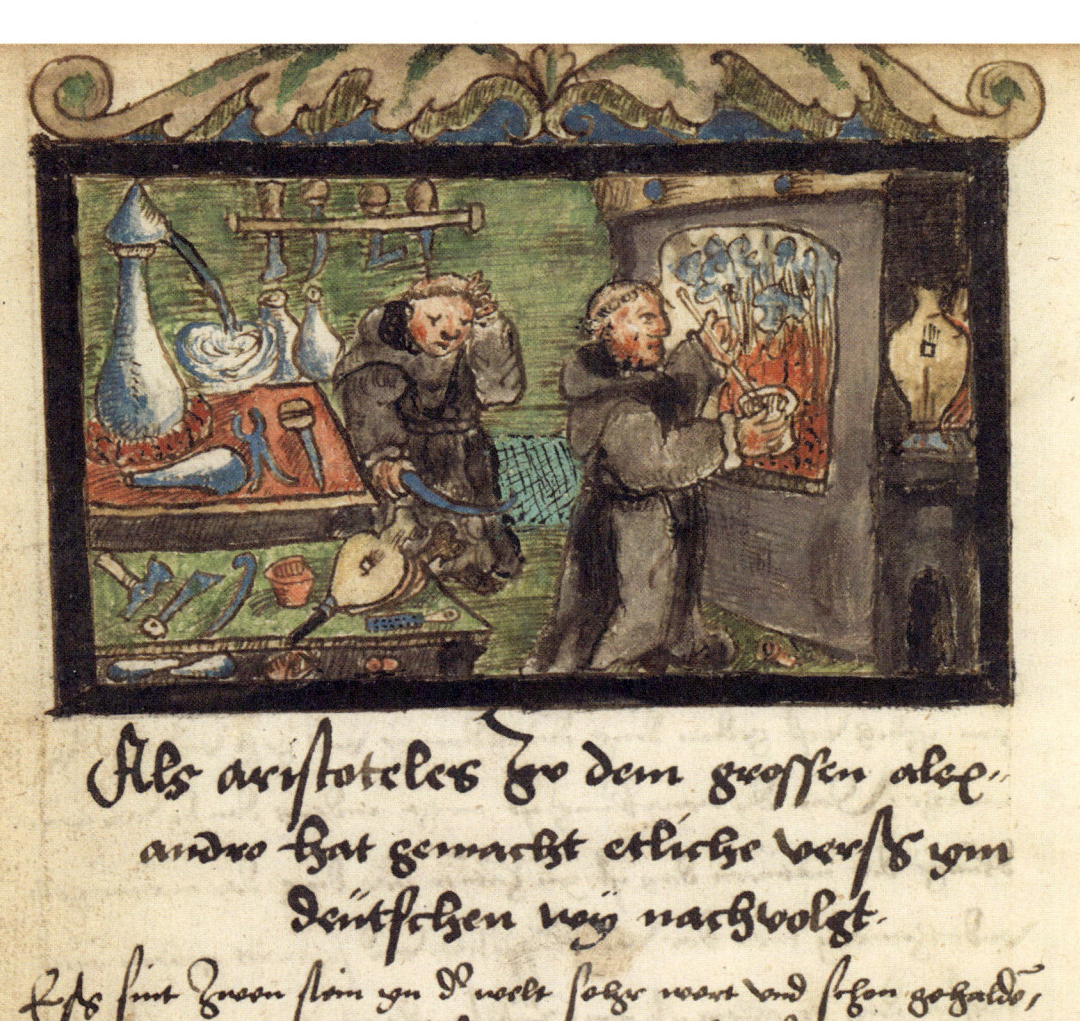

Als aristoteles zo dem groszen alex-
andro hat gemacht etliche verß ym
deütschen vey nachvolgt.

Es sint zween stein yn d'welt sehr werd und schon gehalt=,
der manlich ist rot, der frewlich ist yn sein farb weiß. Ist
sach das ein stein zu dem andern befestigt sey, glauben, wiewol
ein lange zeit dorzu gehort vnd dy kunst das zeit zuvormei=
den vnd es musz solichs lange oder zeit sein, bis dy sünde sterm=
be werden, vnd dy weil ist idermann musz niemand mag zu
wederstohn.

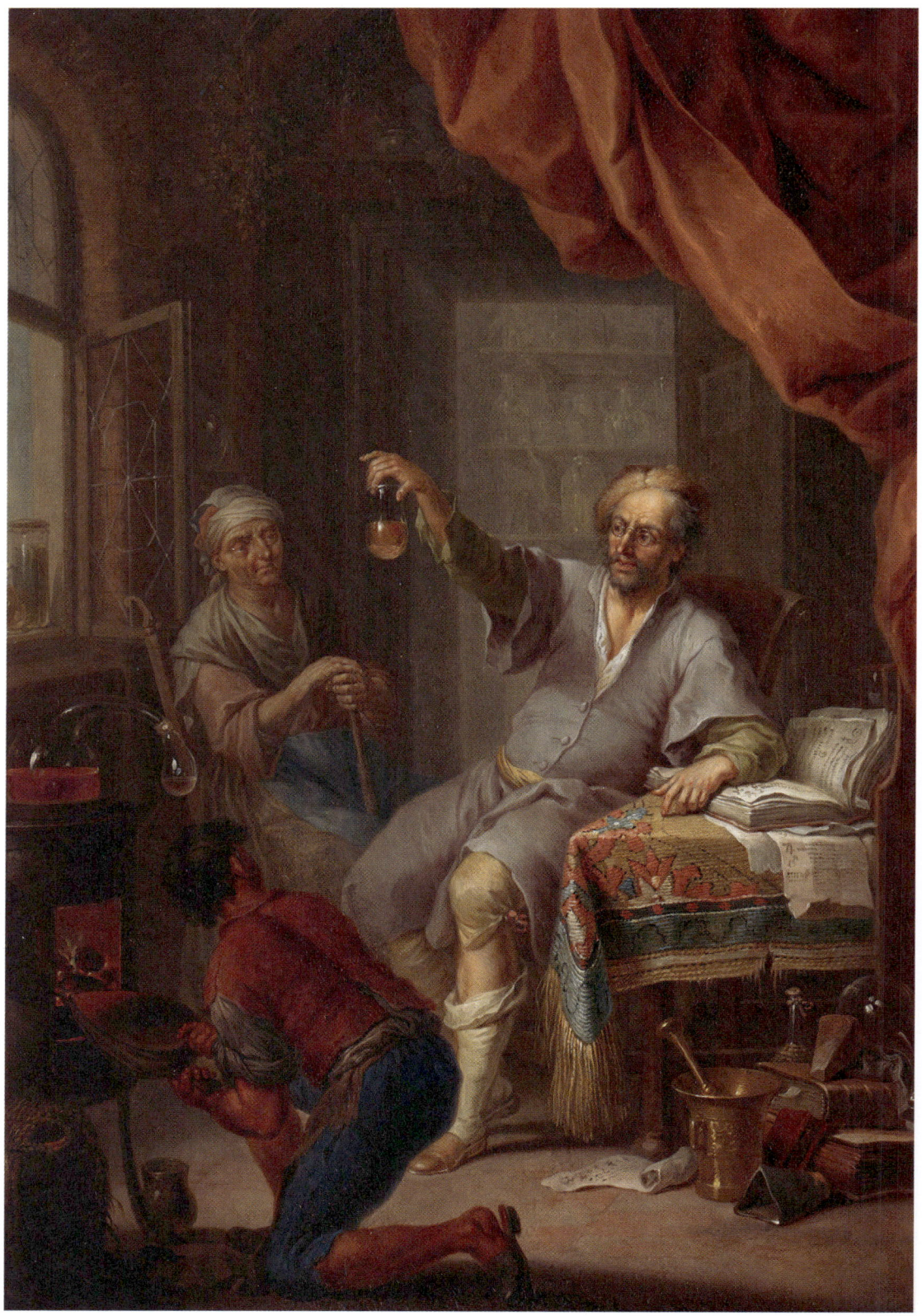

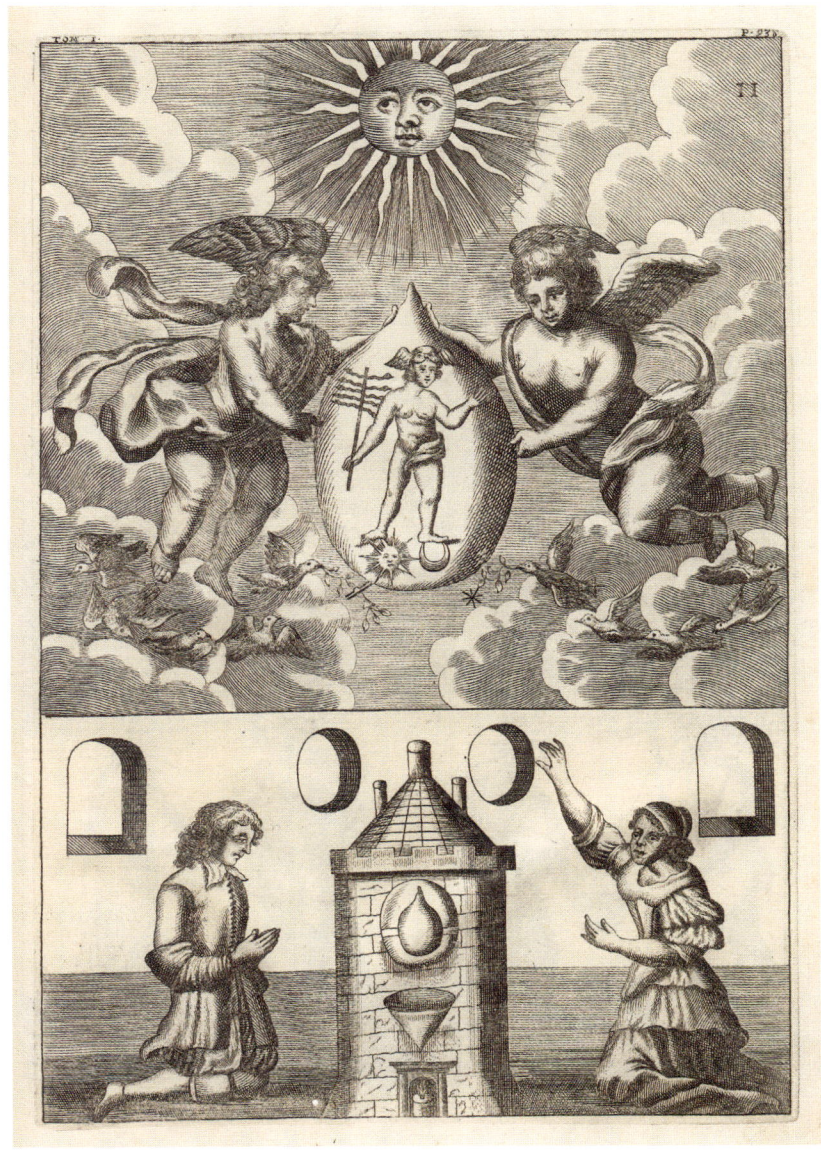

OPPOSITE In Franz Christoph Janneck's 18th-century painting *The Medical Alchemist*, a physician, surrounded by alchemical symbols, conducts a urinoscopy for a patient, while his assistant tends to a distillation at the furnace. It was common at that time for physicians to make a diagnosis by inspecting the patient's urine, assessing its color, cloudiness, and so forth. The Swiss alchemist Paracelsus was contemptuous of such an approach to medicine, saying of doctors "All they can do is to gaze at piss." Instead, he recommended the preparation of specific chemical medicines for specific diseases.

ABOVE The operation of planetary influences on alchemical processes, as depicted in *Bibliotheca chemica curiosa* (Curious chemical library; Geneva, 1702), an extensive volume of alchemical texts edited by Jean Jacques Manget, a physician in Geneva. The illustration here reflects the common belief during the Middle Ages and the Renaissance that mundane substances were associated with celestial bodies: for example, the seven known metals each had a corresponding planetary counterpart.

PROFILE

Johann Joachim Becher

(1635–1682)

The German alchemist Johann Joachim Becher exemplifies how challenging it is to characterize the alchemy of the seventeenth century. From one perspective, Becher looks like the archetypal chrysopoeian charlatan, adept only at persuading princes and nobles to fund his improbable gold-making schemes. Yet in many ways he was a man thoroughly in tune with his times, able to bridge the essentially medieval power structures of the courts and the dawning age of trade and commerce.

Becher was born to Lutheran parents in the German city of Speyer. He claimed rather proudly never to have gone to any school or university, saying that he acquired his learning from knowledgeable experts and from the "light of nature"—from experience. Like Paracelsus, Becher was a restless wanderer, traveling from one court to another seeking patronage and promising marvels. In 1655 he claimed to be mathematician and alchemist to Holy Roman Emperor Ferdinand III in Vienna. By 1660 he was working in Mainz as physician and mathematician to the archbishop; four years later he was in Munich conducting similar duties for Elector Ferdinand Maria, ruler of Bavaria.

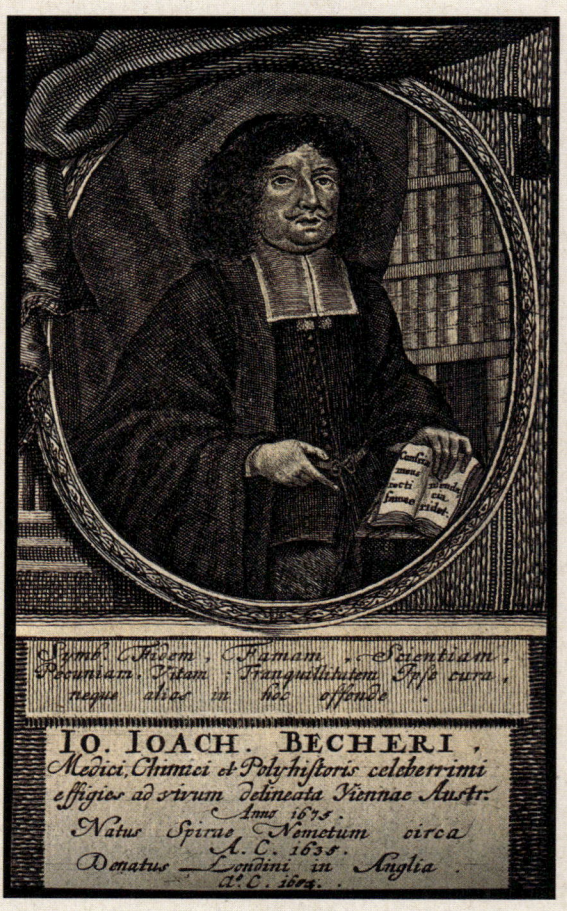

RIGHT Johann Becher, in a line engraving by W. P. Kilian, 1675. Becher was a complex, somewhat mercurial character: a showman and charlatan who duped nobles into believing he could turn metals into gold, but also an entrepreneur who saw that chemical transformations could be an instrument of state power.

OPPOSITE Frontispiece from an edition of Johann Becher's *Physica subterranea* (*Subterranean physics*; Leipzig, 1738). In this treatise on minerals, Becher laid out a new theory of the elements.

138 CHAPTER FIVE

From 1666 Becher styled himself Adviser on Commerce to the Holy Roman Emperor Leopold I, although in the mid-1670s he was forced to wander Europe in search of new patrons. He personifies the early modern "virtuoso," who could make an impression in the courtly world with knowledge of chemistry, mathematics, and the mechanical arts such as the construction of arcane machines: a very practical kind of "magic" in which human art was considered to imitate nature in a manner both productive and entertaining.

He found support in 1678–79 in Holland, where he promised the Dutch government that he could transform sand from the beaches into gold. For this venture, Becher told the rulers that he would need funds of a hundred pounds in silver. Even though Holland, at war with France, desperately needed the wealth that Becher claimed his transmutation experiments would provide, they were not so trusting as to bankroll the scheme without seeing some evidence that it was possible. With the aid of a furnace and a large waterwheel to power his machinery, Becher allegedly carried out a successful act of gold-making in 1679—at first with just a sole witness, but subsequently in the presence of the mayor of Amsterdam and a committee from the Dutch government. "He claims that there will be almost as much net profit in it as in the mines of Hungary," wrote the German mathematician Gottfried Leibniz.

But Becher fell foul of political intrigues in the Dutch court—again like Paracelsus, he was good at making enemies—and he was forced to flee to England, to the dismay of the Dutch authorities who had paid so generously for the trial run. He spent the rest of his life in London,

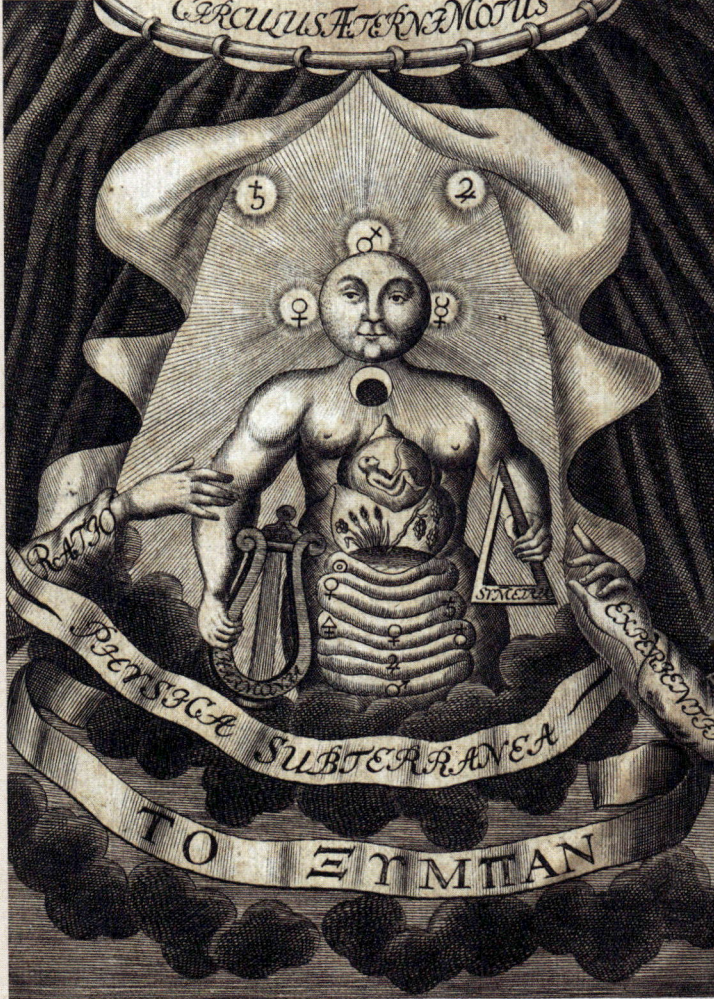

where again he managed to inveigle his way into a position of influence, advising the naval commander Prince Rupert of the Palatinate on mining in Cornwall.

Just as Becher strove to achieve the projection of metals into gold, historian Pamela Smith says that his "whole life was engaged into the projection of his own plan . . . onto the world around him." It was a very modern plan: to turn chymistry into commerce.

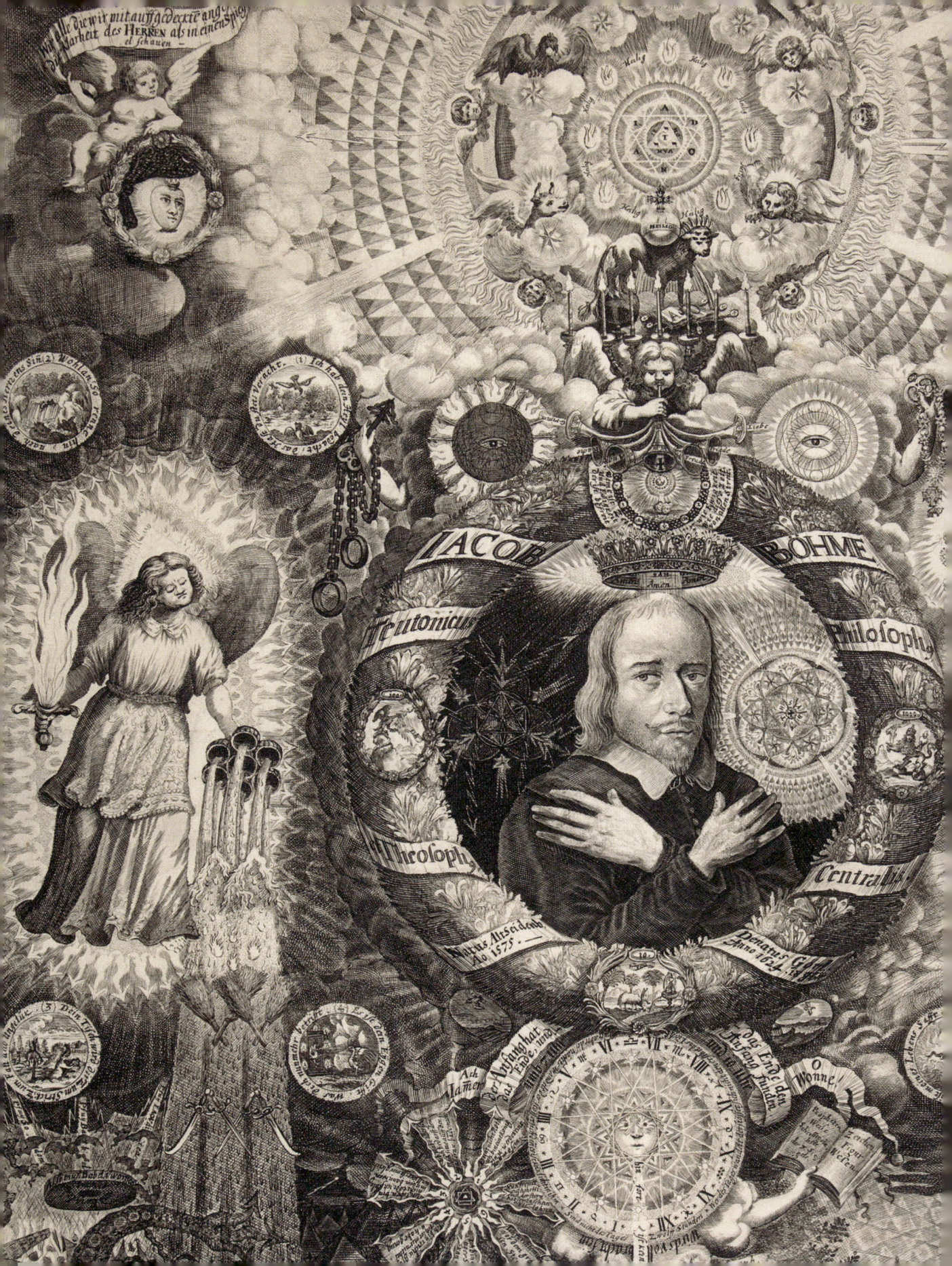

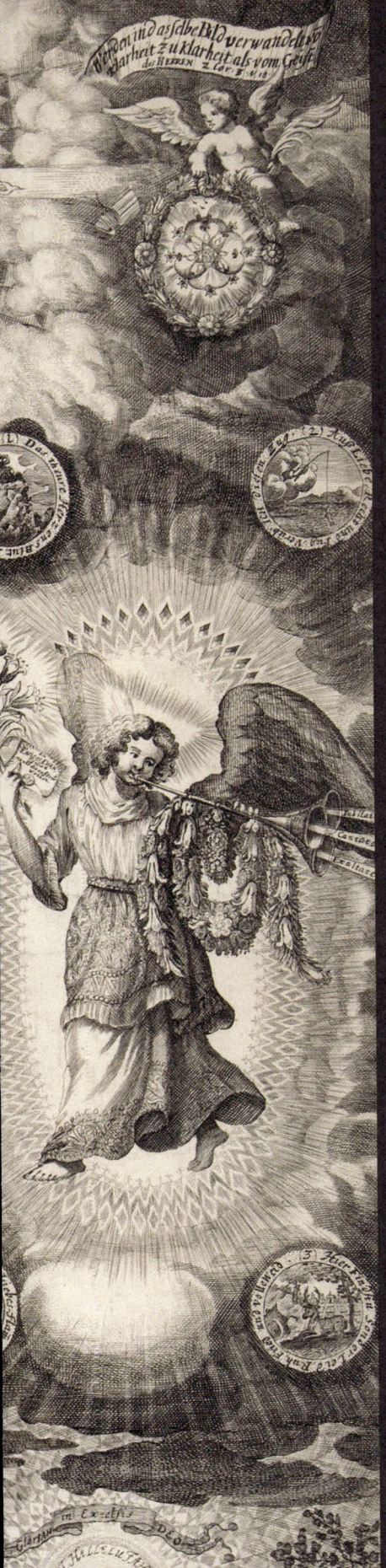

CHAPTER SIX

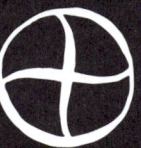

Theater of the World

THE CHEMICAL PHILOSOPHY

LEFT Portrait of the mystic Jacob Boehme by Nicolaus Häublin, after Lucina of Liebenau, 1677. Boehme (1575–1624), a Christian mystic born in a part of Bohemia now in Poland, exemplifies the late phase of alchemy, in which it merged with mystical theological movements that themselves influenced the German Romantics of the 18th and 19th centuries. Boehme drew heavily on the ideas of Paracelsus to inform a position known as theosophy, which held that it was possible to attain direct knowledge of God and his divine purpose and design.

CHAPTER SIX

It is common today to imagine that a scientific cosmology will be found in the abstruse theories of physics: general relativity, quantum theory, string theory. Chemistry, meanwhile, seems a literally mundane science, a grimy, pragmatic affair accompanied by smells and bangs. But there was a time when it was chemistry, not physics, that seemed most likely to furnish a "theory of everything"—a vision of how the universe is composed and arranged. This is precisely what alchemy seemed to some of its adepts to promise: a chemical philosophy of nature, the human body, and all God's works.

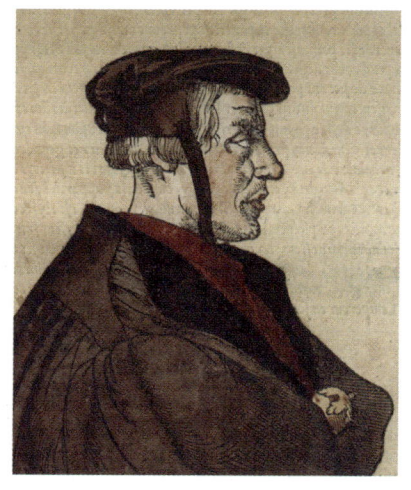

ABOVE Portrait of the renowned natural philosopher Heinrich Cornelius Agrippa, from his *De occulta philosophia libri tres* (Three books of occult philosophy; Cologne, 1533). A German from Cologne, Agrippa was one of the most renowned natural magicians of the Renaissance, and became the subject of demonic rumors; he was one of the inspirations for Goethe's Faust.

The ridicule sometimes heaped on alchemy today (imagine thinking metals could be transmuted into one another!) echoes that often directed at a belief in magic. Indeed, the two practices were allied, but not because both were absurd superstitions—quite the opposite. There was a kind of magic, vigorously asserted and defended during the late Middle Ages and the Renaissance, that was presented as an antidote to superstition, and which played an important role in the development of early science as well as in the foundations of alchemy.

It was called natural magic. The very name insisted on a contrast with the "unnatural" and illicit magic of witchcraft and necromancy, which relied on the assistance of demons. Natural magic, its advocates said, did nothing more than draw on the forces and influences inherent in nature: the principles that cause plants to grow, rain to fall, and, as it was believed at the time, metals to mature in the earth from one kind to another. "Natural magic," wrote Paracelsus, "is the use of true, natural causes to produce rare and unusual effects by methods neither superstitious nor diabolical."

The practitioners of natural magic emphasized that distinction to evade accusations of blasphemy (they were not always successful). It was equally important to them that they not be accused of credulous mysticism, and that natural magic be seen as perfectly rational. The forces of magic were invisible, it is true:

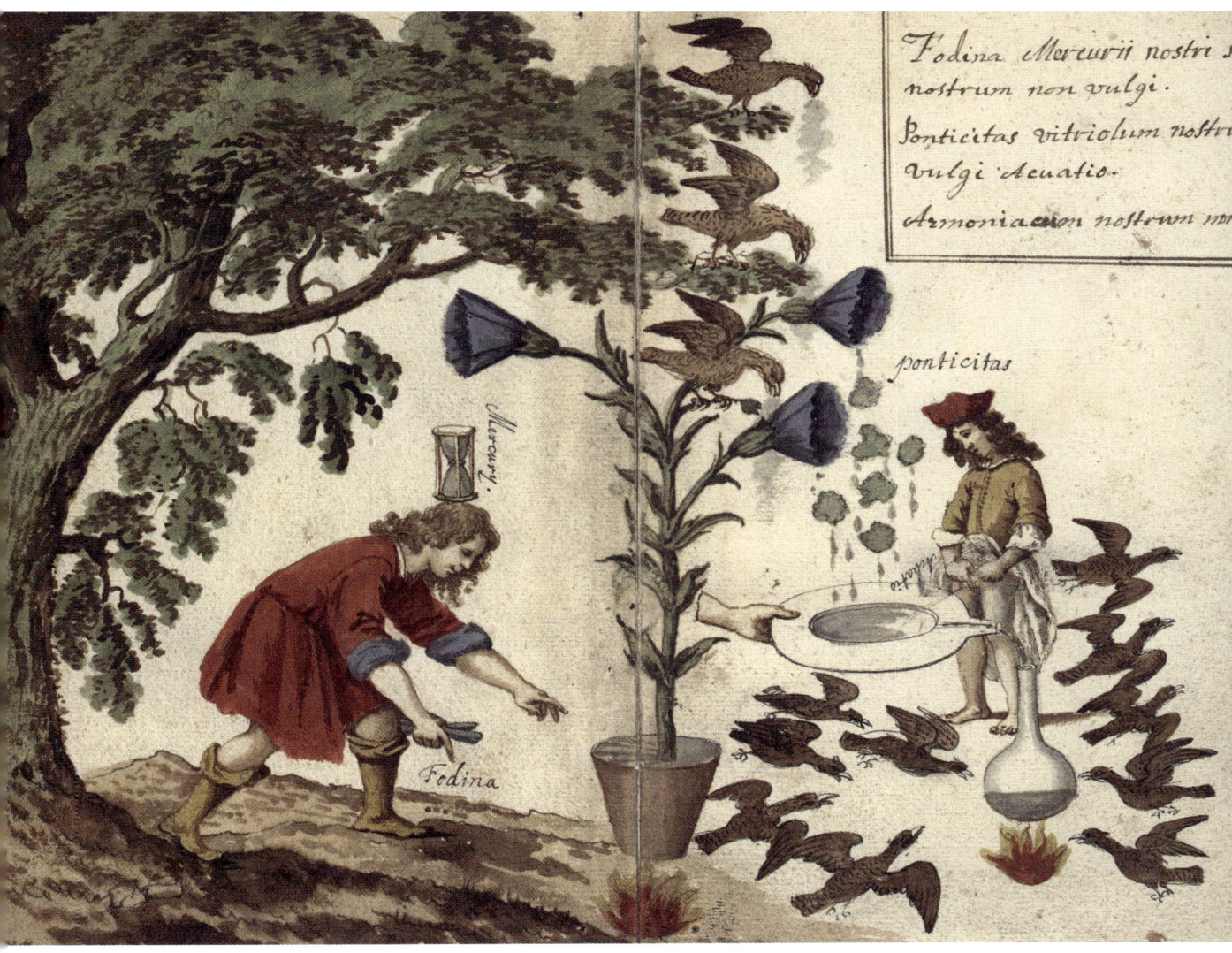

ABOVE "Natural magic," represented in Rabbi Simeon ben Cantara's manuscript *Cabala mineralis* (c. 1675–1700). A boy urinates into a distillation vessel, providing raw matter for the "Secret Fire/Mercurius," symbolized by the blue-flowered plant. This work is thought by some to engage in ambiguous and alchemical word-play: *cabala* seems at first to allude to the Jewish mystical tradition of the Kabbalah, but might also be derived from the female form of the Latin *caballus*, "horse." The word *Ponticitas* above the urinating boy seems to be a neologism alluding to a capacity to build bridges.

they were occult in the literal sense. But what was so fanciful in that? The operation of invisible forces was plain for all to see—for example, in the attraction of iron to a magnet. What the natural magician sought to do was to harness these forces to achieve wondrous things.

But precisely because the forces of natural magic were hidden, they were difficult to discern, comprehend, and command. They were "secrets" of nature. As the German physician Heinrich Cornelius Agrippa, one of the most renowned advocates of natural magic in the sixteenth century, wrote,

RIGHT "The Proportions of Man and Their Occult Numbers," from Agrippa's *De occulta philosophia libri tres* (Three books of occult philosophy; Cologne, 1533). Agrippa's treatise was considered the most important work on natural magic at that time—although, perplexingly, Agrippa also published an apparent refutation of that work.

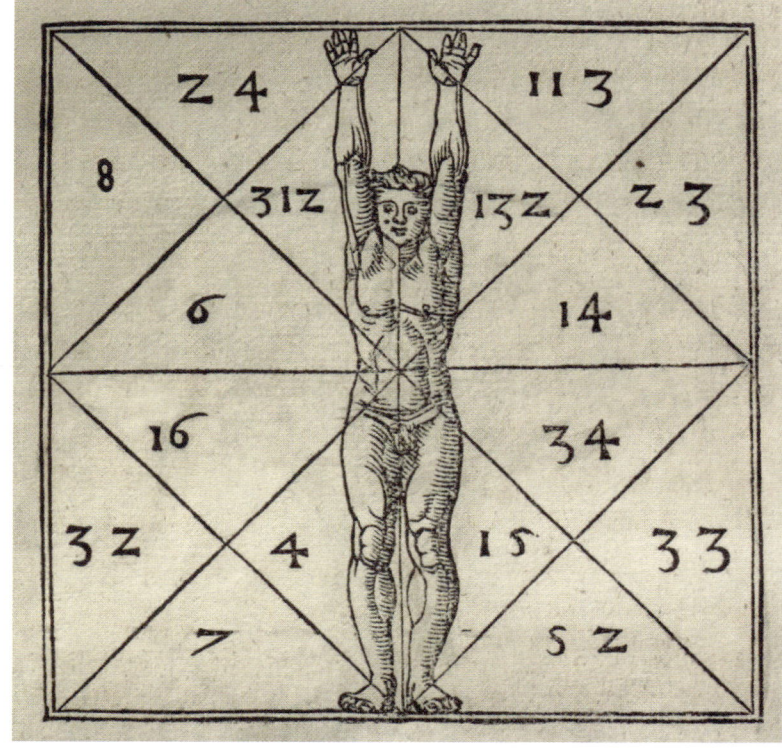

Magic comprises the most profound contemplation of the most secret things, their nature, power, quality, substance, and virtues, as well as the knowledge of the whole nature. It instructs us concerning the differences and similarities among things, from whence it generates its marvellous effects, by uniting the virtues of things.

The magician, then, is merely a servant and scholar of nature, enabling nothing to happen that nature could not also in principle achieve, albeit typically more speedily.

Plato's return

In the late fifteenth century an Italian humanist scholar named Marsilio Ficino translated a collection of Hellenistic manuscripts called the *Corpus hermeticum*, probably written in the second and third centuries B.C. but attributed to the legendary magus Hermes Trismegistus. The *Corpus* touched on topics including astrology, medicine, botany, theology, and alchemy. Ficino also translated Plato's classic work *Timaeus* and the writings of the third-century Greek philosopher Plotinus, architect of the revival at that time of Platonic ideas in the movement that became known as Neoplatonism. This philosophy postulated a hidden cosmic order uniting the macrocosm and the microcosm: "As above, so below," the central precept of the *Emerald Tablet* (see page 29). Ficino himself established the Platonic Academy in Florence to expound Neoplatonic philosophy.

This tradition held that there are "correspondences" between objects and events in the heavens and those found on Earth. The seven known celestial bodies—the Sun and Moon, Mercury, Venus, Mars, Jupiter, and Saturn—each had an associated metal (respectively, gold, silver, mercury, copper, iron, tin, and lead). Correspondences also existed between the properties of metals and herbs and the principles governing the human body, so that,

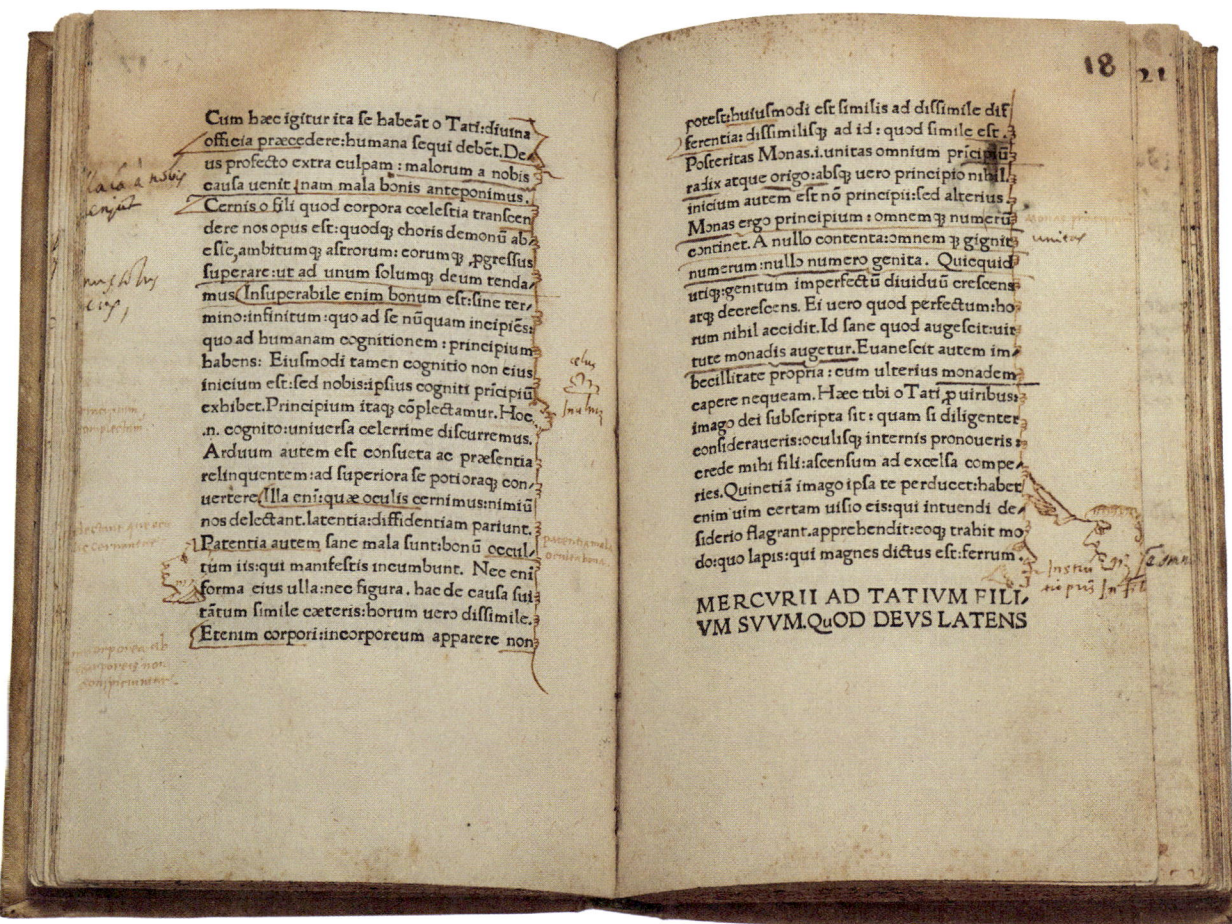

BELOW The first printed edition, with annotations, of Marsilio Ficino's *Pimander*, or *De potestate et sapientia dei* (On the power and wisdom of God; Treviso, 1471). It comprises translations from Greek of the *Corpus hermeticum* attributed to Hermes Trismegistus; Ficino made the translation at the request of the Florentine Cosimo de' Medici. It triggered a flourishing of interest in the tradition known as Neoplatonism.

> Astrology had long been as controversial as alchemy, and for much the same reason—not because it was inherently absurd but because it was abused by charlatans.

for example, nature contained clues about the medicinal virtues of its components. Plants with heart-shaped leaves might help to cure diseases of the heart, and so on, while illnesses associated with an excess of fluid (such as the swelling known as dropsy) needed to be treated with substances with a sun-like (and thus "drying") yellow complexion, such as iron oxides or sulfur. Paracelsus suggested that these "signatures" were clues left by God that the natural magician could recognize and interpret to devise cures.

The natural magician knew how to manipulate this divine web of forces and correspondences to achieve wonders. Ficino's friend and protégé Pico della Mirandola wrote: "As the farmer weds his elms to vines, even so does the magus wed earth to heaven, that is, he weds lower things to the endowments and powers of higher things." Some of this secret knowledge might be found by careful observation and experimentation; some might be deduced from numerological interpretations of the Bible and other works, including the branch of Jewish mysticism called Kabbalah that (some claimed) God had revealed to Moses.

Neoplatonism seemed to offer a justification for astrology, the belief that the positions of the stars and heavenly bodies influenced events on Earth, so that astronomical observations could be used to cast horoscopes and foretell the future. Astrology had long been as controversial as alchemy, and for much the same reason—not because it was inherently absurd but because it was abused by charlatans. Pico della Mirandola and Ficino both wrote attacks on astrology, yet they remained convinced of its core principles. Within the framework of natural magic, it was supposed that the heavenly bodies exert real (but invisible) forces on earthly objects—including the human body, thereby influencing our health. After all, did not the sun bake moisture from the fields, and, many believed, cause diseases by making the air humid? Did the rise and fall of the tides not follow the waxing and waning of the moon? More broadly, cosmic events reflected terrestrial ones: the appearance of a comet was commonly thought to be an omen, perhaps foreshadowing social unrest or pestilence. (Such traumatic events were common enough during the Renaissance to find matches with any cometary apparition.) Astrological prognostication was thus seen as no different from being able to predict the occurrence of eclipses, which astronomers could do with impressive reliability.

Alchemy as a world model

Neoplatonism and natural magic embedded alchemy within a broader mode of thought and helped to establish it as a rational process that reflected the way all of nature worked. For the chemical philosophers, the transformations of chemistry seemed a better basis for understanding the natural world than the mathematical and logical approach favored by the Aristotelian tradition. Rather than trying to develop explanations based on deductions from first principles, it was preferable to be guided by "experience"—by careful observation and experiment. This approach meant that the chemical philosophy could turn its attention to the study of phenomena that did not obviously permit a quantitative, mathematical formulation: meteorology, geology, biology, botany, metallurgy. In this way, the picture of an alchemical cosmos contained within it the seeds of later scientific methodology, while broadening the scope of natural philosophy itself.

Paracelsus argued that the human body was itself a microcosm of this chemical macrocosm, with its own sources of internal heat and circulation of fluids. He wrote that we each have inside us a kind of inner alchemist he called the *archeus*, which separates the good from the bad in what we ingest and uses the former to make flesh and blood while we expel the latter as waste. This might be seen as the first intimation of our biochemical nature, operating through the same principles as the rest of chemistry.

Neoplatonic notions of invisible forces, influences, and emanations, in contrast to the "mechanistic" picture that relied on physical contact between objects, could accommodate phenomena such as magnetism, which had long seemed magical in its ability to act across empty space. The English philosopher William Gilbert proposed at the start of the seventeenth century that the Earth itself is a giant magnet, explaining how the compass works, while magnetic action at a distance provided the model used by Isaac Newton for his theory of gravity in the 1680s.

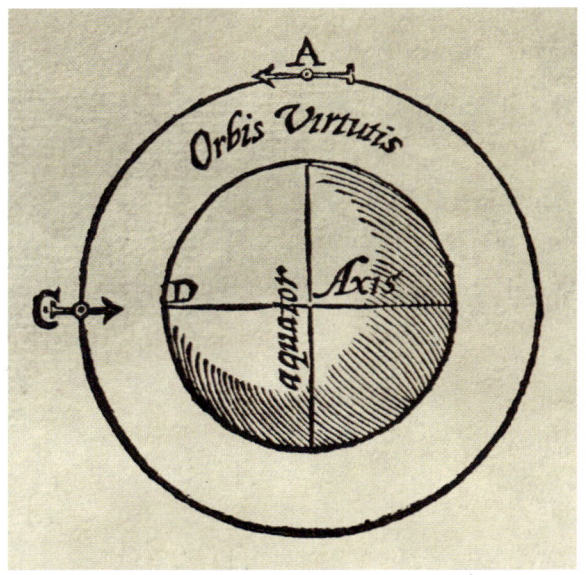

RIGHT Illustration from William Gilbert's treatise on the magnetic Earth, *De magnete* (London, 1600), showing *Orbis virtutis*, or the magnetic field of force. Gilbert was the first to propose that the Earth itself might be a giant magnet, explaining why compass needles always point to the poles.

The chemical philosophers of the late sixteenth and early seventeenth century regarded all of nature as a kind of chemical laboratory. Thus, the Earth was like a furnace, producing volcanic heat that expels groundwater via mountain springs and streams in a kind of distillation, and driving the fermentation and transformation of metals in veins of ore. Clouds were said to be formed by evaporation and condensation (as indeed they are!), again like distillation in the alchemical alembic. Lightning and thunder were compared to the detonation of gunpowder, caused by a kind of "sulfur" and "niter" (the chemical ingredients of gunpowder) in the air.

This vision of the Earth as a giant alchemical furnace was articulated by the German Jesuit and polymath Athanasius Kircher in his 1665 book *Mundus subterraneus* (The subterranean world). The book showed the Earth with a fire at its core connected by conduits to smaller conflagrations deep underground, linked in turn to flaming volcanoes on the surface: a remarkable presentiment of the current view in the earth sciences that the planet has an incandescent core of iron and that globs of molten mantle feed magma to volcanoes.

Kircher is the kind of figure who is hard to fit into the conventional history of science that posits a "Scientific Revolution" in the seventeenth century based on Newton's mathematical mechanics. While not exactly an alchemist himself, Kircher shared

BELOW Hand-colored illustration from Athanasius Kircher's *Mundus subterraneus* (The subterranean world; Amsterdam, 1665) showing the Earth with fire at its center. Kircher, a German Jesuit, was a polymath who wrote on a wide range of topics, from music to geology and Egyptology. His speculations about the source of volcanism might be considered to foreshadow the modern understanding of the planet's inner structure.

the spirit of the chemical philosophy, working on geology, biology (where his ideas hinted at the evolution of species), epidemiology, pure mathematics, and magnetism, as well as the study of Egyptian hieroglyphics, the theory of music, and interpretation of the Bible. Kircher speculated that the planets are other worlds and the stars other suns.

Many historical accounts of the Scientific Revolution tend to obscure just how different seventeenth-century "science" really was from that of today. Take the English physician Robert Fludd, whose enthusiasm for the Hermetic philosophy that included

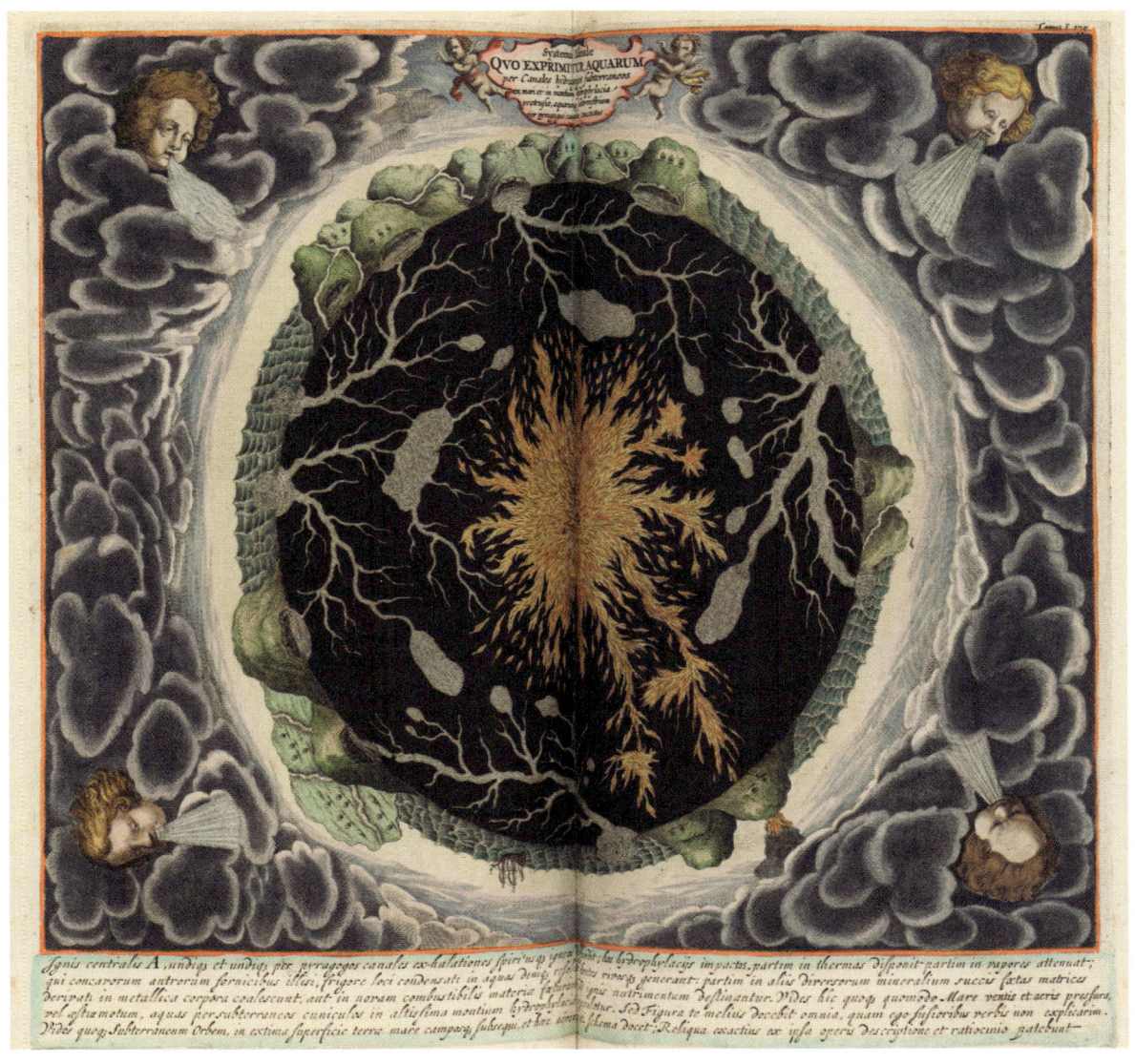

The chemical philosophy also had a strong religious dimension. "Experience" for the chemical philosopher did not just mean experiment and observation but also divine revelation.

natural magic and alchemy makes him sound like a throwback to a more superstitious age. This was not really so: Fludd was a respected member of the Royal College of Physicians, and, in his belief that blind adherence to the precepts of classical authorities such as Aristotle and Galen needed to be replaced by a new teaching that placed emphasis on observation and experience, he was very much in tune with his times.

However tempting it might be to distinguish mathematically inclined "proto-scientists" like the German astronomer Johannes Kepler from alleged mystics like Kircher and Fludd, that simply won't do. Kepler disapproved of Fludd, saying that he offered explanations that were "enigmatic" and "emblematic" and not based on mathematics like his own, while Fludd responded that Kepler was one of those mathematicians who "concern themselves with quantitative shadows." But the natural philosophy of the two men was not really so different: both believed in a kind of mathematical cosmic harmony and in a vital *spiritus mundi* that animated the world.

Alchemy and religion

The chemical philosophy had broader ambitions than explaining the natural world—or perhaps we should rather say that such an explanation was expected to have a somewhat different character from the naturalistic accounts that scientists seek today. For the chemical philosophy also had a strong religious dimension. "Experience" for the chemical philosopher did not just mean experiment and observation but also divine revelation. This religious inflection is one reason why it does not quite suffice to say that alchemy was merely the forerunner of modern chemistry.

Paracelsus and his later followers frequently justified their ideas using biblical references: he himself argued that his notion of the *tria prima* of sulfur, mercury, and salt makes sense because "God made everything from three." The Creation itself was envisaged as an alchemical process of separation—of water from earth, or, as the English clergyman and writer Thomas Tymme wrote in the

OPPOSITE Engraving from the English philosopher Robert Fludd's *Tomus secundus ... de supernaturali, naturali, praeternaturali et contranaturali microcosmi historia* (Second volume ... of the supernatural, natural, supernatural and contranatural history of the microcosm; Oppenheim, 1619) illustrating Fludd's belief in a deep harmony that governed both the cosmos and the human body.

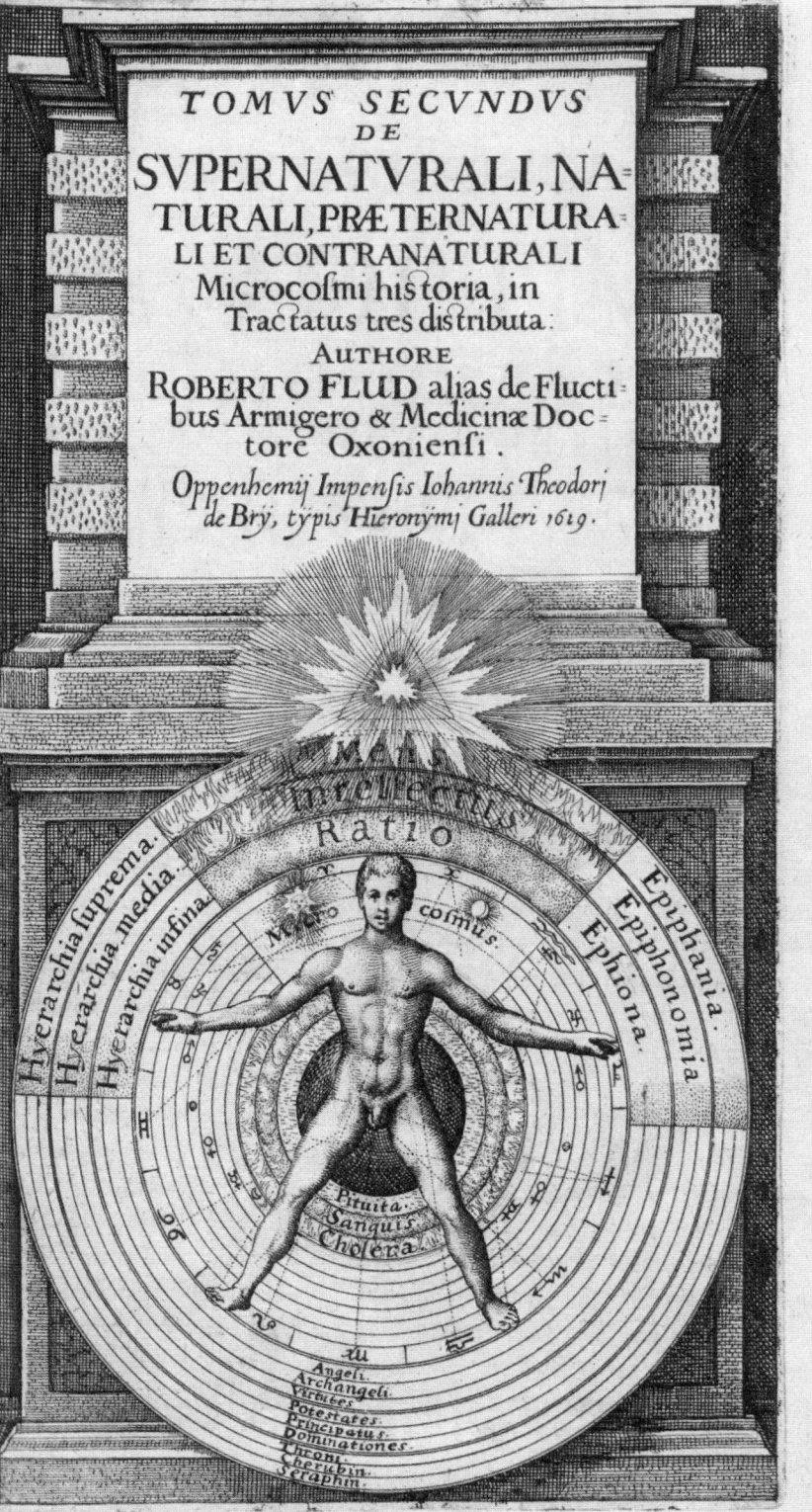

TOMVS SECVNDVS
DE
SVPERNATVRALI, NA=
TVRALI, PRÆTERNATVRA=
LI ET CONTRANATVRALI
Microcosmi historia, in
Tractatus tres distributa:
AUTHORE
ROBERTO FLUD alias de Flucti=
bus Armigero & Medicinæ Doc=
tore Oxoniensi.

Oppenhemij Impensis Iohannis Theodori
de Bry, typis Hieronymi Galleri 1619.

early seventeenth century, a matter of "Halchymicall Extraction, Separation, Sublimation, and Conjunction."

This Christian aspect of alchemy was particularly prominent in the works of the German physician and alchemist Heinrich Khunrath, a follower of Paracelsus who worked for a time in the court of Rudolf II in Prague. An illustration in Khunrath's key work *Amphitheatrum sapientiae aeternae* (Amphitheater of eternal wisdom; 1595) shows the alchemist kneeling in prayer in his laboratory, hoping for enlightenment from God. Khunrath firmly believed that true adepts were *theodidaktoi*: taught by God through visions and dreams. He considered the processes of purification and perfection pursued by alchemists to be analogous to the Christian goal of purifying the spirit, and he compared the raising

BELOW The Paracelsuian *tria prima* of sulfur, mercury, and salt, depicted in a three-volume alchemical manuscript produced in Germany in 1738, *Clavis artis* (The key of art), attributed to the Persian Zoroaster (Zarathustra). Paracelsus augmented the older idea that all metals are composed of sulfur and mercury, adding salt—said to engender solidity—so that the scheme could encompass all substances, including the fabric of the human body.

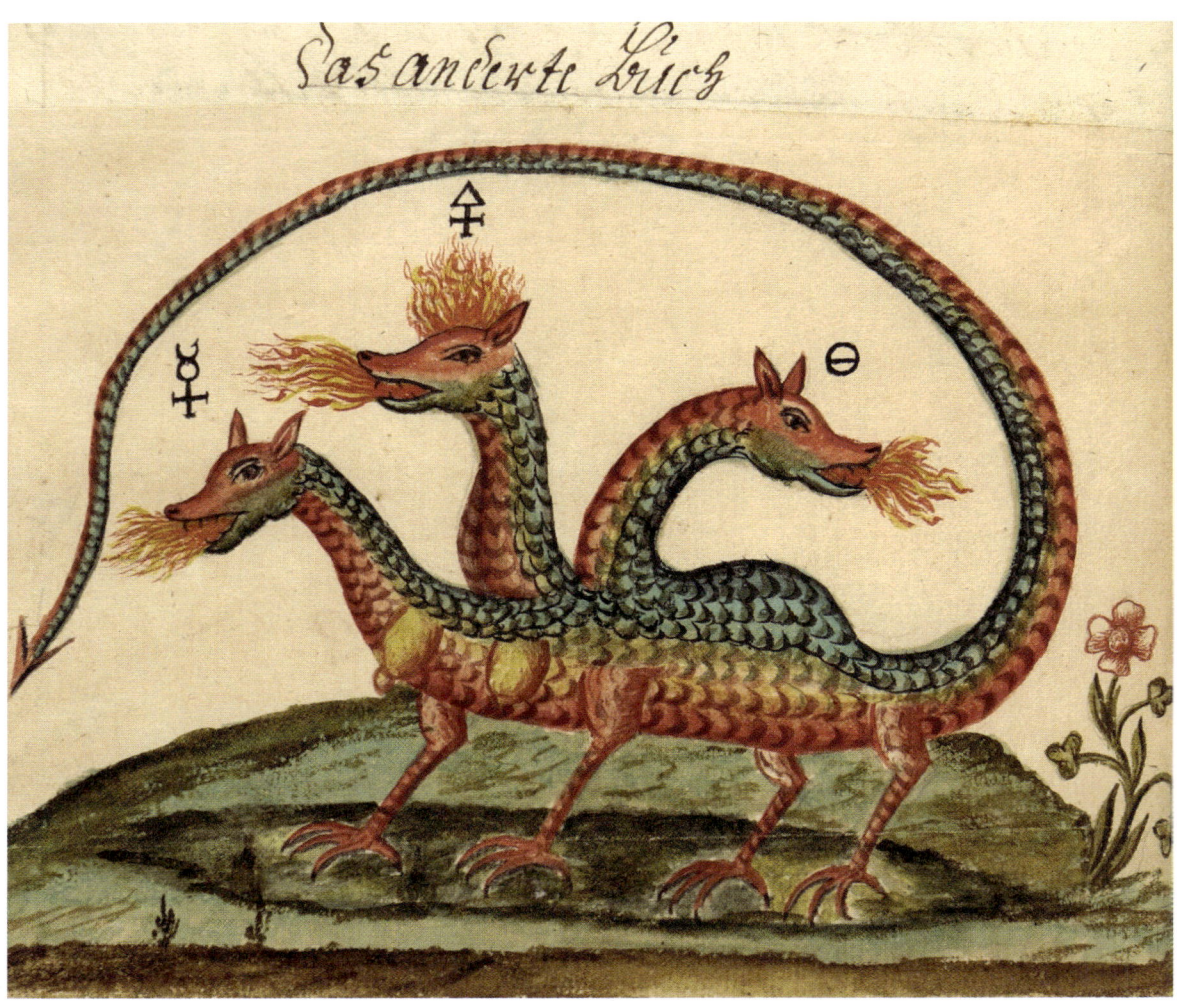

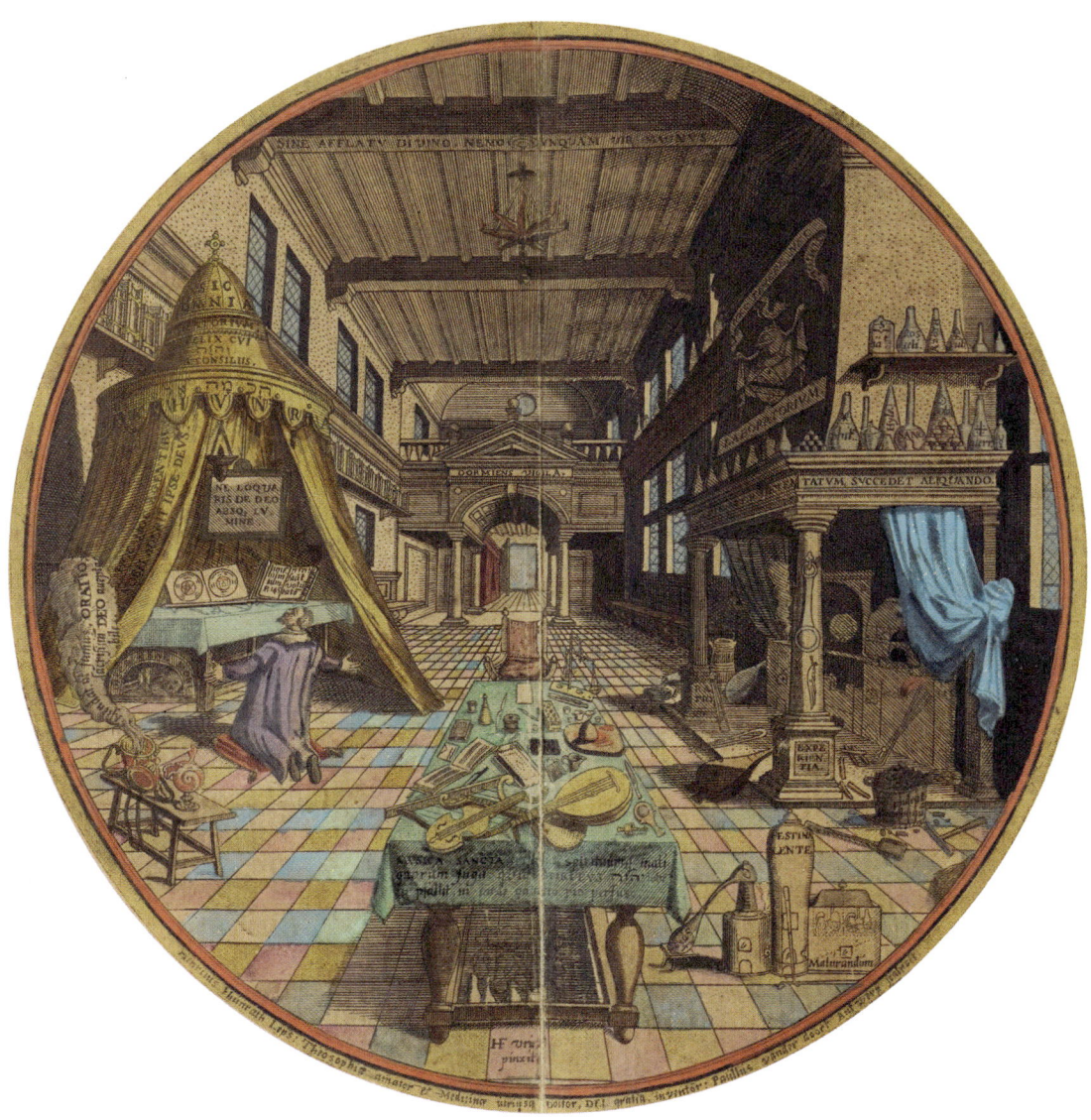

ABOVE An illustration in Heinrich Khunrath's key work *Amphitheatrum sapientiae aeternae* (Amphitheater of eternal wisdom; 1595) shows an alchemist kneeling in prayer in his laboratory, hoping for enlightenment from God. The image reflects Khunrath's conviction that alchemy was as much a spiritual as a practical discipline, requiring religious devotion from its practitioners.

of base metals to an elevated state by the philosopher's stone to the redemption of humankind offered by Christ.

To some of its advocates, then, alchemy seemed less like a practical art performed in smoky laboratories and more like an allegory for something else. The German scholar Michael Maier, physician to Rudolf II, was one of these. His most celebrated book *Atalanta fugiens* (Atalanta fleeing; 1617) contained beautiful engravings that displayed alchemical processes in highly symbolic and sometimes rather obscure forms, along with alchemical poems and even a piece of alchemical music composed by Maier. The book was certainly

THEATER OF THE WORLD: THE CHEMICAL PHILOSOPHY

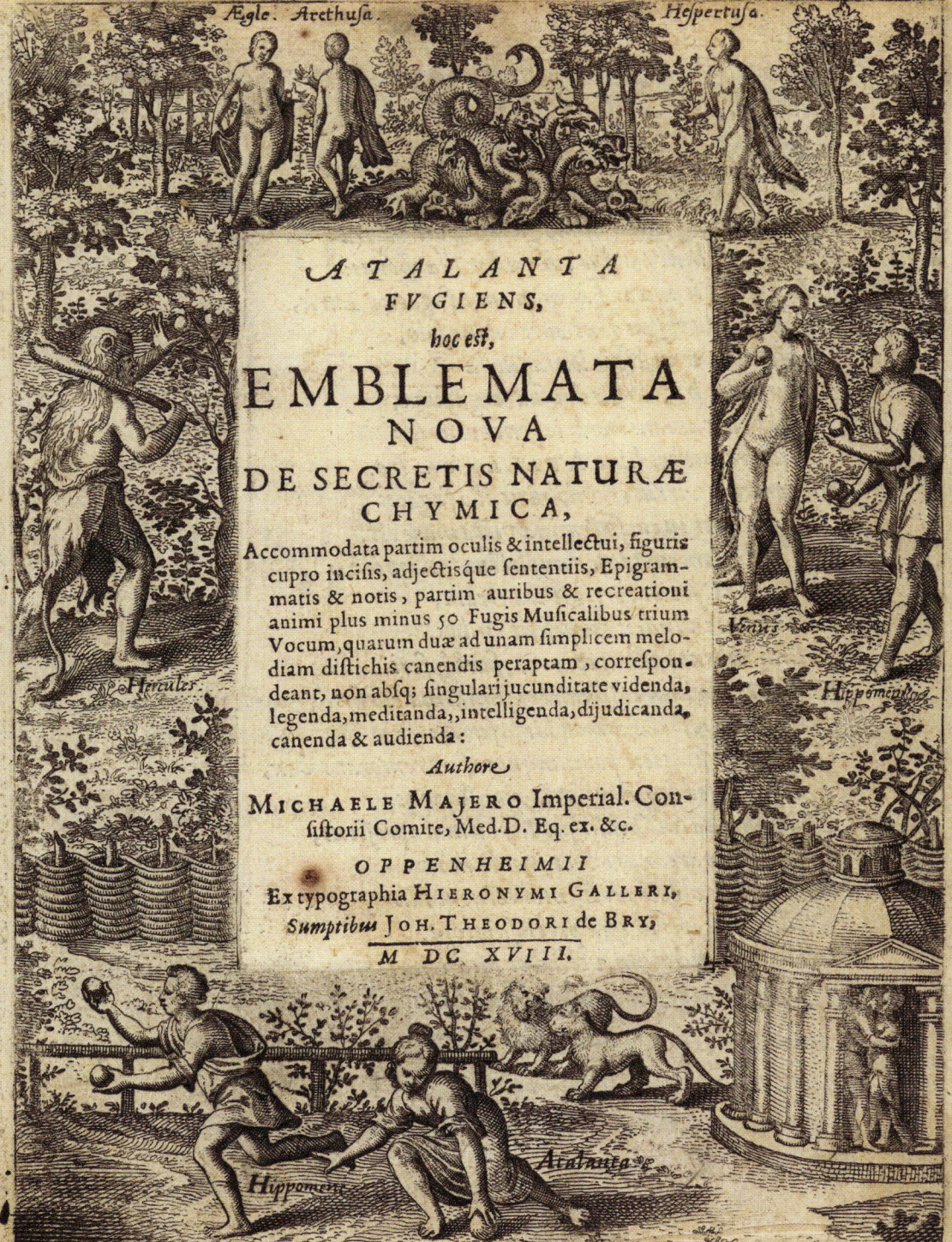

ATALANTA
FVGIENS,
hoc est,
EMBLEMATA
NOVA
DE SECRETIS NATURÆ
CHYMICA,
Accommodata partim oculis & intellectui, figuris cupro incisis, adjectisque sententiis, Epigrammatis & notis, partim auribus & recreationi animi plus minus 50 Fugis Musicalibus trium Vocum, quarum duæ ad unam simplicem melodiam distichis canendis peraptam, correspondeant, non absq; singulari jucunditate videnda, legenda, meditanda, intelligenda, dijudicanda, canenda & audienda:

Authore
MICHAELE MAJERO Imperial. Consistorii Comite, Med. D. Eq. ex. &c.

OPPENHEIMII
Ex typographia Hieronymi Galleri,
Sumptibus Joh. Theodori de Bry,
M DC XVIII.

OPPOSITE Title page of Michael Maier's *Atalanta fugiens* (Atalanta fleeing; Oppenheim, 1618), a phantasmagorical compendium of alchemical ideas, expressed partly through emblematic engravings, often alluding to Greek myth, that depicted the quest for the philosopher's stone.

BELOW "His suckling nurse is the earth": Epigram 2 and facing music by Michael Maier, in a hand-colored copy of *Atalanta fugiens* (Atalanta fleeing; Oppenheim, 1618). In this engraving, Earth suckles the offspring of Wise Men while the infant Jupiter is fed by the goat Amalthea, and Romulus and Remus by a she-wolf.

no practical manual; rather, in layers of dense allusion it sought to align alchemical thought with broader currents in humanistic scholarship. Maier wrote that "God has hidden infinite secrets in nature . . . chymical secrets are not the least of these, but are, after the investigation of divine things, the first and most precious of all."

On a visit to England after Rudolf's death, Maier probably met his kindred spirit Robert Fludd. Fludd considered alchemy to be a sort of practical theology: one could comprehend the Creation and God's works by studying chemical transformations in the laboratory. To some this seemed profoundly impious. The French mathematician and Catholic priest Marin Mersenne, a supporter of René Descartes's mechanical philosophy, regarded Fludd as a dangerous heretic, while Mersenne's colleague Pierre Gassendi accused him and his ilk of wanting to make "alchemy the sole religion, the alchemist the sole religious person, and the *tirocinium* [textbook] of alchemy the sole Catechism of the Faith."

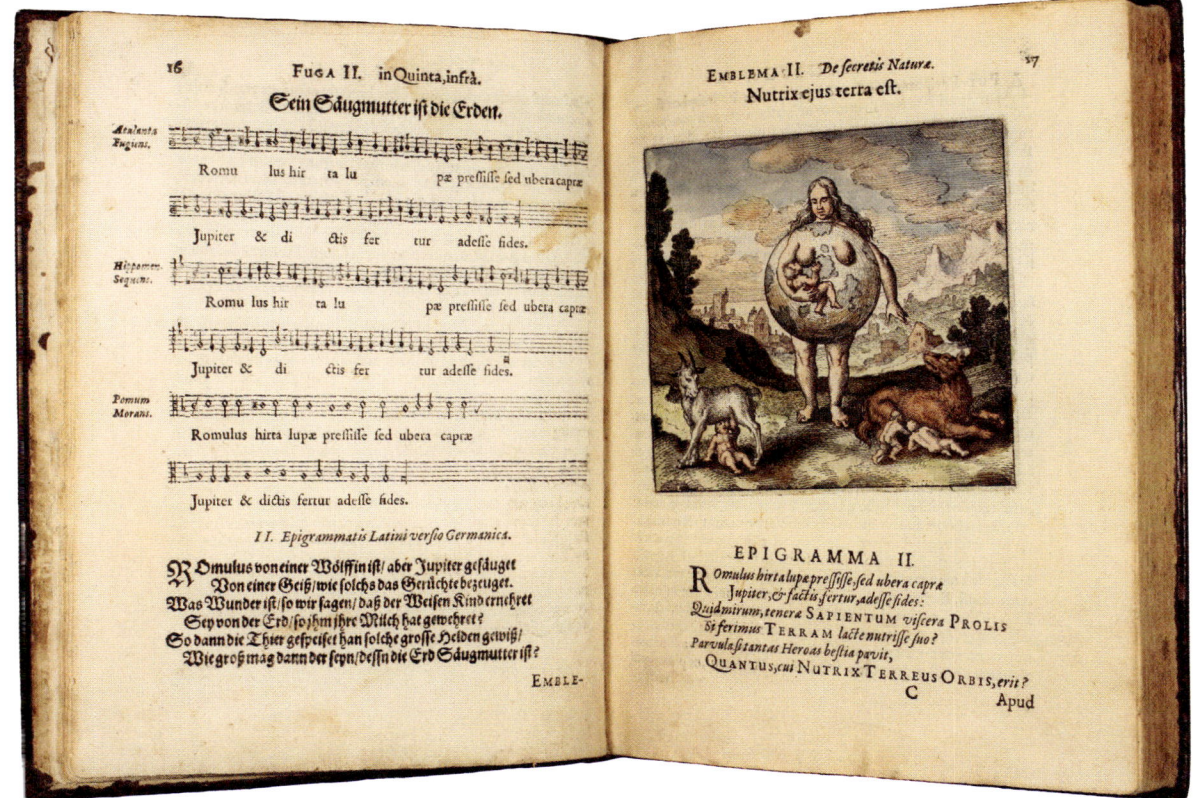

THEATER OF THE WORLD: THE CHEMICAL PHILOSOPHY

Van Helmont was eager to separate what seemed valuable in alchemy from what he deemed its more fanciful elements.

RIGHT Woodcut from the title page of Jan van Helmont's *Ortus medicinae* (Rise of medicine; Amsterdam, 1648). The printer's mark shows Minerva (goddess of wisdom) holding a banner bearing the legend "Ne extra oleas" (Nothing beyond the olive)—meaning that one should stay within the bounds of wisdom. Van Helmont was a key figure in the transition from alchemy to chymistry, and he was keen to separate what seemed useful in the subject from what he deemed superstitious and obsolete.

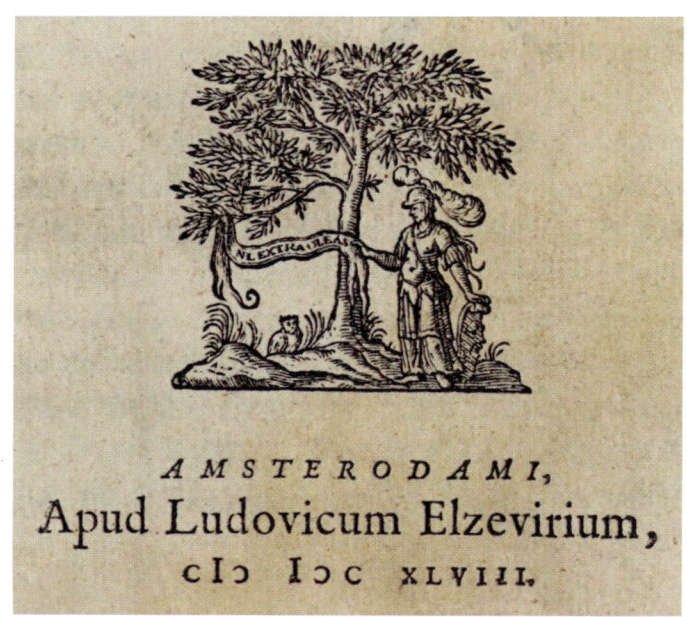

In his condemnation of Fludd, Mersenne sought to enlist the support of the Flemish physician Jan Baptista van Helmont—who responded that Fludd was not worth the effort, not because he was an alchemist but because he was a bad one, with little real learning. In truth, van Helmont shared many of the same beliefs. Despite arguing that we didn't need theologians to understand nature, nor alchemists and physicians to understand God, he too considered the Creation itself to be an alchemical process, and felt that the obsolete natural philosophy of the ancients, based on mathematical logic, needed to be replaced with one grounded in experience. He endorsed the Paracelsian *tria prima* and the idea of a *spiritus vitae* that quickened creatures into life: he even tried to distil that vital essence from blood.

Van Helmont was himself no stranger to religious condemnation. When he ridiculed the Jesuits over an issue connected to a "magnetic cure", he was summoned before the Inquisition and briefly imprisoned before being put under house arrest in the Flemish town of Vilvorde, and his great work the *Ortus medicinae* (Rise of medicine; 1648) could not be published until four years after his death. Van Helmont was eager to separate what seemed valuable in alchemy from what he deemed its more fanciful elements. He stressed the importance of careful measurement, and in his most famous experiment he "proved" that plants (and by extension, all things) are ultimately made from water alone by measuring the mass of a willow tree as it grew in a pot over several years, supplied only (or so it seemed to van Helmont, who could not know about photosynthesis) with rainwater.

Some historians regard van Helmont's willow-tree experiment as a landmark in the evolution of quantitative science. Yet the Flemish physician's work illustrates how intimately blended early modern science, alchemy, and theology were. That mixture can be found to some degree in most of the natural philosophers of the Renaissance and early modern times, most famously Robert Boyle and Isaac Newton. They have been described as "Janus-faced," looking both forward and back—but that imposes modern distinctions that would have been meaningless to them. They and their contemporaries sought a unified view of all creation, but they looked for it in different places than we do today.

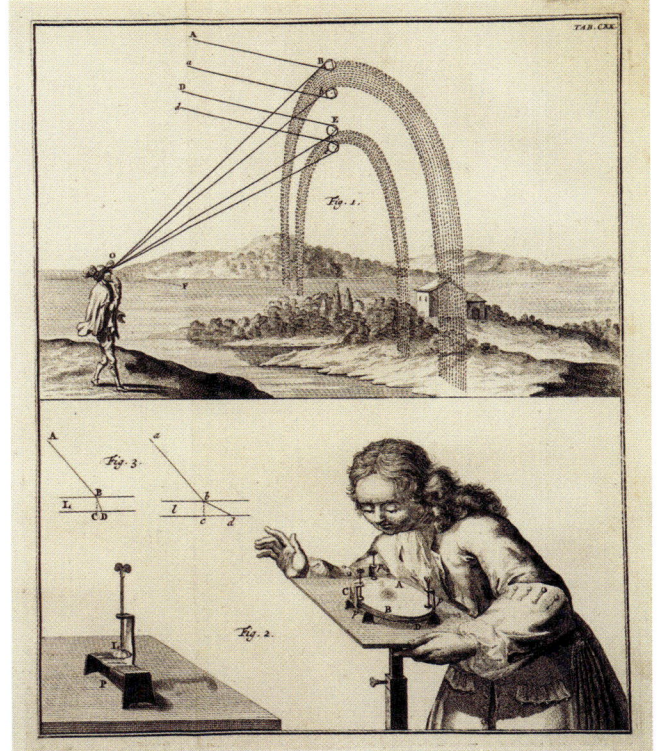

RIGHT Like Robert Boyle, Isaac Newton has been depicted as "Janus-faced," looking both forward to the modern scientific age and backward to the mystical art of alchemy. Such a view, however, misrepresents both Newton—who was very much a man of his time—and alchemy, which was by no means moribund in the 17th century. This illustration from *Physices elementa mathematica* (Mathematical elements of natural philosophy, 1742), an introduction to Newton's theorem, by his friend Willem Jacob 's Gravesande, shows Newton's theory of light and color, and how his ideas explained the colors of the rainbow.

PROFILE

Heinrich Khunrath

(c. 1560–1605)

The German alchemist Heinrich Khunrath personifies one version of what the Hermetic art had come to represent by the end of the sixteenth century: a fundamentally religious quest into the deepest secrets of nature, as revealed through chemical transformations.

Born in Dresden, Saxony, Khunrath gained his medical doctorate from the University of Basel in 1588. He knew John Dee (see page 173) and spent some time at the "Hermetic court" of Rudolf II in Prague. A proselytizer of Lutheran theology, he was influenced by the ideas of Paracelsus and took an interest in the healing powers of plants and mineral salts. He considered the goal of alchemy was to make the philosopher's stone, "the magnificent object and subject of all wonderfulness in the Heavens and on Earth"—a gift of God that could only be created with divine inspiration.

For Khunrath the stone furnished potential evidence of God's power in nature, making the quest to manufacture it a moral act of devotion. The elaborate titles of his works, including (as translated in English) *Of the Secret, External, and Visible Fire of the Mages and Philosophers* (1608) and *Physico-Chemical Testament of the Natural, Triune, Wonderful and Miraculous, Most Secret Universal Chaos of Physico-Chemists* (1598), give a good indication of this revelatory, almost rapturous approach to alchemy.

In an image that appeared first in the posthumous 1609 version of Khunrath's *Amphitheatrum sapientiae aeternae* (Amphitheater of eternal wisdom; 1595, see page 153), he is shown in his workspace, which combines a Hermetic temple on the left with a laboratory on the right: the realms of God and of Nature that, in Khunrath's view, must be united to realize the miraculous transformations of alchemy. Khunrath kneels in prayer before an altar-like table that holds mystical texts, beneath a canopy decorated with Latin and Hebrew inscriptions, testifying to his enthusiasm for the Kabbalah. Opposite this holy shrine is a curtained alcove with apparatus, including a furnace, bellows, and tongs, and shelves stocked with vessels for alchemical work. On a table between altar and furnace lie a viol, lutes, and other musical instruments, hinting at the idea of a mathematical cosmic harmony governing the relationships between all things.

Despite the dark suspicions of some of his contemporaries, Khunrath's piety could not be doubted. He warned people against intellectual arrogance, saying that no true knowledge could come "without the inspiration, aid and guidance of God." He argued that such knowledge might be found both in the writings and laboratory demonstrations of true adepts and by direct revelation from God or angels—for example, through dreams or visions.

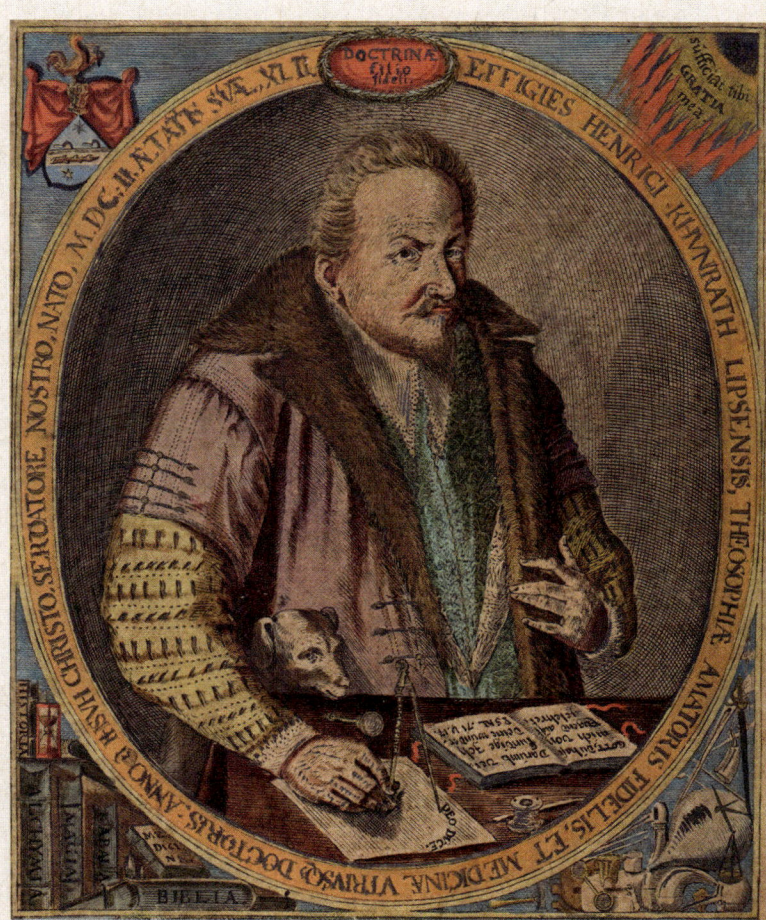

RIGHT Portrait of Heinrich Khunrath from his *Amphitheatrum sapientiae aeternae* (Amphitheater of eternal wisdom; Hanau, 1609).

To skeptics, this was a kind of magic that verged on necromancy. The German chymist Andreas Libavius attacked Khunrath and other Paracelsians for their allegedly diabolical enthusiasms, and the seventeenth-century French philosopher Marin Mersenne accused Khunrath of being "most devoted to the wicked and illicit arts, insolent to nature, injurious to men, and blasphemous to God." Khunrath denied that his work had anything to do with the "Devil's magicians or sorcerers," and defended alchemists against the accusation that they were prying into forbidden secrets by saying that God wishes all the mysteries of nature to be known.

Khunrath's divine alchemy might look today as though it sacrificed the practical for the mystical; he has been denounced as an "example of alchemy's spiritual extremists," and historian John Read alleges that he exerted "no influence upon the progress of alchemy towards chemistry." But, more recently, historian of science Peter Forshaw argues that in fact Khunrath's alchemy "spans the whole spectrum." It is only in retrospect that his theological emphasis seems at odds with the emergence of chemistry as a purely practical affair. Khunrath serves as a reminder that at the dawn of the so-called Age of Reason, alchemy retained both aspects.

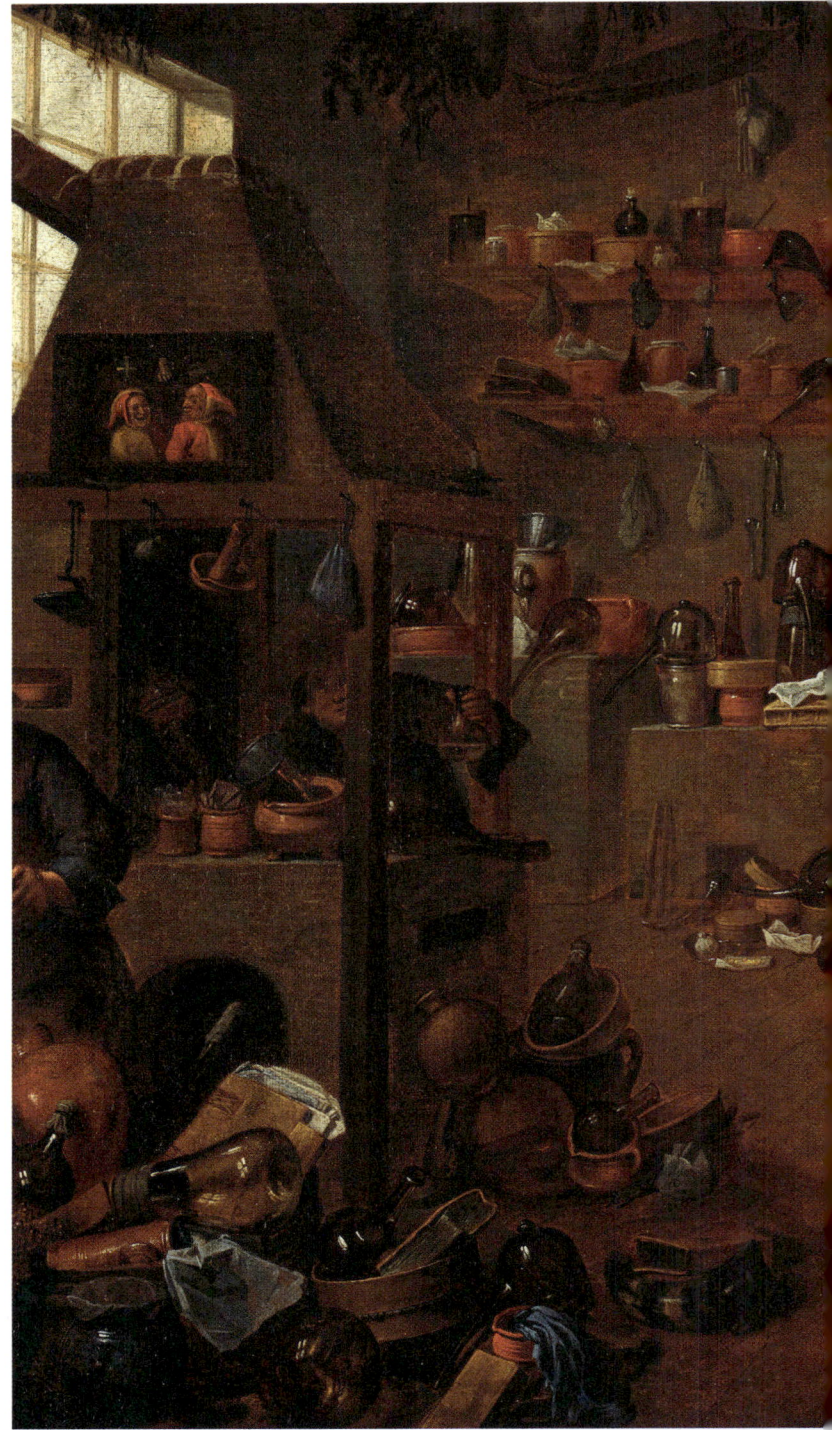

RIGHT *A Savant in His Cabinet, Surrounded by Chemical and Other Apparatus, Examining a Flask*, by Mattheus van Helmont (not a known relative of the famous Flemish physician), before 1674. Such depictions of alchemists were a popular theme for Netherlandish artists of this time, in part because of the opportunities they presented for demonstrations of an artist's skill in rendering gleaming glassware and ceramic glazes. Notice again the stylized gesture of the alchemist holding aloft a flask of colored liquid (perhaps urine?).

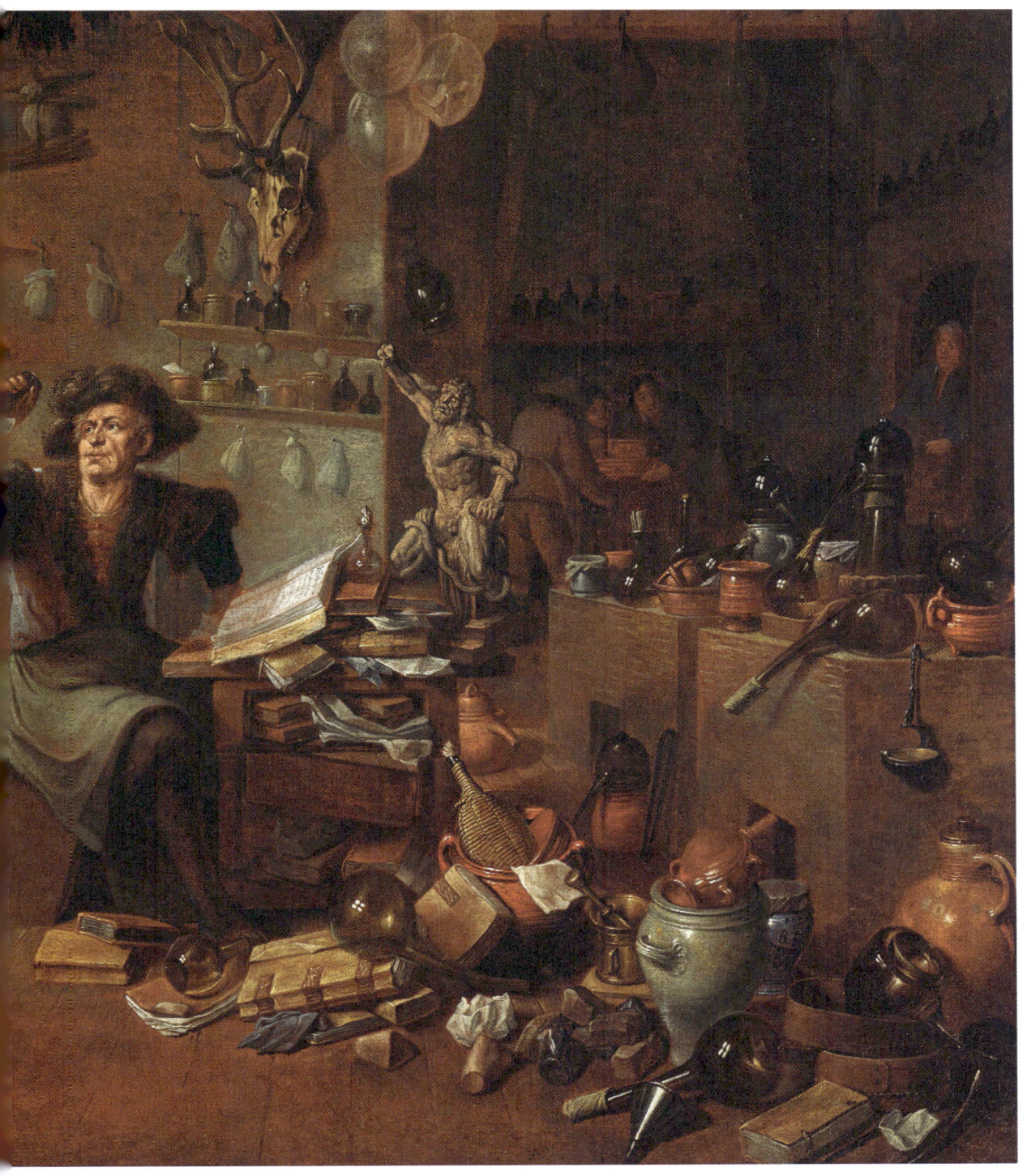

THEATER OF THE WORLD: THE CHEMICAL PHILOSOPHY

DISCURSUS XXV.
DRACO non moritur, nisi cum fratre & sorore sua interficiatur, qui sunt Sol & Luna.

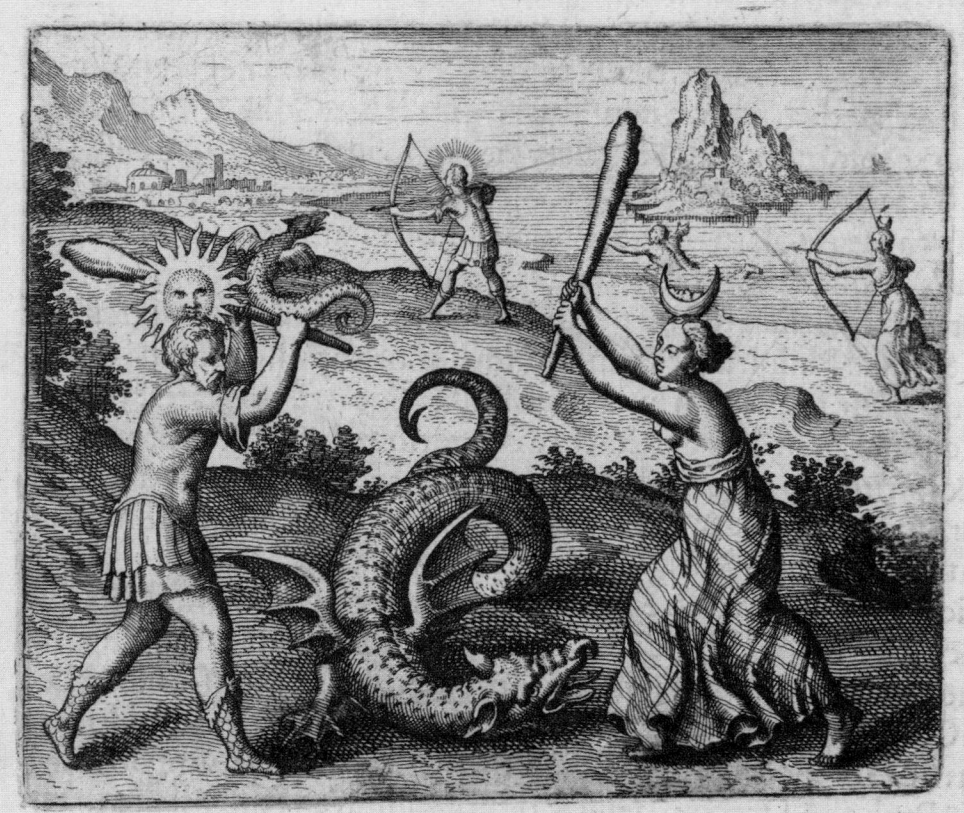

EPIGRAMMA XXV.

Exiguæ est non artis opus, stravisse Draconem
    Funere, ne serpat mox redivivus humo.
Frater & ipsa soror juncti simul illius ora
    Fuste premunt, nec res fert aliena necem.
Phœbus ei frater, soror est at Cynthia, Python
    Illâ, ast Orion hac cecidêre manu.

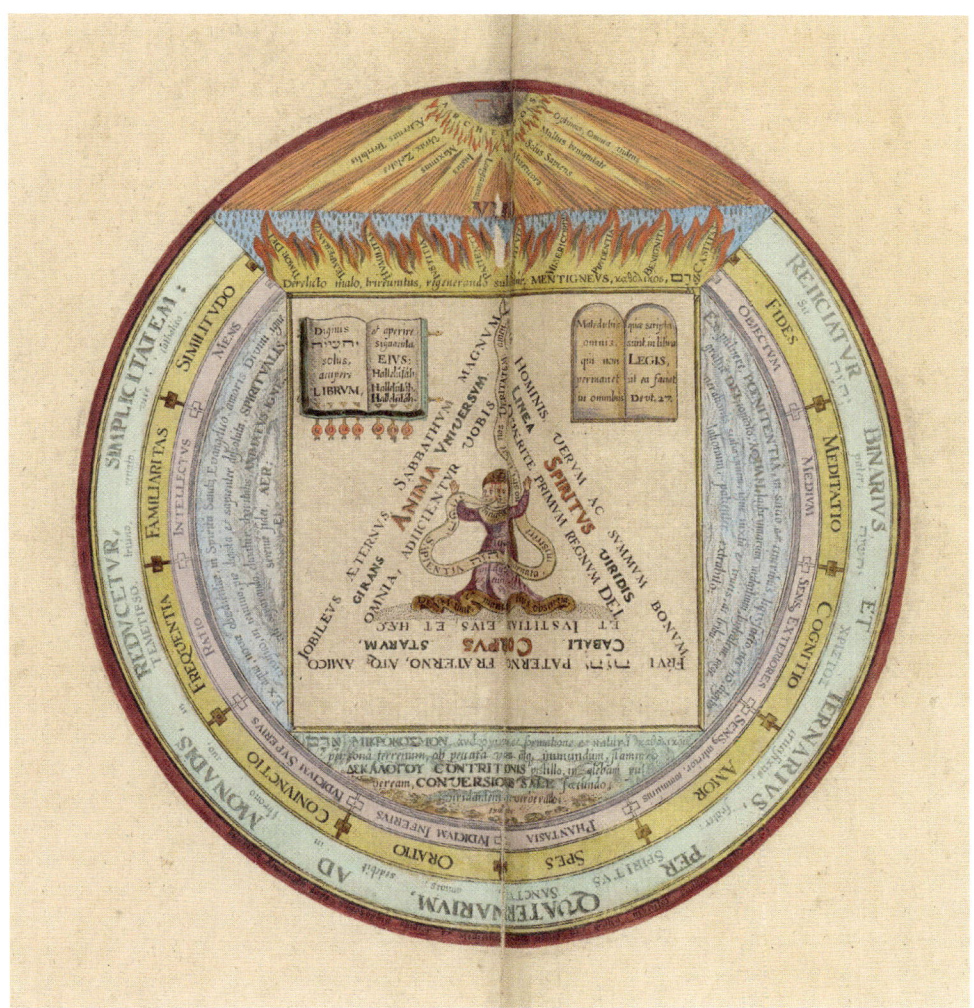

OPPOSITE "The dragon dies, unless he is slain with his brother and sister, who are the Sun and the Moon," Epigram XXV from Michael Maier's *Atalanta fugiens* (Atalanta fleeing; Oppenheim, 1618). The epigram explains that the dragon represents "philosophical mercury," which can only be coagulated with Sol (gold) and Luna (silver)—a very stylized account of the formation of amalgams of these metals.

ABOVE The conjunction of soul (*anima*), body (*corpus*), and spirit (*spiritus*): an illustration from the German alchemist Heinrich Khunrath's *Amphitheatrum sapientiae aeternae* (Amphitheater of eternal wisdom; Hanau, 1609). This text is Khunrath's best-known work, and its copperplate engravings are among the most elaborate and beautiful of that age, some of them densely inscribed with text and laden with symbolism. This complex diagram encodes, among other things, Khunrath's numerological conviction that are three things that constitute the world: a "trinitarian" structure that he discerns in Christian, Hebrew, and Greek theological/philosophical systems, reflected too in the Paracelsian *tria prima*.

THEATER OF THE WORLD: THE CHEMICAL PHILOSOPHY

RIGHT An allegory of Distillation (left) and Mercury workers (right), from the *Book of Alchemical Formulas* compiled by Claudio de Domenico Celentano di Valle Nove and dedicated to Gonzalo Pérez, a powerful aristocrat in Spanish-ruled Naples (1606). Little is known of this book's author, but the work itself is a slightly eccentric collection of heavily allegorical images, like that on the left, alongside rather straightforward practical illustrations of techniques, like that on the right.

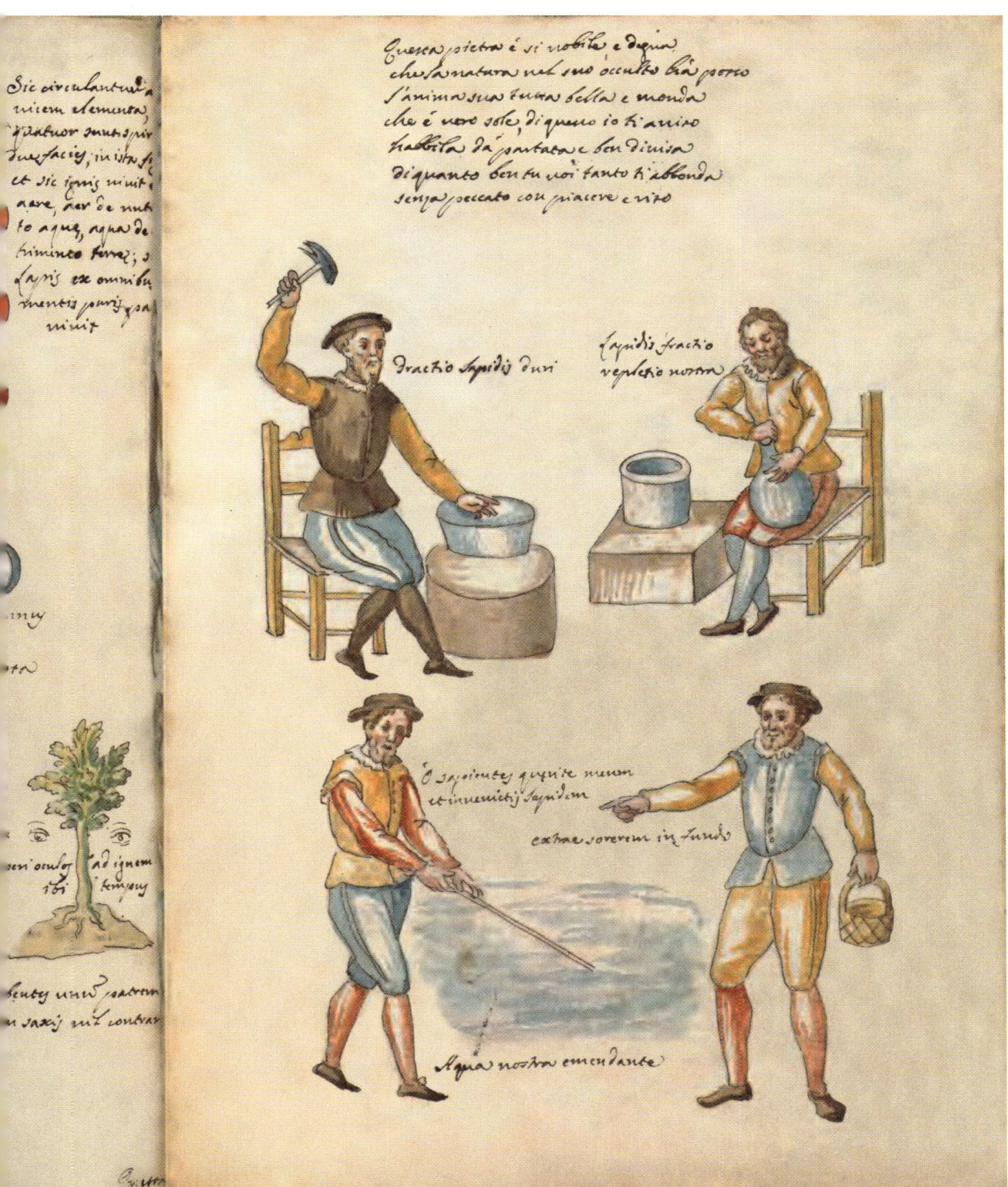

THEATER OF THE WORLD: THE CHEMICAL PHILOSOPHY

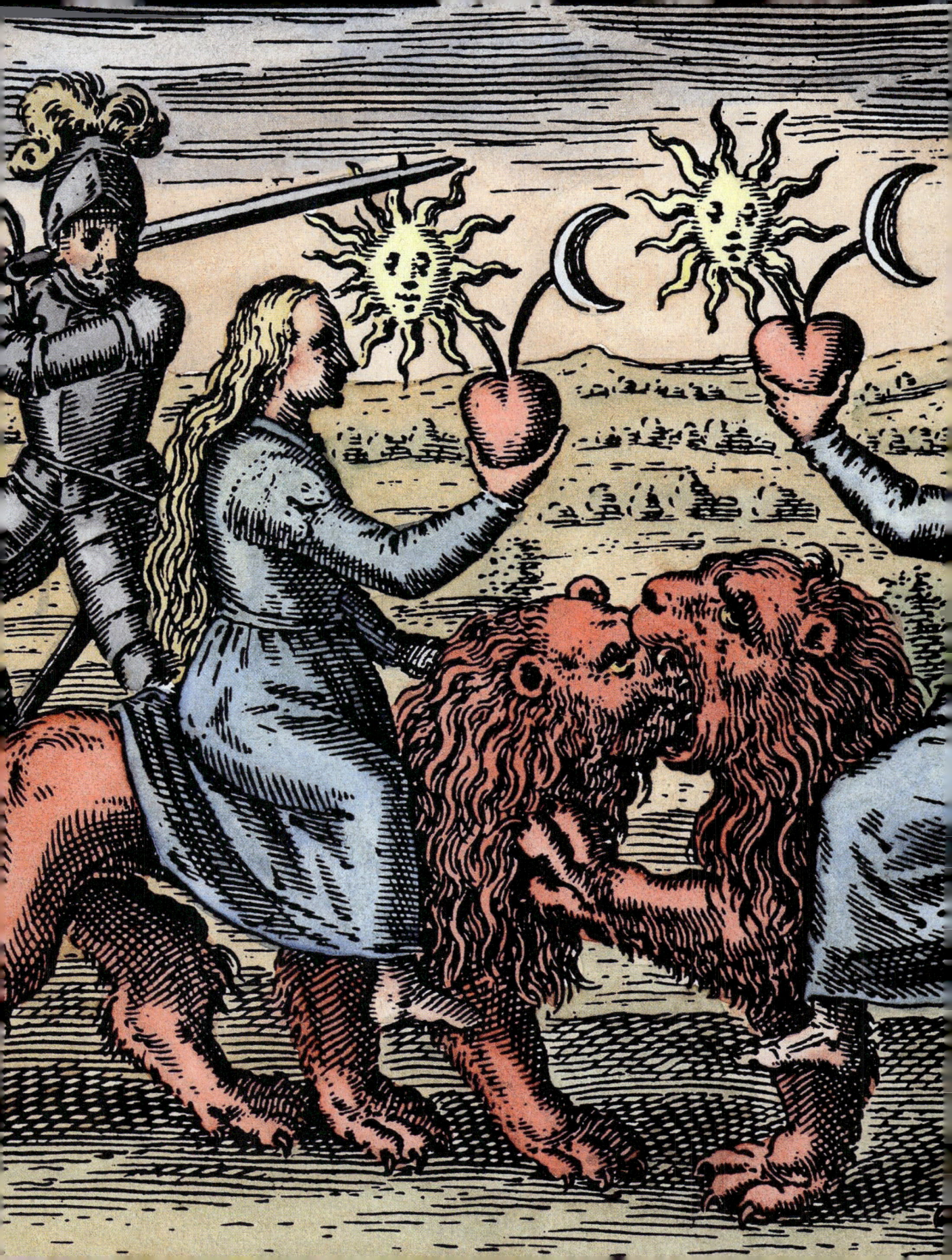

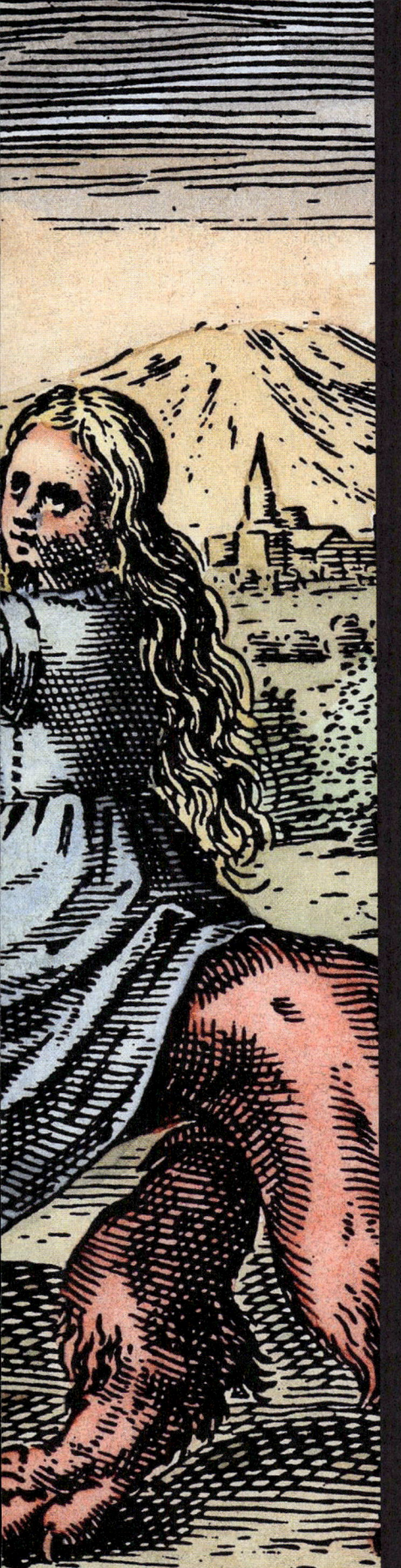

CHAPTER SEVEN

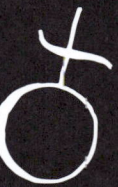

Alchemical Wars

CONTROVERSIES OF ALCHEMY

LEFT Detail of Key XI from Basil Valentine's *The Twelve Keys of Basil Valentine*, in *Musaeum hermeticum* (Hermetic museum; Frankfurt, 1678). Basil Valentine was a pseudonym for a 16th-century writer (or group of writers), and his advocacy of the element antimony as a medical cure provoked furious controversy between traditional physicians and those who supported the new "chemical medicine" or iatrochemistry.

CHAPTER SEVEN

Alchemy has always had a mixed reception from rulers, intellectuals, and society in general. Kings and emperors feared the effect that gold-making might have on the monetary system. The public image of alchemy, meanwhile, has always been tinged with notoriety, and there is a long history of alchemists being the object of ridicule and satire, their grand claims contrasting with the grimy realities of their work. In addition to this, because of the challenge alchemy posed to traditional ideas in natural philosophy and medicine, it has often become entangled with political and religious controversies. Alchemy has always been, sometimes literally, a battleground.

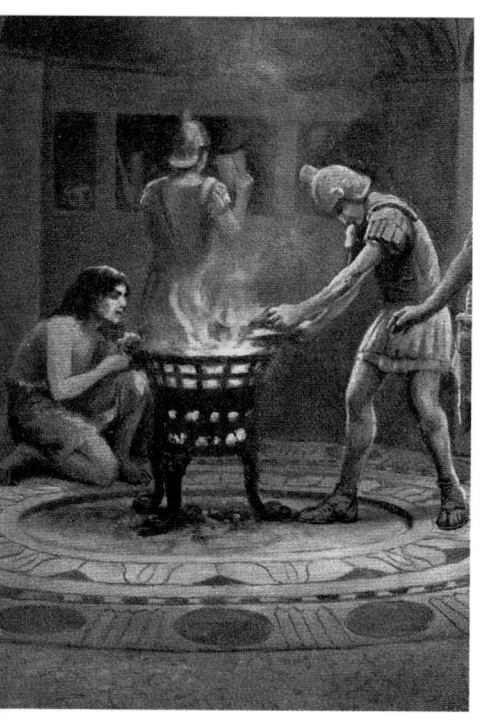

ABOVE Roman soldiers burning alchemical books, as imagined in *Hutchinson's History of the Nations* (1920). There is little evidence for this episode, but it reflects a distrust of alchemy by rulers who feared the impact of alchemical gold on their nation's currency.

If alchemy was really (among other things) a regular aspect of the history of science, why hasn't it generally been depicted as such? Why have we tended to consign alchemy to the dustbin of superstitious nonsense, or, alternatively, to suggest that it was never about scientific understanding anyway?

A part of the answer is that early chemists (often called chymists by historians today) found it expedient to assert the respectability of their discipline by stripping it of all that seemed obsolete or otiose—and that detritus became misidentified as alchemy's sole content. Alchemy has doubtless also suffered from the long-standing prejudice in the Western intellectual world that denigrates the pragmatic, the manual, the applied: it looked too grubby an affair to be taken seriously in the academies.

But there was more than this to alchemy's poor reputation. It had always been tarnished with a patina of disrepute: there has never really been a time when alchemy lacked detractors who denounced it as fraudulent or corrupt. Even astrology has not suffered so persistently from such attacks.

If someone tells you they can make gold, then even if you have no reason to suppose that this is an impossible thing you would be a fool to take their word for it and buy their services or their manufactured yellow metal. When alchemy was suspected of being riddled with frauds and charlatans, it was with some justification. At the same time, there seemed something illicit, even diabolical, about gold-making.

Alchemy was also perceived as a threat if it *did* work. If today someone were to discover a real process by which iron and lead could be turned into gold, the value of the natural precious metal would plummet. How much worse that would be in times when a country's currency was yoked to its gold reserves and when money was made from gold itself. In one of the earliest-known official measures taken against alchemy, the Roman emperor Diocletian in the third century A.D. is said to have ordered the burning of alchemical books in Egypt for the sake of social stability. Having suppressed an uprising there in 290, he feared that alchemical gold might be used to finance another rebellion. Diocletian also oversaw the issuing of new coinage after the currency had become debased

RIGHT "Multipliers must cease": a woodcut showing alchemical "multipliers" at work, from the English antiquarian Elias Ashmole's compilation of alchemical texts *Theatrum chemicum britannicum* (London, 1652). This illustration accompanies *The Ordinall of Alchimy* by Thomas Norton, a poetical work of the 15th century, and depicts several athanora furnaces. Alchemists were multipliers because they could allegedly transform tiny amounts of gold to prodigious quantities using the philosopher's stone.

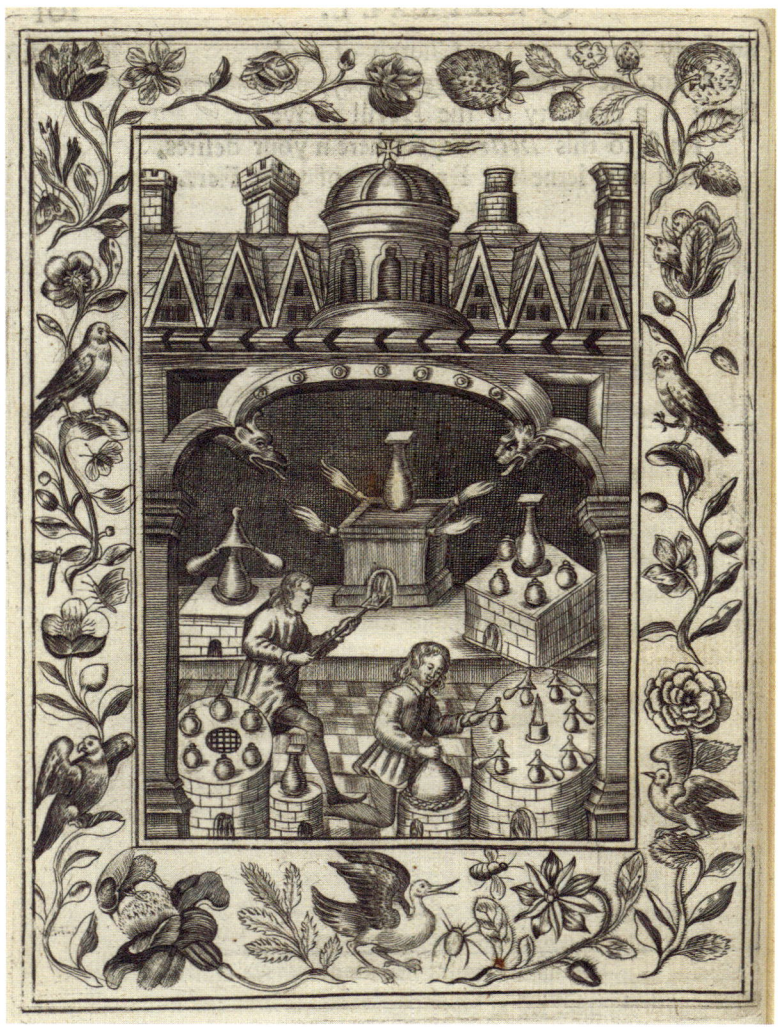

> While emperors and kings were troubled by alchemy's threat to their power and authority, many intellectuals and artists regarded it with mockery.

by alloying precious metals with others—another reason he did not welcome a possible influx of alchemical gold. That was the motive, too, for an edict by the Chinese Han emperor Jingdi in 144 B.C. forbidding the manufacture of "false yellow gold."

Official bans on alchemy recurred in Europe from time to time. In 1317 Pope John XXII issued a papal decree that threatened to levy a fine on anyone found to have produced or sold false alchemically produced gold. However much of it they had made or sold, they would have to pay the same weight in real gold. Here, too, the charge was not that all alchemy was impotent, but rather that alchemists might be making substandard gold. "The impoverished alchemists promise riches that they do not deliver," the edict began:

To such an extent does their damned and damnable temerity go that they stamp upon the base metal the characters of public money for believing eyes, and it is only in this way that they deceive the ignorant populace as to the alchemical fire of the furnace.

The papal edict evidently did not suppress alchemy, but it meant that alchemists were best advised to conduct their business in secrecy. A similar ban—the Act Against Multipliers—was imposed in 1404 in England by Henry IV, again because of the fear that a weakening of the currency might foment rebellions, which were a constant danger for the king after he ascended the throne in 1399. Yet when, four decades later, Henry VI was in urgent need of money to finance his wars against France and subsequently against the House of York, he began to issue licenses to alchemists in the hope that they could fill his coffers. The 1404 Act was not repealed, however, until 1689, partly on the appeal of Robert Boyle.

The alchemical charlatan

While emperors and kings were troubled by alchemy's threat to their power and authority, many intellectuals and artists regarded it with mockery. Again, their skepticism often came not from a conviction that transmutation was inherently impossible, but from a sense that many either were wasting their time in fruitless pursuit of it, beggaring themselves in the process, or were deceiving others for their own profit. In "The Canon's Yeoman's Tale," Geoffrey Chaucer in the fourteenth century shows us both archetypes. The yeoman explains that he and his master (the canon) look so shabby because the canon has impoverished himself in his efforts to make

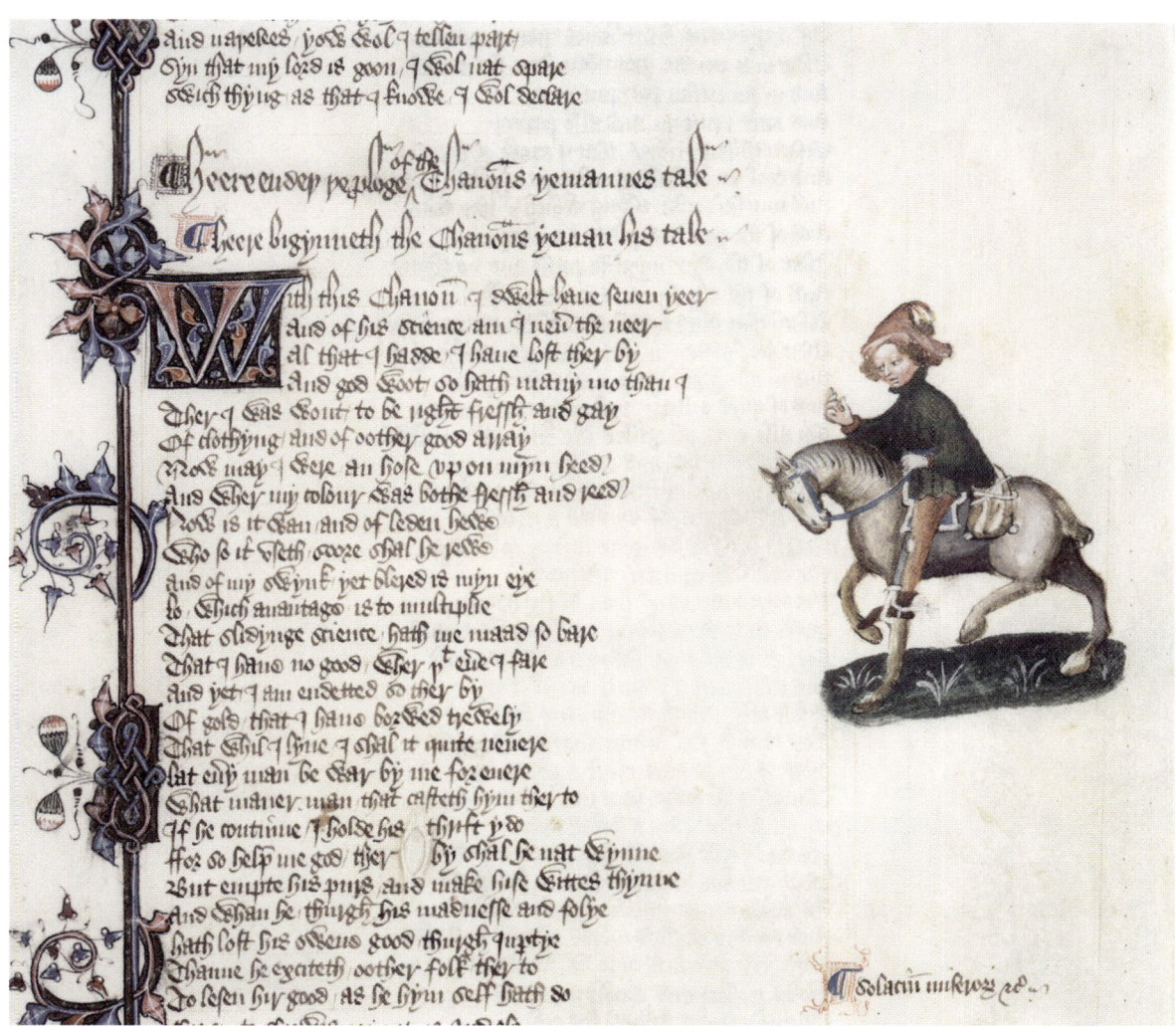

ABOVE The Canon's Yeoman, a detail from Geoffrey Chaucer's *Canterbury Tales* (c. 1340–1400). The yeoman recounts his labors as an assistant to two alchemists —one incompetent, the other deceitful— and advises that alchemy is a fool's quest.

gold. After serving him for seven years, says the yeoman, "That slippery science has made me so bare, that I've no goods, wherever I may fare."

The yeoman adds that he once worked for another alchemist who, rather than pursue so hopeless a quest, resorted to trickery and deceit. This individual performed a faked transmutation before a priest using some worthless powder that he then sold to the cleric for a hefty sum. Alchemy, the yeoman concludes, is a fool's quest: "all the coin he spends therein goes out, and is but lost." Earlier in that century, Dante made his feelings about alchemists plain: in his *Inferno* he consigned them to the eighth circle of hell in punishment for their deceitfulness.

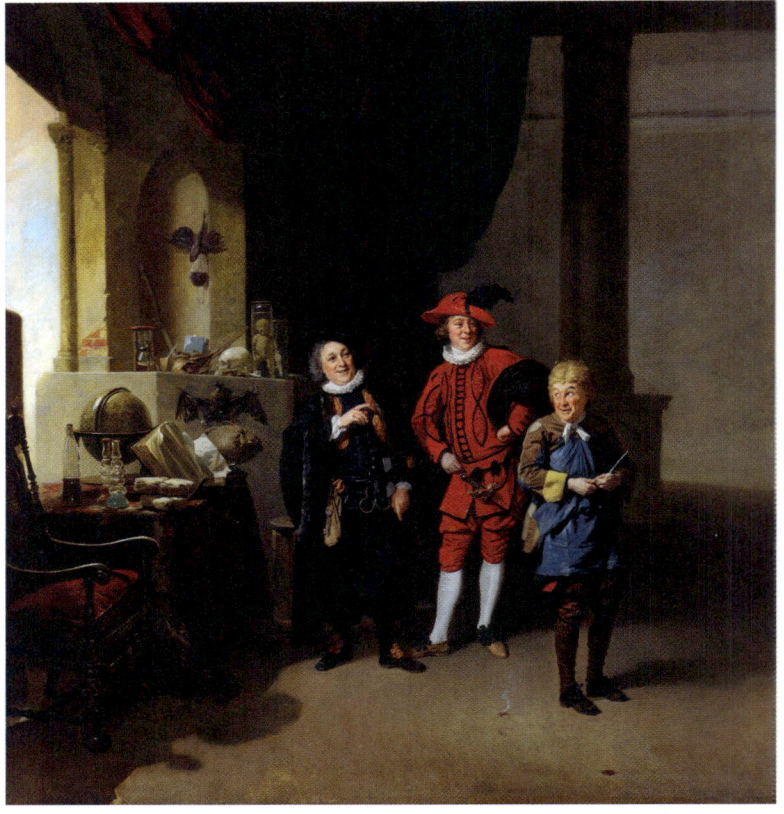

RIGHT Wily tricksters Subtle (left), posing as an alchemist and astrologer, and Captain Face (center), being visited by Abel Drugger (right) in Ben Jonson's 1612 play *The Alchemist*, in a painting by Johann Zoffany, c. 1769–70. Subtle is the archetypal alchemical charlatan who swindles his generally rich, noble, and credulous clients out of their money.

There was still plenty of skepticism about alchemy during the Renaissance. The Italian humanist writer Francesco Filelfo wrote: "Those who think that by spoiling and corrupting copper, silver or gold can be made, seem to me stupid fools." The most celebrated depiction of Hermetic deceit is Ben Jonson's 1610 play *The Alchemist*. The title character is the wily trickster Subtle, who, together with his assistant Face, sets about swindling aristocratic dupes. The joke is not really at the expense of alchemy—we are never in any doubt that Subtle is quite unable to perform transmutation—but of those made gullible by greed.

The Italian humanist writer Francesco Filelfo wrote: "Those who think that by spoiling and corrupting copper, silver or gold can be made, seem to me stupid fools."

All the same, Jonson skewers the overblown jargon of alchemy as Subtle seeks to blind his client with "science": "Your sun, your moon, your firmament, your adrop, your lato, azoch, zernich, chibrit, heautarit." In this regard, *The Alchemist* is in the tradition of later satires on the "experimental philosophy," such as Thomas Shadwell's *The Virtuoso* and Jonathan Swift's *Gulliver's Travels*, which ridiculed its interest in seemingly trivial phenomena and its theories that seemed to defy common sense.

The alchemical court

Some historians suspect Jonson modelled Subtle on the English physician, astrologer, and alchemist Simon Forman, who, despite being repeatedly fined and imprisoned for practicing medicine without a license, grew rich on his reputation for wondrous cures. Others suggest the character is based on John Dee, mathematician, magus, and sometime court astrologer and alchemist for Elizabeth I.

Dee's interest in alchemy was shared by the queen herself, who hoped like many rulers that it might augment the royal revenue. As the son of a court official for Henry VIII, Dee was well connected to noble families at the center of the Tudor Reformation. He personifies the flourishing of "occult" philosophy—natural magic,

BELOW An ink and wax impression of a ring made by Simon Forman, inscribed with magical formulae, from a 1598 manuscript of magical notes by Elias Ashmole. Forman was a notorious physician, astrologer, and alchemist of London who was repeatedly in trouble with the authorities for unlicensed practice of medicine.

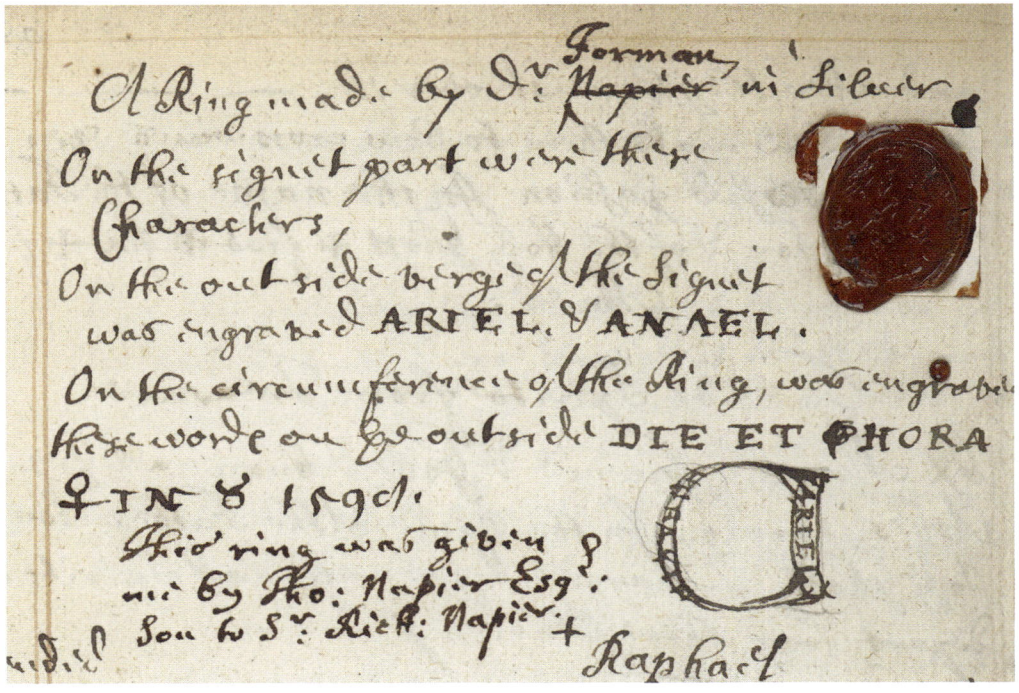

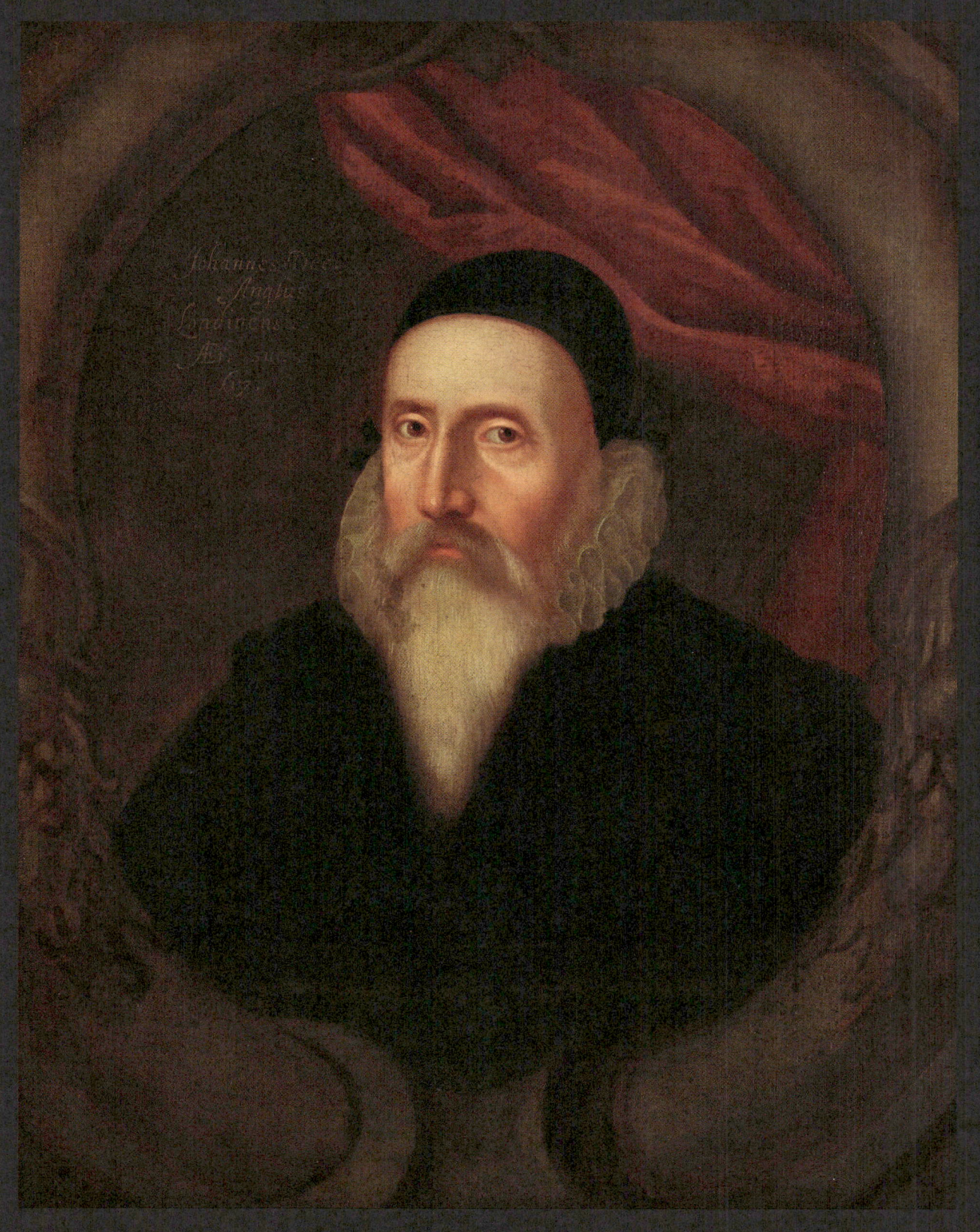

OPPOSITE Portrait of John Dee by an unknown artist, c. 1594. Dee was a mathematician, astronomer, antiquarian, and alchemist who advised for some time in the court of Elizabeth I. The preface he wrote to an English translation of Euclid's *Elements* advocating for the importance of mathematics was highly influential.

BELOW Magical items in a set belonging to John Dee used for divining: a leather-covered wooden case, three wax discs engraved with magical figures and names, a gold disc engraved with a diagram of a vision by his assistant Edward Kelley (made in Kraków, 1584, see page 189), and a crystal ball. Kelley claimed to be able to commune with angels using the crystal ball.

alchemy, astrology—in the sixteenth century. Deeply influenced by Neoplatonism and Kabbalism, Dee believed that the innermost secrets of the world are mathematical and geometrical.

He sought to improve methods of navigation and thereby help to expand Elizabeth's nascent British empire. But Dee lacked the skills to navigate the intrigues of the royal court, and in 1583 he left England in the retinue of a Polish prince, accompanied by an assistant named Edward Kelley who claimed to possess occult powers and to be able to use Dee's crystal ball to converse with angels. Dee and Kelley made their way to the court of Rudolf II in Prague, where they succeeded only in slighting the emperor and making powerful enemies. Accused of black arts, the pair were banished from the city. Dee returned to England but struggled to regain royal favor, eking out his days as an impoverished college warden.

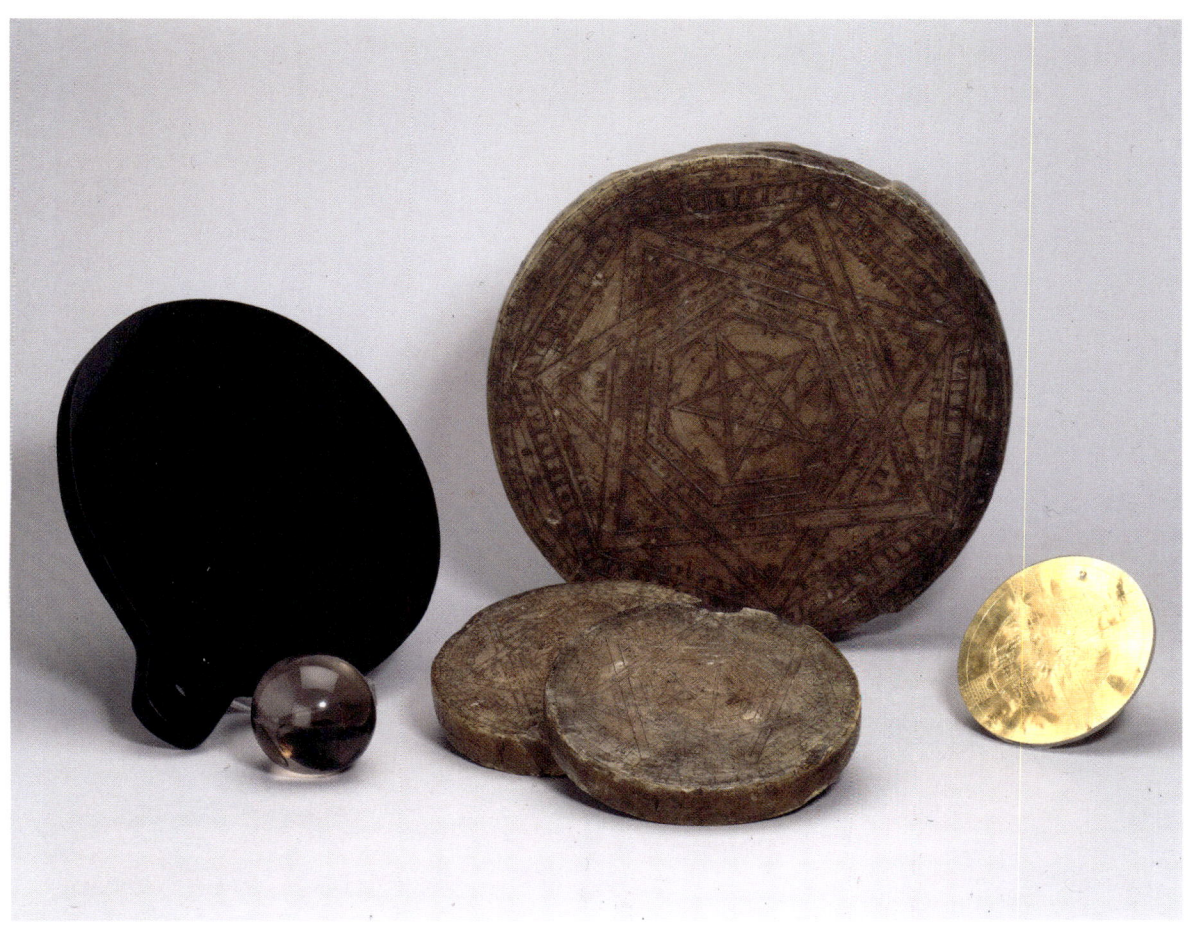

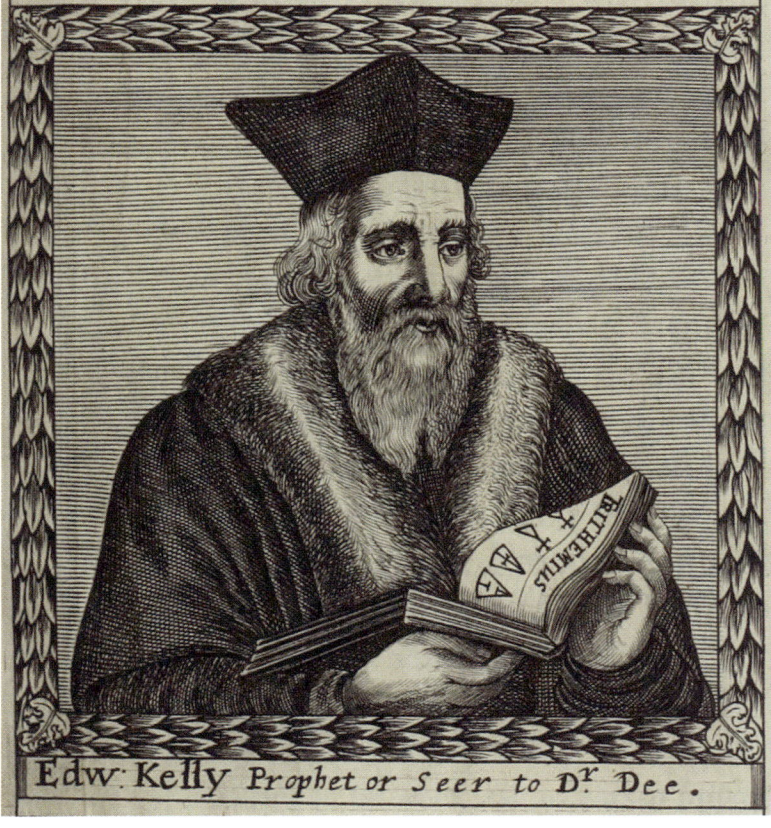

RIGHT Edward Kelley, sometime assistant to John Dee, holding a book by the German Benedictine abbot Johannes Trithemius, a renowned 15th-century occultist, in an engraving in *A True & Faithful Relation of What Passed for Many Yeers between Dr. John Dee . . . and Some Spirits* (London, 1659).

In journeying from Elizabethan England to Rudolf II's Prague, Dee and Kelley were perhaps unconsciously borne along by a deep current of the times. According to the British historian Robert Evans, "Both the unmarried Emperor and the [English] Virgin Queen were widely regarded as figures prophetic of significant change in their own day." Alchemy and other occult arts were generally thought to herald—and, indeed, to encode—incipient political and religious transformation.

Surrounding himself with gold-makers, wizards, and stargazers instead of ambassadors and princes, the Holy Roman Emperor seemed to many in the Catholic church to be more interested in probing the hidden secrets of nature than in defending the faith against the threat of the Reformation. Educated in the austere Madrid court of the Habsburgs, Rudolf apparently did not relish his throne; maladroit and depressive, he was rumored to be an atheist, or mad, or both. He sponsored scholars in the arts and sciences, from the pioneering astronomers Tycho Brahe and Johannes Kepler

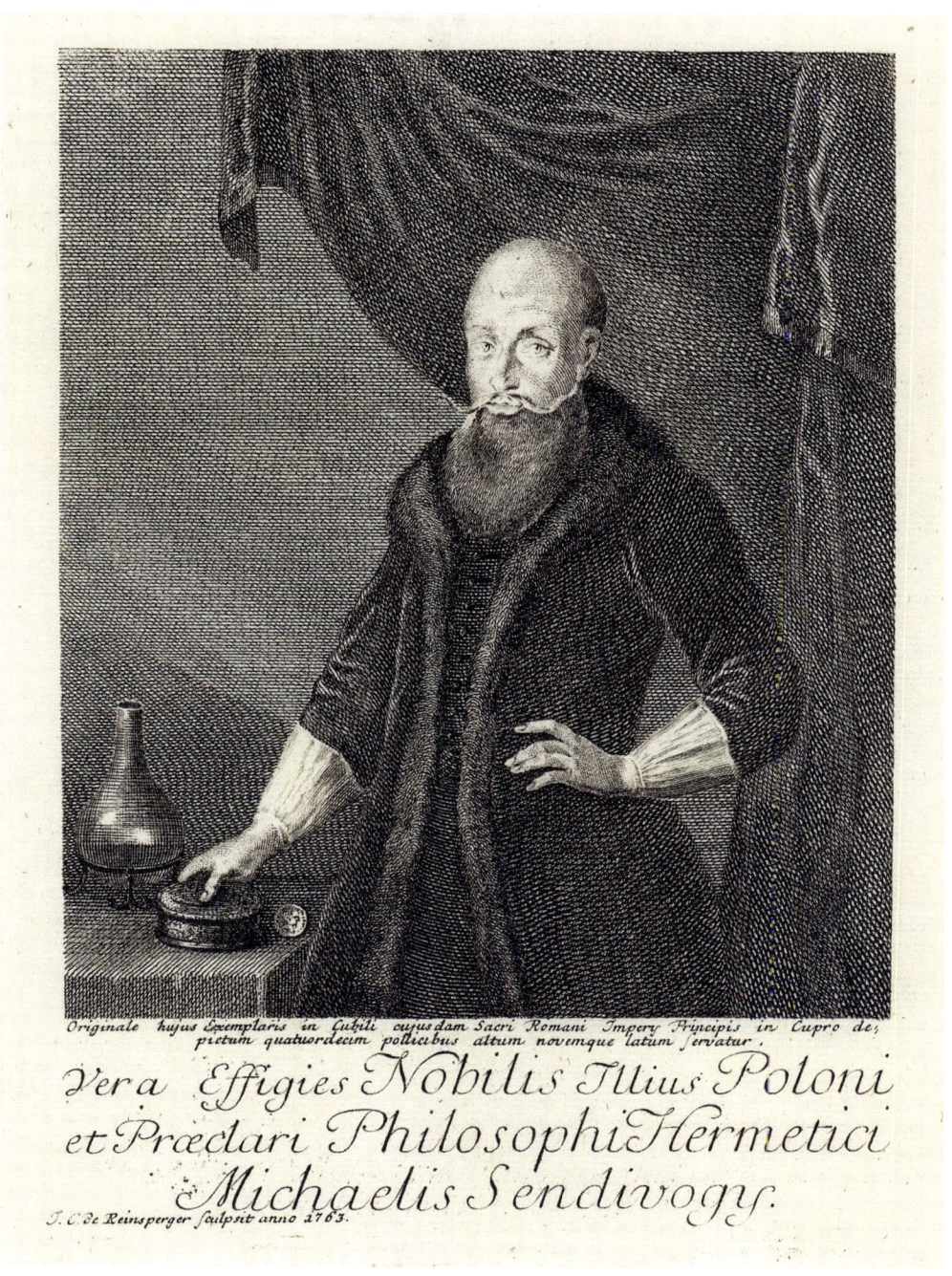

OPPOSITE The 17th-century Polish alchemist Michael Sendivogius (Michał Sędziwój), engraved in 1746 after an unknown oil painting. This image was reproduced as the frontispiece to a 1766 edition of his most famous work, *Novum lumen chemicum* (translated as *A New Light of Alchymie*), published in Nuremberg.

to the painter, designer, and engineer Giuseppe Arcimboldo. Among the influential alchemists who sought Rudolf's patronage were Michael Maier, Heinrich Khunrath (see page 158), the German Oswald Croll, and the Pole Michael Sendivogius.

But Rudolf was not destined to pursue his esoteric passions in peace. The religious tolerance he displayed in Prague, where a Protestant like Kepler could secure court employment and Jews were safe, did not please the papacy. In 1600 the papal ambassador in Bohemia told Pope Clement VIII that it was widely thought that the emperor "has been bewitched and is in league with the devil." In 1611 Rudolf's ambitious brother Matthias led an army against him and took control of Bohemia, Austria, and Hungary, permitting Rudolf to remain a puppet emperor until his death the following year. When Matthias died in 1618, the fragile peace dissolved, and the new emperor, Ferdinand II of Styria, found himself facing a revolt in Bohemia. The rebels offered the crown to Frederick V, Elector Palatine of the Rhine and head of the Protestant Union of German states, who was married to Elizabeth, daughter of King James I of England and Scotland.

Frederick and Elizabeth were the archetypal Hermetic king and queen, figureheads of a utopian humanistic liberalism. They "ruled" Bohemia only during the winter of 1619–20—dubbed the Winter King and Queen, like an allegory plucked straight from an alchemy text—before the Protestant forces were crushed by the Habsburg army. That conflict precipitated the Thirty Years' War between the Protestant German states and the Holy Roman Empire that devastated Germany and Bohemia for decades.

The Rosicrucian Enlightenment

The visionary, even revolutionary, themes that Michael Maier discerned in alchemy informed his book *Themis aurea* (Golden Themis), subtitled *The Laws of the Fraternity of the Rosie Cross* when published in English in 1656. It was one of several works of that febrile time that spoke of the Order of the Rosy Cross, a secret society of reformers later known as the Rosicrucians. In Rosicrucianism we can see alchemy spinning off into a counter-cultural movement in which Hermeticism was allied to an almost eschatological view of religious reform. Historian Frances Yates has called this moment in the early seventeenth century the Rosicrucian Enlightenment: a period intermediate between the Renaissance and the Scientific Revolution. "It is," she wrote, "a phase in which the

BELOW Michael Maier's *Themis aurea* (Golden Themis), subtitled *The Laws of the Fraternity of the Rosie Cross*, published in Latin in Frankfurt in 1618, and in English in 1656. Maier's book is one of several works of the early 17th century making reference to the Rosicrucian fraternity, a fictitious brotherhood of adepts with secret knowledge.

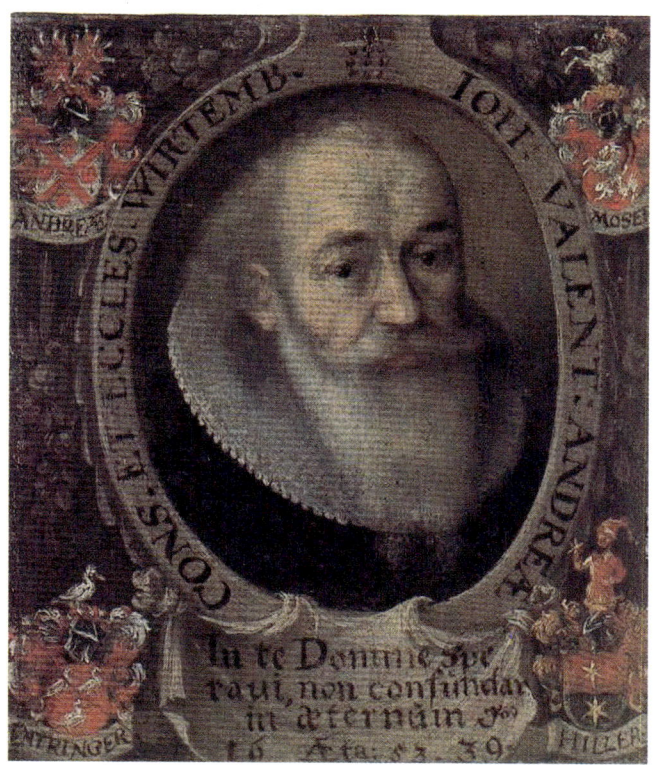

RIGHT Portrait of Johann Valentin Andreae of Herrenberg, a member of the German secret society called the Order of the Inseparables, founded in 1577. Andreae was the likely author of at least some of the foundational texts of the Rosicrucian order, which espoused an alchemically inflected utopian vision of society.

Renaissance Hermetic-Cabalist tradition has received the influx of another Hermetic tradition, that of alchemy."

Whether such a well-defined era can truly be identified is now questioned by many historians. Yet it is surely true that the disintegration of Rudolf's court and the terrible war that followed was a strange but formative prelude to the Scientific Revolution. The Thirty Years' War drove to England scholarly refugees such as the Prussian Samuel Hartlib, an educational reformer and one of the instigators of the Royal Society in London. Rosicrucianism was a product of that tumultuous time.

Some alleged that the spiritual founder of the Rosicrucians was Paracelsus himself. One of the figures most clearly linked to its murky beginnings was the German Paracelsian alchemist and theologian Johann Valentin Andreae of Herrenberg, a member of the utopian Protestant movement that also included Hartlib. Andreae belonged to another shadowy German society, the Order of the Inseparables, founded in 1577, which was interested in mining and alchemy. He is probably the author of at least the first of the two texts published anonymously to launch the Rosicrucian movement: the *Fama fraternitatis* (Report of the brotherhood; 1614) and the *Confessio fraternitatis* (Confession of the brotherhood; 1615). They told of one Christian Rosencreutz, a Dutchman of the fifteenth century who purportedly gained secret knowledge in the Middle East that was now being transmitted through a secret network of adepts. The manifestos called for these individuals to declare themselves

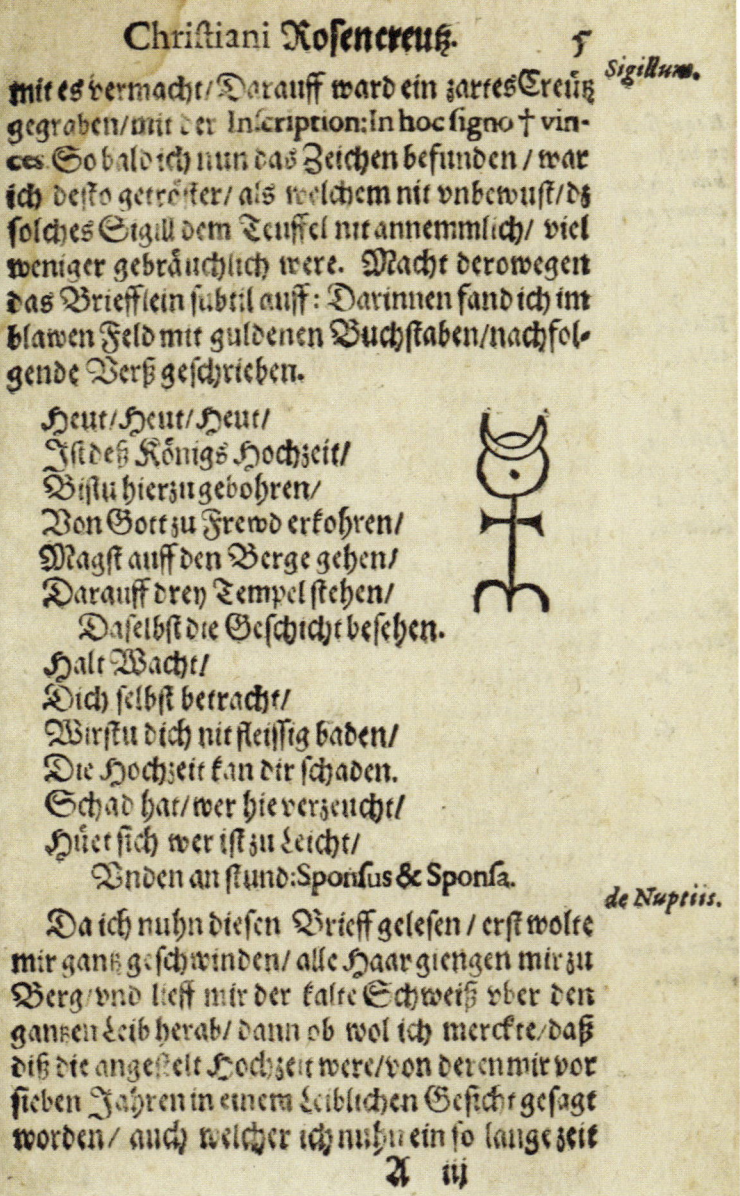

RIGHT A symbol of alchemical union called the Monas Hieroglyphica, invented by John Dee. The symbol is depicted in *Chymische Hochzeit Christiani Rosencreutz* (The chymical wedding of Christian Rosencreutz; Strasbourg, 1616), attributed to Johann Valentin Andreae.

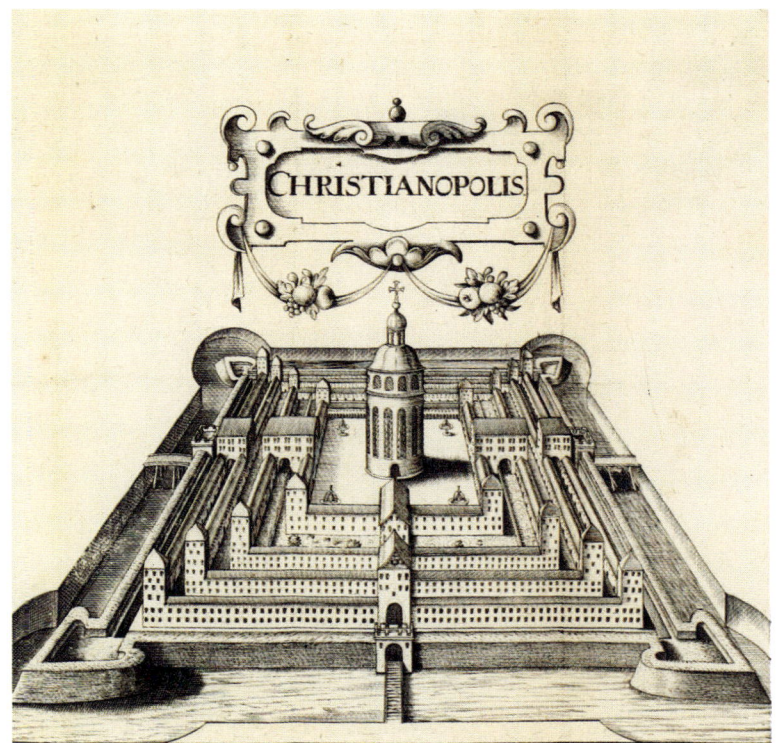
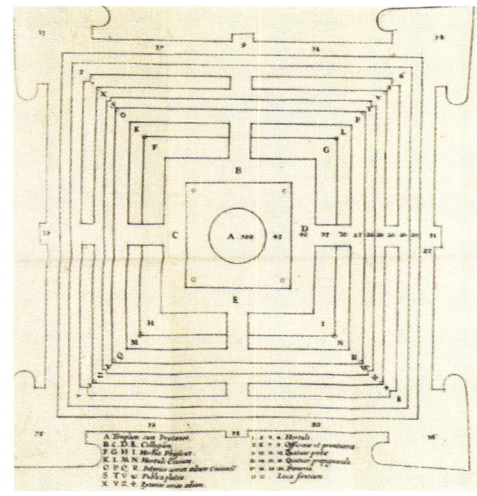

ABOVE AND RIGHT An idealized vision of the utopian city-state of Christianopolis, which was envisaged by Johann Valentin Andreae in his book *Reipublicae Christianopolitanae descriptio* (Description of the Christianopolitan republic; Strasbourg, 1619). Christianopolis represented an ideal Protestant society where humble Christian piety was respected and lavish displays of wealth were rejected.

and be united in the Brotherhood of the Rosy Cross. The society's first secretary is said to have been the German salt manufacturer Johann Thölde, the likely author of influential alchemical works attributed to the fictitious "Basil Valentine" (see page 186).

The link with alchemy was particularly explicit in the third Rosicrucian tract, published anonymously (again probably the work of Andreae) in 1616 and titled *The Chymical Wedding of Christian Rosencreutz*. It describes Rosencreutz's journey to a fabulous castle where he assists in the wedding of a king and queen: the classic allegory of the formation of the philosopher's stone.

The Rosicrucian movement was, in modern terms, a classic conspiracy theory. There is no indication that the network mentioned in the *Fama* and *Confessio* even existed, although there was no shortage of people ready to declare themselves a part of it or to petition for admission. Many of these enthusiasts were sympathetic to the vision of an alchemically inspired Protestant utopia, a vision depicted by Andreae in his 1619 book *Reipublicae Christianopolitanae descriptio* (Description of the Christianopolitan republic), which is believed to have inspired Francis Bacon's science-

ALCHEMICAL WARS: CONTROVERSIES OF ALCHEMY

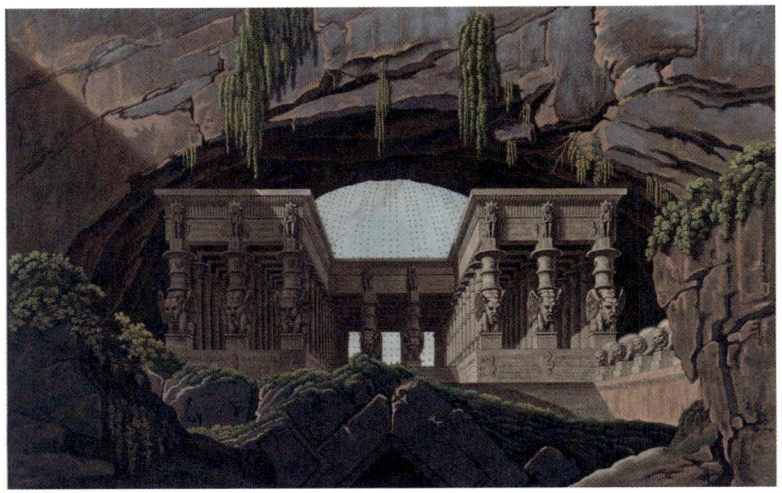

RIGHT A fabulous temple designed for *The Magic Flute*, act 1, scene 1, by Karl Friedrich Schinkel, 1847–49. Mozart's opera has been interpreted as an alchemical parable, and is replete with symbolism relating to Rosicrucianism and the Chymical Wedding.

based theocratic utopia in his *New Atlantis* (1626). When Robert Fludd learned of the Rosicrucian order after meeting Maier in London, he wrote a defense of the proposal to replace a university curriculum based on Aristotle and Galen with one that drew on alchemy, natural magic, and Paracelsian chemical medicine.

A secret society united by privileged knowledge and special codes of conduct sounds reminiscent of Freemasonry—and in the eighteenth century the Freemasons indeed absorbed and adapted Rosicrucianism. The Masonic Rosicrucian Society, founded in Scotland, established colleges and lodges in Britain and the United States that still exist today. It had close connections to the esoteric occult revival of the late nineteenth and early twentieth centuries (discussed in Chapter 9).

The antimony wars

Rosicrucianism shows how the rich imagery and allegory of alchemy and the chemical philosophy lent itself to spheres far removed from the manufacture of useful and precious materials. Given how alchemy became associated with challenges to old traditions and authorities, it is not surprising that it was implicated in the religious and political struggles of the Renaissance and early modern period. As well as the conflict in Germany, those tensions exploded, less bloodily but no less bitterly, in France during the heyday of Paracelsian iatrochemistry in the late sixteenth century.

At that time, the medical faculty of the University of Paris remained stolidly allied to Galenic medicine. When the Norman physician Roch le Bailiff began to gain favor at the royal court, they were alarmed—for he was not just a Paracelsian but also a Protestant. In 1578 the faculty succeeded in getting le Bailiff expelled from the city and banned from practicing medicine. But their relief was short-lived, for in 1589 the prince of Navarre—a Huguenot (adherent of the French Protestant church)—was crowned King Henri IV. The new king was compelled to become a Catholic on taking the throne, but he brought to the court a retinue with unwelcome ideas about both medicine and religion. Henri's royal doctor was Jean Ribit, a Huguenot and Paracelsian, and he was

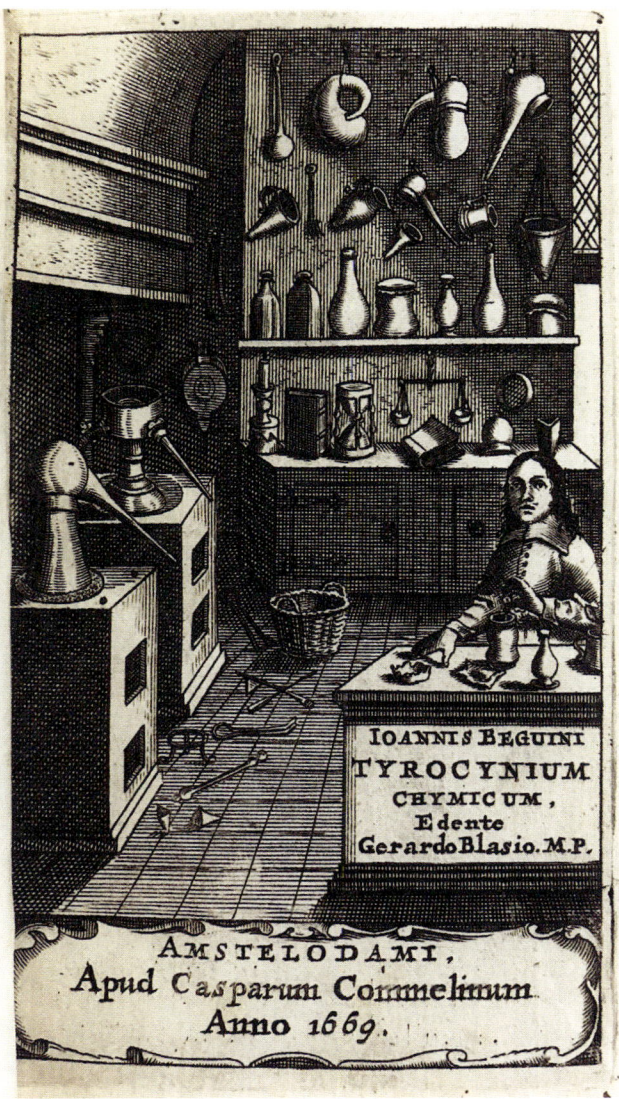

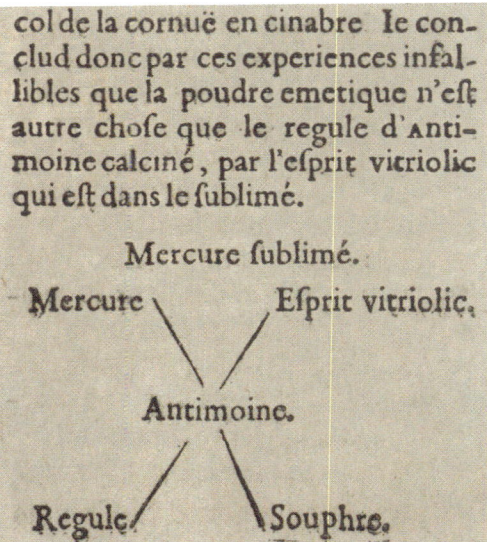

ABOVE Jean Béguin's 1610 *Tyrocinium chymicium* (loosely meaning "introductory essays on chemistry") was a stepping stone in the transition from alchemy to chemistry. His *Elemens de chymie* (Elements of chemistry; Paris, 1615), contained the first ever diagrammatic representation of chemical relationships. Here antimony is related to mercury, sulfur, spirit of vitriol (sulfuric acid), and alloys (*regule*), for example with tin or lead. The latter were sometimes used in casting.

LEFT The Huguenot and Paracelsian chymist Jean Béguin in his iatrochemical laboratory in Paris, from an edition of his *Tyrocinium chymicum* (Amsterdam, 1659).

ministered to also by the like-minded physicians Joseph Duchesne and Théodore Turquet de Mayerne.

The head of the conservative Paris faculty, Jean Riolan, responded with a tract praising Hippocrates and Galen and denouncing Mayerne and Duchesne personally. This power struggle rumbled on until the king was assassinated by a Catholic zealot in 1610, whereupon Mayerne was forced to flee to England. That was not the end of the matter, however. In 1604 Mayerne and Ribit helped the Huguenot and Paracelsian chymist Jean Béguin establish an iatrochemical laboratory in Paris, and Béguin's 1610 textbook, *Tyrocinium chymicum* (loosely meaning "introductory essays on chemistry"), became influential in chemical teaching. By liberating practical Paracelsian chymistry from its more speculative and mystical aspects, Béguin's book was a stepping stone in the transition from alchemy to chemistry.

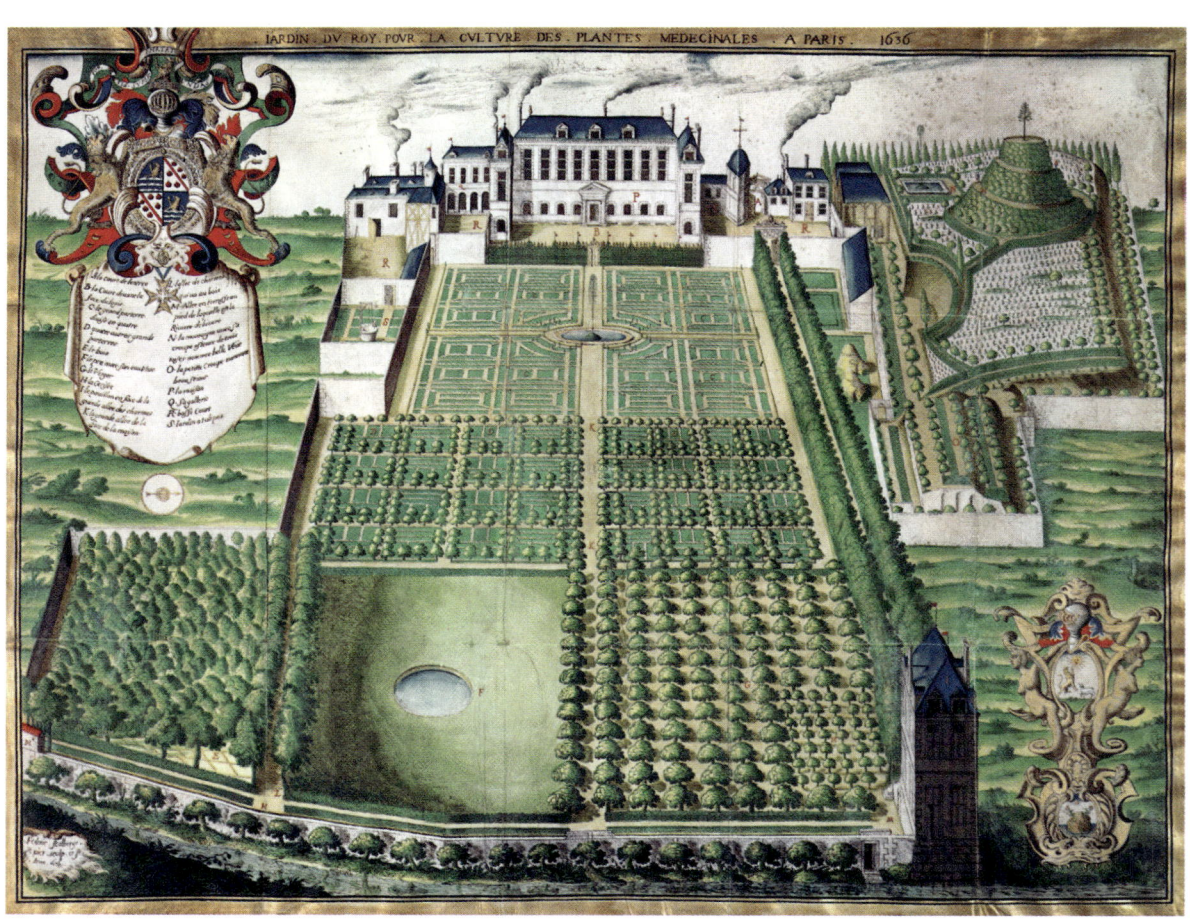

BELOW The French king's medicinal plant garden in Paris, known as the Jardin du Roi, depicted here in a 1636 engraving after Federic Scalberge, was a place of horticultural and chymical research and experimentation. It was dominated in the early 17th century by Paracelsian iatrochemistry.

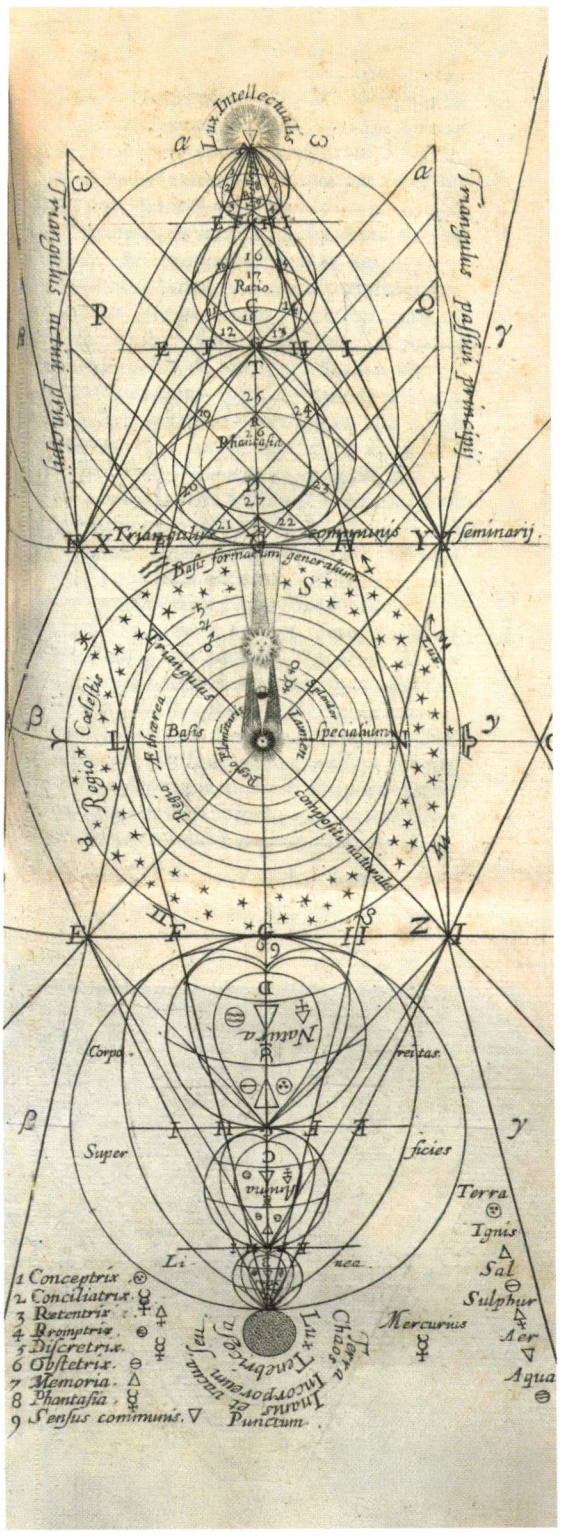

LEFT A chart of "the whole corporeal and incorporeal nature," from William Davidson's *Philosophia pyrotechnia* (Paris, 1640). It was written in 1633–35, a few years before Davidson, a Scottish Paracelsian chymist, was appointed *intendant* of the Jardin Royal des Plantes, Paris. In later life, Davidson became a physician to kings of Poland.

That process was abetted by the establishment of the Jardin du Roi, an herb garden in Paris, in 1635. Although the garden was originally Riolan's idea, it was administered by the royal physician Guy de la Brosse, who was favorably inclined towards iatrochemistry. In 1647 the Jardin appointed as professor of chemistry a Scotsman, the Protestant and Paracelsian William Davidson. He was especially enthusiastic about a Paracelsian cure that used antimony, which ingredient was championed for medical use in Basil Valentine's 1604 tract *Triumph-wagen antimonii* (The triumphal chariot of antimony).

The traditionalist Paris faculty had been campaigning against the use of antimony for years, and now, led by the aging Riolan, they hounded Davidson out of his post in 1651. It was a rather Pyrrhic victory, for iatrochemistry (sometimes called spagyric chemistry after the Greek terms for the alchemical process of separation and recombination) was inexorably on the rise. In a gambit that has often proved effective in science, the argument was won via the textbook. A string of influential texts by French iatrochemists, such as Nicaise Lefebvre's *Cours de chimie* (Chemistry course; 1660) and Nicolas Lemery's book of the same title in 1675, secured the place of the "new" chemistry and of chemical medicine. As these titles attest, the transition from alchemy to chemistry—and in the process, the stripping away of philosophical baggage from chemical experimentation—was by that time well underway.

ALCHEMICAL WARS: CONTROVERSIES OF ALCHEMY

PROFILE

Basil Valentine

(allegedly fifteenth century)

> "Antimony, you affirm, is a poison. Therefore let everyone beware of using it. But this conclusion is not logical, Sir Doctor, Magister, or Baccalaureas; it is not logical, Sir Doctor, however much you may plume yourself on your red cap... Antimony can be so freed of its poison by our Spagyric Art as to become a most salutary Medicine."

This bombastic and provocative speech sounds very much like the Swiss alchemist Paracelsus, but in fact it appears in a book published six decades after his death, titled *Triumph-wagen antimonii* (The triumphal chariot of antimony; 1604) and apparently written by the alchemist Basil Valentine. He was long alleged to be a fifteenth-century German monk from Erfurt who became an alchemical adept, but there is no evidence at all of Valentine's existence; his name alone—*basileos valens* is a bastardized blend of Latin and Greek meaning "valiant king"—suggests that the character is wholly fictitious. Most historians now believe that the real Basil Valentine—or at least, the author of the first few books bearing his name—was a sixteenth-century German salt-maker from Hesse named Johann Thölde, who published them.

The Triumphal Chariot was an influential text, championed in the French "antimony wars" (see page 182) by those who believed that this semi-metal could indeed be made into a potent medicine. Basil Valentine himself attests to the toxicity of the substance if it is not processed by "spagyric art," spawning a later, spurious etymology: *anti-monachos*, the nemesis of monks. It was alleged that Valentine once tried to use it to fatten up his sickly monastic colleagues, having noticed that it had that effect on pigs. But when he added antimony secretly to the monks' food, they become violently ill, and some died. The notion that this poison could be turned into a cure by alchemy is characteristic of Paracelsian iatrochemistry. A modern reconstruction of the process Valentine describes, conducted by historian of science Lawrence Principe, suggests that it might have produced the harmless and slightly sweet-tasting (as Valentine attests) compound iron acetate, having no antimony in it at all.

The character invented by Thölde (if indeed it was he) proved very popular. As works attributed to Valentine became highly influential in the alchemy and chymistry of the early seventeenth century, legends accrued around the apocryphal Benedictine monk. Some say that his *Last Will and Testament* was concealed within the altar of the abbey church until revealed when a pillar was destroyed in a lightning strike.

These works exemplify the alchemy of those times, being written in the highly allegorical language that often encrypts practical chemical manipulations. On their own they read as dreamlike poetry, as in this passage from Valentine's *Von dem grossen Stein der Uralten* (Of the great Stone of the ancients):

Take the ravenous gray wolf that on account of his name is subjected to bellicose Mars, but by birth is a child of old Saturn, and that lives in the valleys and mountains of the world and is possessed of great hunger. Throw the king's body before him that he may have his nourishment from it . . .

The gray wolf is probably antimony ore (the mineral stibnite), then thought to be related to metallic lead, which was identified with the planet Saturn. The king is gold; the account describes the result of mixing these substances in the heat of a furnace. And so the account continues, at least partially comprehensible and even reproducible to the modern chemist who knows how to translate it. The allegorical imagery was enhanced by elegant woodcuts in the 1618 edition of *The Twelve Keys of Basil Valentine* published by Michael Maier—a step-by-step account of how to make the philosopher's stone that was studied closely, and probably somewhat despairingly, by Robert Boyle and Isaac Newton.

BELOW Imaginary portrait of Basil Valentine, depicted alongside Hermes Trismegistus, in Basil Valentine's *Révélation des mystères des teintures essentieles* (Revelation of the mysteries of essential tinctures; Paris, 1646).

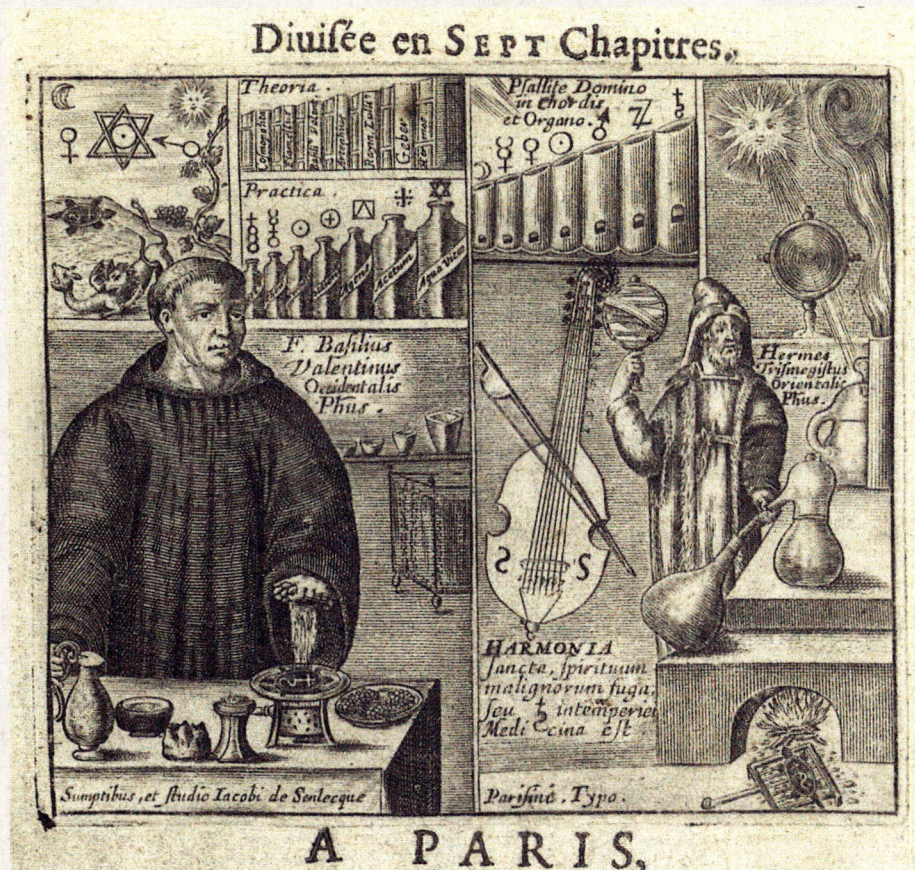

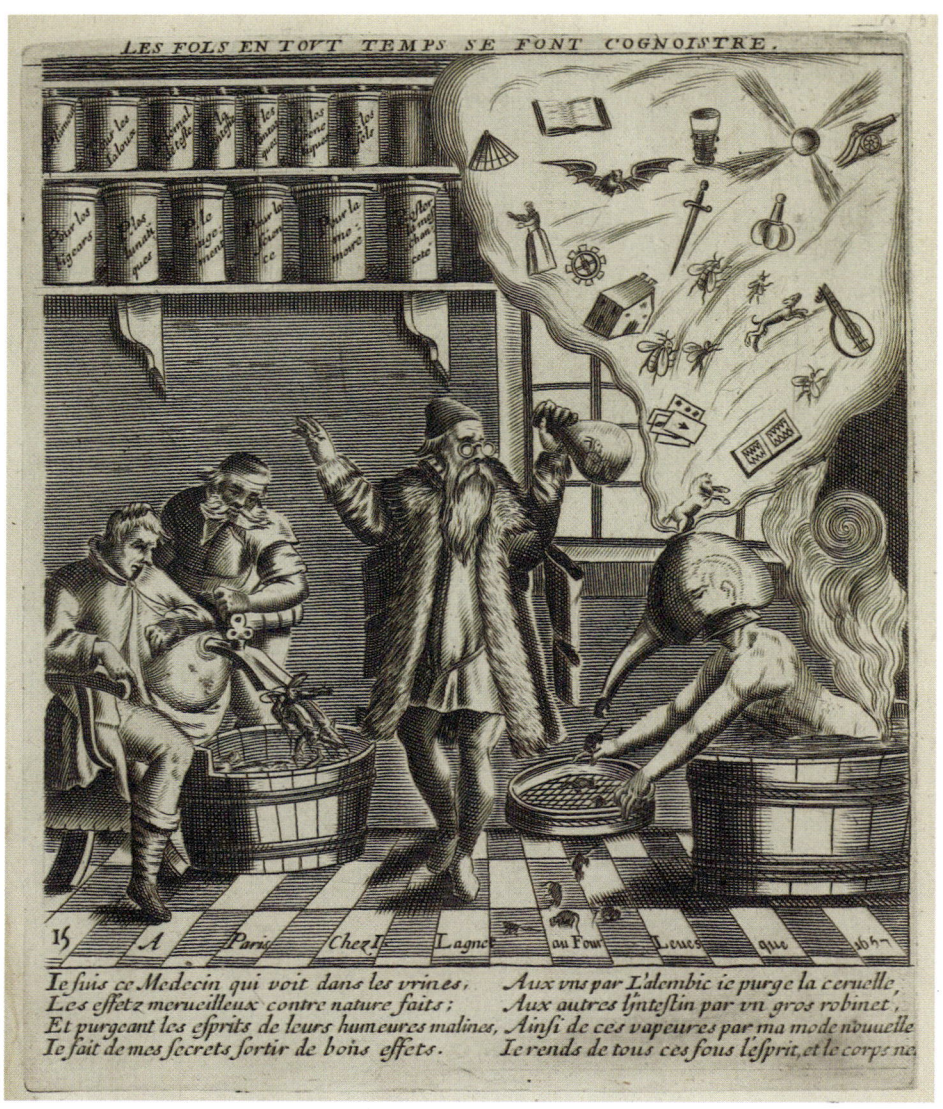

ABOVE Alchemy's poor reputation is highlighted in this satirical engraving by the Flemish artist Theodor de Bry, titled *Doctor of Fools*, 1657. The figure in the tub wears an ambix on his head, and is apparently distilling mice. The alchemical doctor brandishes a flask, inside which is a tiny creature— an homunculus depicted as an "narn" (fool), mocking the alchemical claim to be able to create artificial life in the laboratory. The shelves are stocked with fake remedies for all manner of ailments.

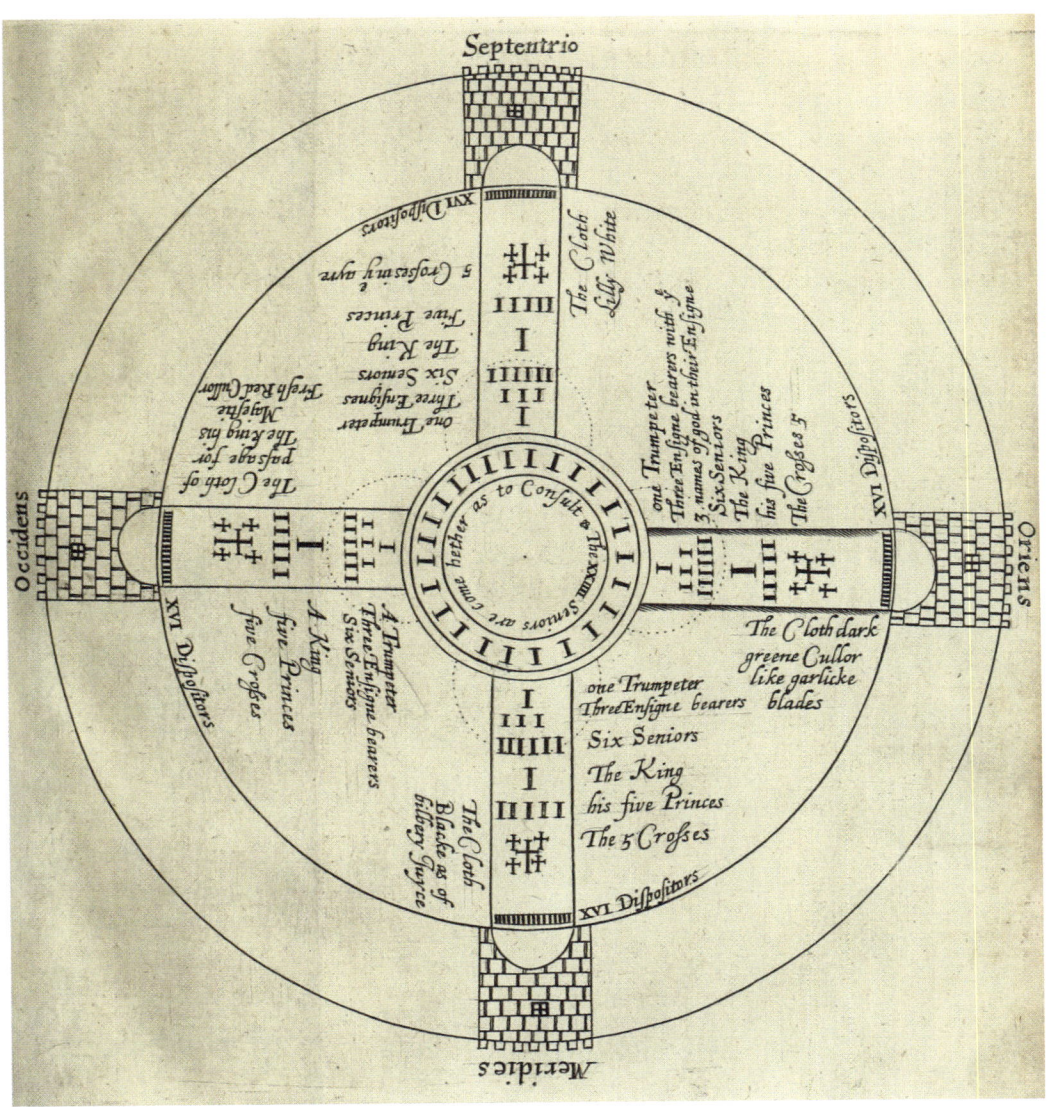

ABOVE Edward Kelley's Four Castles Vision, in a diagram from the Elizabethan magus John Dee's *A True & Faithful Relation of What Passed for Many Yeers between Dr. John Dee . . . and Some Spirits* (London, 1659). Kelley worked with Dee for several years as a scryer, able to communicate with angels via "shew stones" or mirrors. The relationship was constantly fractious, and the two men fell out for good after traveling through Europe to the court of Rudolf II. Kelley claimed to know the secret of alchemical transmutation, achieved using a red tincture. He is widely regarded now as a fraudster, although the line between deceit and delusion is often hard to determine in such cases.

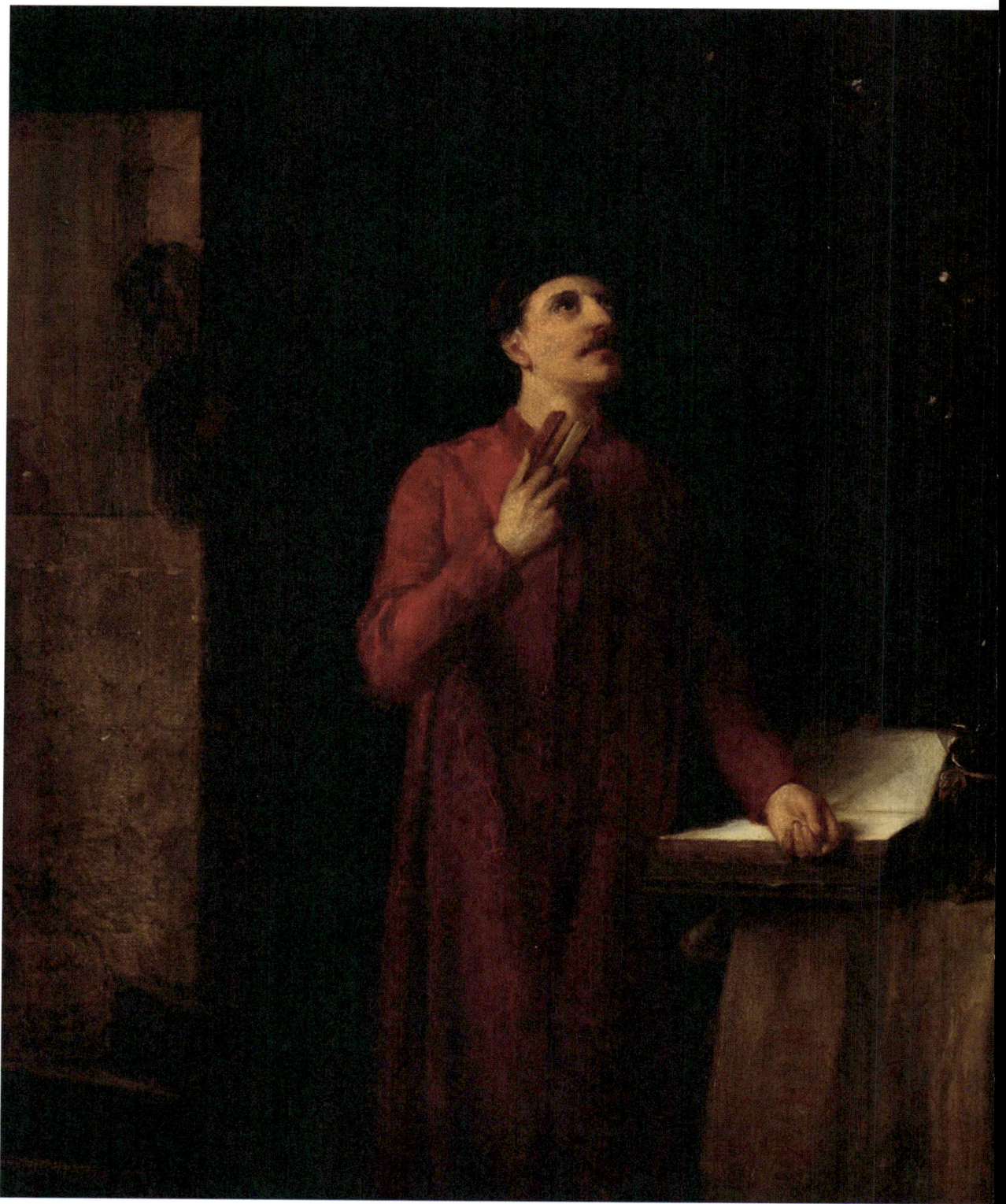

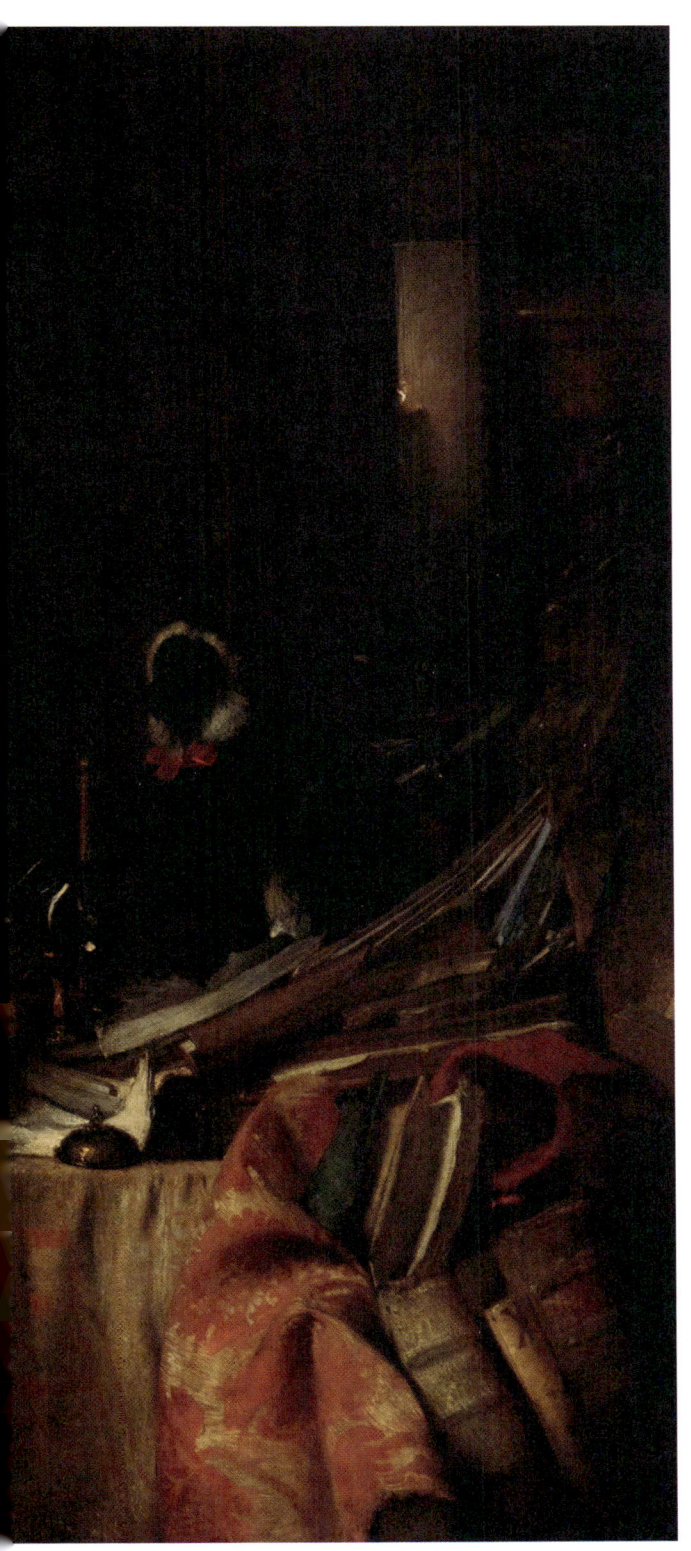

LEFT The 19th-century Scottish academy painter William Fettes Douglas had a strong interest in alchemy and mysticism. This portrait from c. 1872 depicts an unknown Rosicrucian apparently consulting tomes of ancient lore. Douglas also painted *The Alchemist* (1855), in which the title character appears to be inspecting a flask of urine, and *The Spell* (1864), depicting a necromancer conjuring with a skull.

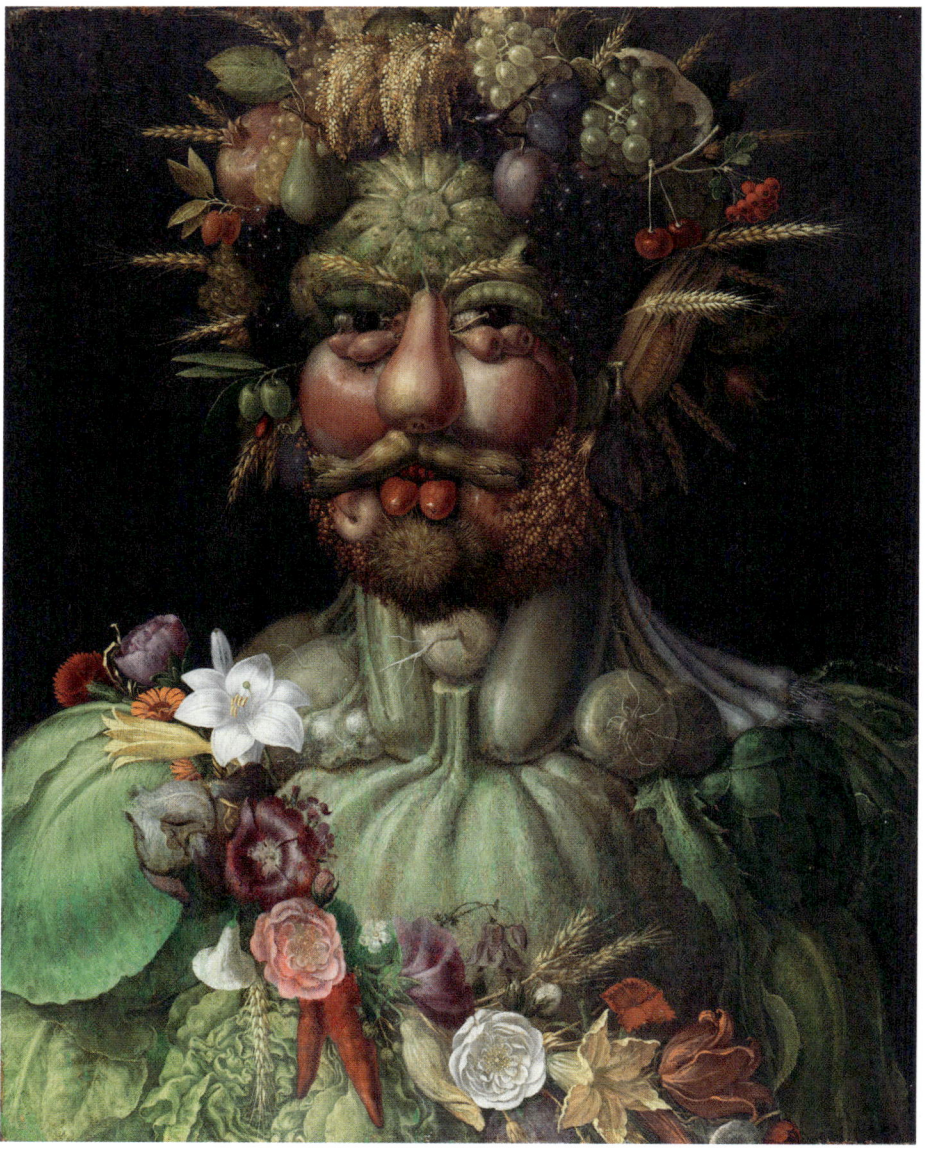

ABOVE Emperor Rudolf II as Vertumnus, the Roman god of the seasons, painted by Rudolf's court artist Giuseppe Arcimboldo in 1591. An imperial allegory, the whimsical portrait also reflects Rudolf's renowned fascination with puzzles, the bizarre, and the occult. In their transformative configuration, Arcimboldo's composite portraits reflect the ethos of alchemical doctrine and spirituality then so prevalent at the Habsburg court. Originally in the Imperial collection at Prague, it is thought to have later been gifted to Skokloster Castle by Queen Christina of Sweden.

OPPOSITE The tree of knowledge of good and evil, taken from *Geheime Figuren der Rosenkreuzer* (Secret symbols of the Rosicrucians; 1785). The book draws on alchemical and mystical imagery from earlier works, albeit seemingly without any clear connection to the then-extent "official" Rosicrucian movement; some consider it an attempt to cash in on a public enthusiasm for mysticism with a set of lavish and gloriously colored prints.

Der Baum der Erkenntniß Gutes und Böses.

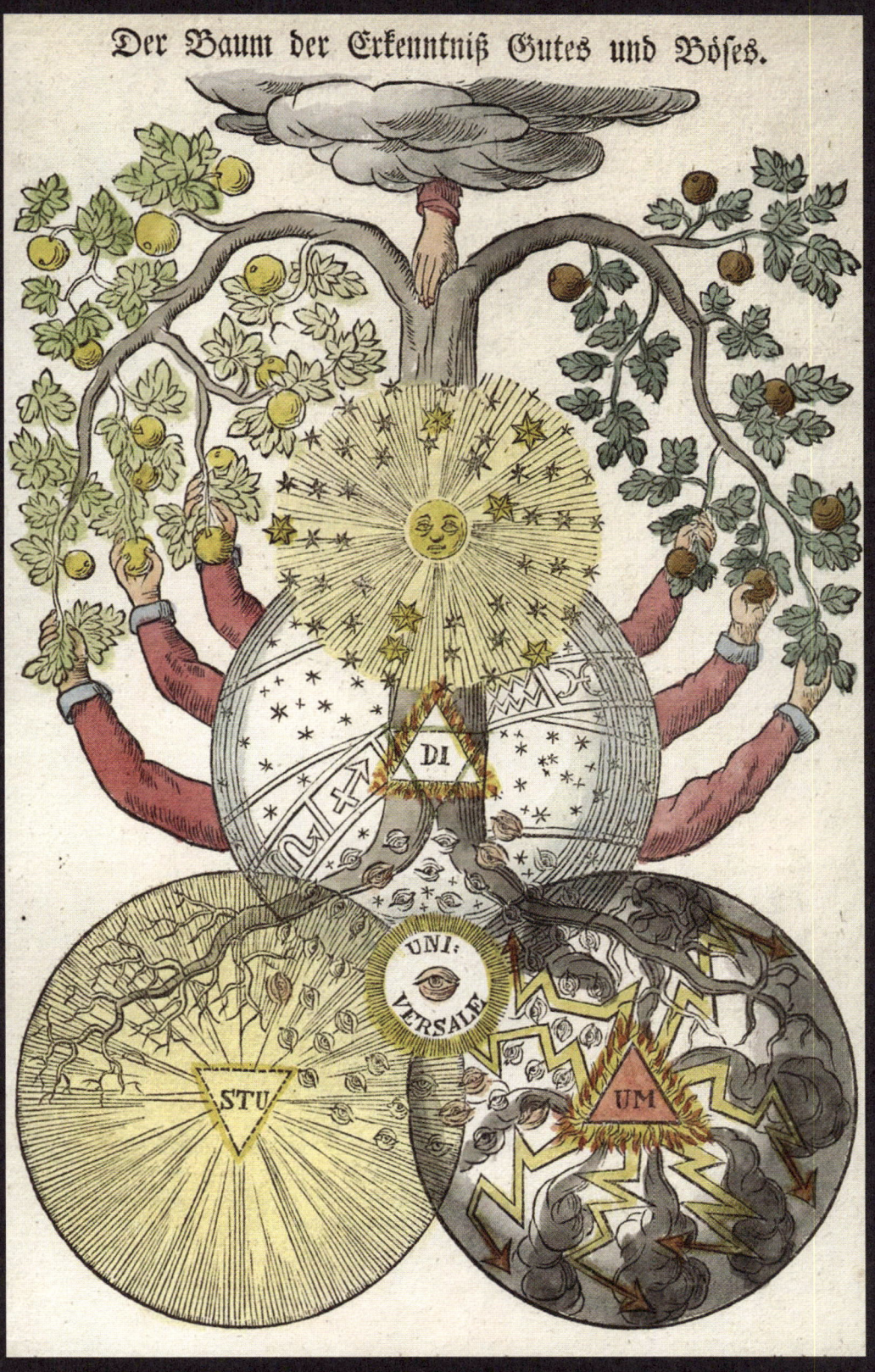

PROFILE

Christina of Sweden

(1626–1689)

Female alchemists are rare, no doubt mostly because—as in art, science, music, and literature—women were generally denied the opportunities for such learning. One remarkable woman for whom this was not true was the seventeenth-century queen of Sweden, who not only attracted and supported some of the leading chemists of her times but conducted her own theoretical and practical research into the Great Work.

In a pivotal conflict of the Thirty Years' War at Lützen in 1632, forces from Sweden and the German Protestant states narrowly defeated the Imperial army. Among the many fatalities was the Swedish king Gustav Adolphus, leaving his six-year-old daughter Christina on the throne.

The king had stipulated that his daughter should be educated as thoroughly as a boy while the realm was administered by the Swedish Privy Council. After being made queen in 1644 Christina became a patron of the arts, making Stockholm a hub for literature, music, and painting, and she hosted gatherings of eminent humanist scholars such as René Descartes. Her personal library contained thousands of volumes, including works on the Hermetic arts, Neoplatonic philosophy, and the Kabbalah, some of them looted from the treasures collected by Rudolf II in Prague after Christina's army conquered the city in 1648. Among these were works by Geber, Pseudo-Arnald of Villanova, and Ramon Pseudo-Lull, as well as more recent chymical texts by Andreas Libavius, the Paracelsian Oswald Croll, and Nicolas Lemery.

Alchemy was central to these interests. It is, Christina wrote, "a royal and furthermore, a divine science. It has fallen into disrepute among those who think they know everything, but know nothing." She seems to have read widely on the topic, and also to have conducted her own experiments, some of them in pursuit of the philosopher's stone at the urging of the Swedish alchemist Johannes Franck, who visited her court in 1651.

Christina abdicated the throne to her cousin in 1654 and converted to Catholicism. She traveled to Rome, where she was welcomed at the Vatican. Taking up residence in a grand palace, she established it as an institution for art and literature that later became the Academy of Arcadia. (The palace now houses the Accademia Nazionale dei Lincei, of which Galileo had been a member.) Among those associated with the academy were the astronomer Giovanni Cassini and the Jesuit polymath Athanasius Kircher.

In Rome Christina was free to pursue her intellectual and Hermetic interests. In 1667 she worked alongside the Italian alchemist

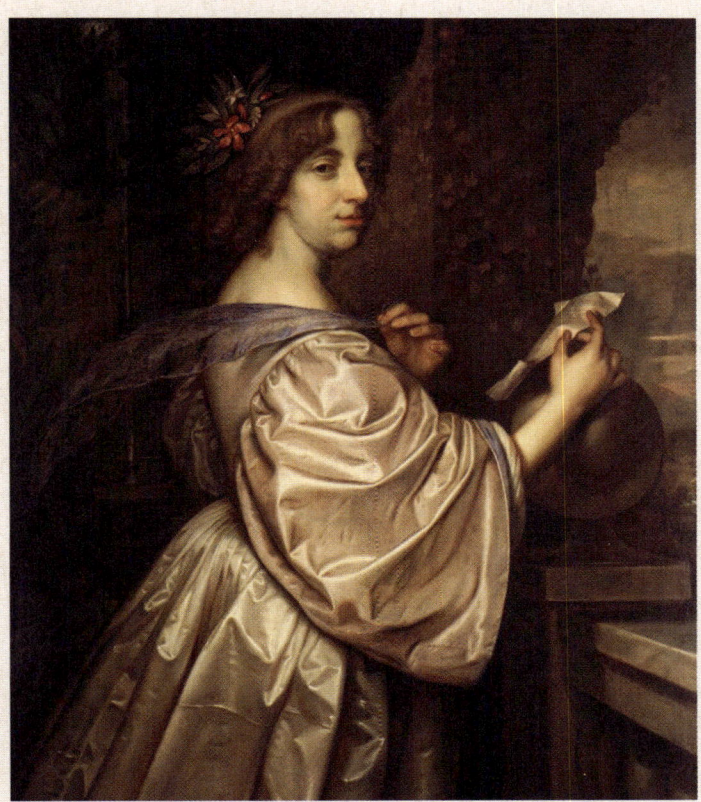

RIGHT Queen Christina of Sweden, in a portrait by David Beck, 1650. Christina's interest in alchemy seems to have been profound and informed. As well as attracting alchemists to her court in Sweden, and later (after abdicating the throne) to her home in Rome, she seems to have engaged in laboratory experiments herself.

and prophet Giuseppe Francesco Borri, until being advised to sever the partnership because Borri was considered a potential heretic whom the Inquisition wished to interrogate. She corresponded with the distinguished German chymist Johannes Glauber and was fascinated to hear of Hennig Brandt's discovery of phosphorus (see page 128). For a time she was the owner of the mysterious Voynich manuscript, acquired from Rudolf II's alchemical collection, which Kircher later tried unsuccessfully to decrypt.

In her laboratory studies Christina allegedly took on a female assistant named Sibylla, and in 1670 she appointed the alchemist Pietro Antonio Bandiera to run her experiments. There seems little doubt that she had hands-on experience: in a surviving document from that time a drawing of alchemical equipment is accompanied by a note written by her, in which she wonders how long a particular material needed to be heated in the furnace for the process of calcination to take place.

Christina died of a bacterial infection in 1689. She remained celibate, and her sexuality has been a matter of much debate. It was said that she "walked like a man, sat and rode like a man, and could eat and swear like the roughest soldiers," although she wrote at the end of her life that she was "neither Male nor Hermaphrodite, as some People in the World have pass'd me for." In Antwerp in 1654–55 she is said to have met and made a strong impression on the English noblewoman Margaret Cavendish, who followed Christina's example of defying the gender stereotypes of the age by delving deeply into natural philosophy.

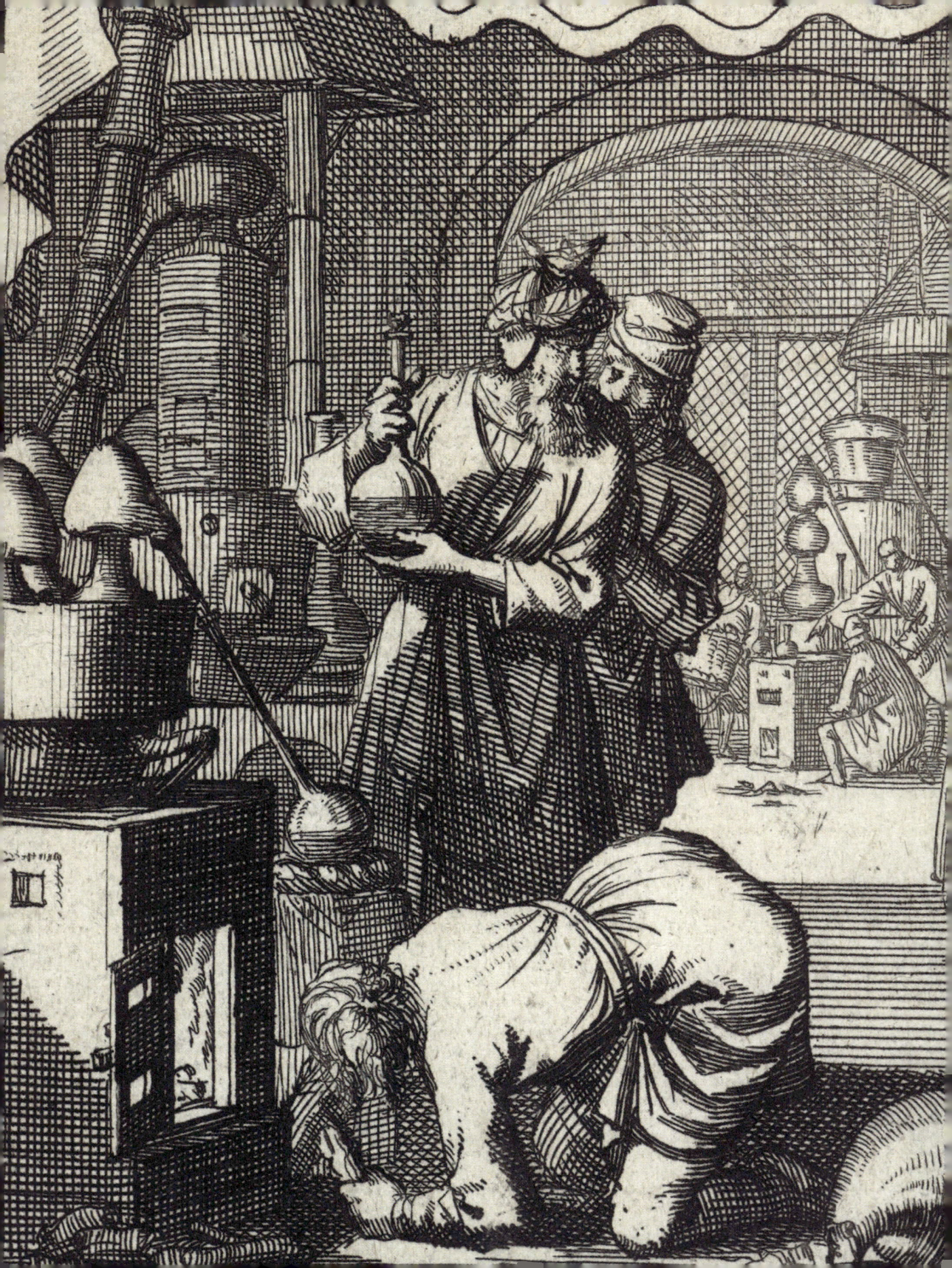

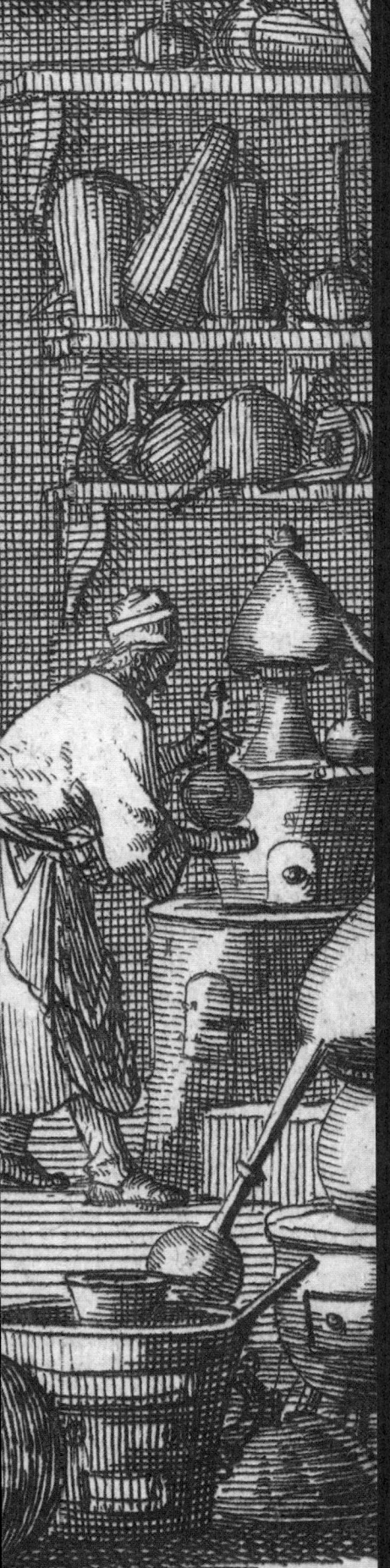

CHAPTER EIGHT

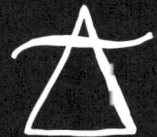

Crucibles
FROM ALCHEMY TO CHEMISTRY

LEFT Glass and pottery distillation vessels of all types feature in Dutch engraver Jan Luyken's *Figures in a Laboratory*, 1693. In contrast to some satirical images from around this time showing alchemical workplaces as disorderly, dark, and smoky, this image depicts a place of orderly and systematic chemical research.

CHAPTER EIGHT

Today, alchemists are recognized as having made profound and systematic contributions to the development of chemistry, rather than merely stumbling by accident over useful methods and materials in their search for gold. Some of the key ideas in the emergence of chemistry have their origins in alchemical thought and practice. And contrary to the old notion that alchemy was disowned and ridiculed by the "experimental philosophers" of the seventeenth century, it is now clear that those scholars continued to explore alchemical ideas even as they laid the foundations of modern chemistry.

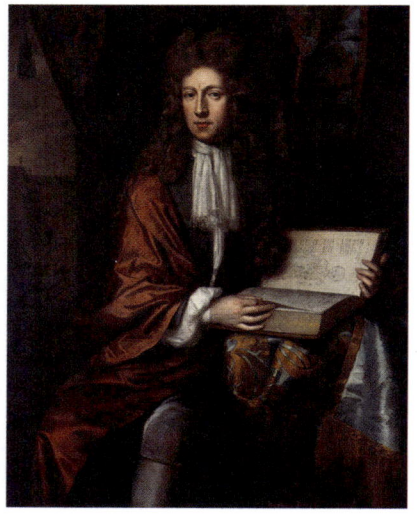

ABOVE A painting of Robert Boyle (attributed to Johann Kerseboom, date unknown), the 17th-century Anglo-Irish experimental philosopher, is sometimes portrayed as the "man who killed alchemy." That reputation is misplaced, however. Boyle's famous work *The Sceptical Chymist* (1661) denounced the cryptic and fraudulent habits of some alchemists, but Boyle himself never doubted that the transmutation of metals was possible.

In popular legend, Robert Boyle is the "man who killed alchemy." The Belgian-American chemist and historian George Sarton, sometimes considered one of the founders of the modern discipline of the history of science, called Boyle "one of the best prototypes of the modern man of science," in contrast to alchemists, who were "fools or knaves" (or both). The American historian of magic and early science Lynn Thorndike denied, despite abundant evidence to the contrary, that Boyle had any interest in gold-making or even believed it possible. The goal of many such "histories" was to find a path linking Boyle to the pioneering chemists of the late eighteenth century: historian Marie Boas Hall claimed in the 1960s that Boyle was "preparing the way" for French chemist Antoine Lavoisier's reassuringly modern ideas while getting rid of the mystical trash of alchemy.

It's now clear that such narratives—examples of what today's historians call Whig history, which prunes and interprets the past to create a sense of inexorable progress towards modernity—are fictions. We've seen already that Boyle was very much interested in alchemy and eagerly sought the philosopher's stone that could transmute metals into gold.

Central to the old view of Boyle is his 1661 book *The Sceptical Chymist*, which Boas Hall called a "withering blast" against alchemy. Here Boyle is credited with introducing the modern concept of a chemical element. Neither of those claims stands up to scrutiny.

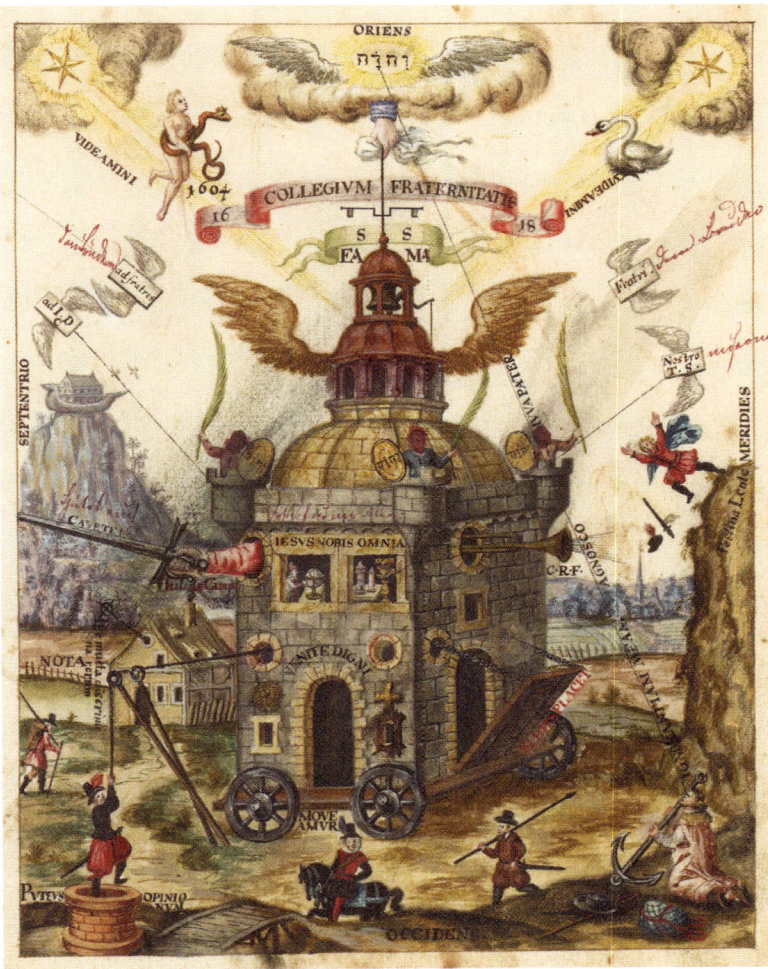

RIGHT Emblematic image of a Rosicrucian college, from *Speculum sophicum Rhodostauroticum (Mirror of the Wisdom of the Rosy Cross)*, a 1618 work by Theophilus Schweighardt, probably the pseudonym of the German alchemist and Rosicrucian Daniel Mögling, the court physician and astronomer of Philip III, Landgrave of Hesse-Butzbach. The book's images are packed with symbolism that can be hard to decipher. Rosicrucians often styled themselves as an "invisible college"—a term used by the Prussian exile Samuel Hartlib—steeped in the tradition of occult secrets, for the group of natural philosophers who convened informally in London in the 1640s. Robert Boyle was a prominent member of this circle, which became one of the progenitors of the Royal Society.

In defining an element as a substance that can't be broken down into simpler ones, Boyle wasn't really saying anything new or controversial—and besides, he seems to doubt whether any such substances truly exist. And like many of his contemporaries, Boyle was not trying to discredit alchemy but rather to separate what is good and useful in it from what is vague or dishonest. He endorsed an old distinction between true adepts who know the secret of chrysopoeia and vulgar cheats or untutored laborers such as dyers and distillers who don't understand what they do (Boyle puts himself in the former category, naturally). *The Sceptical Chymist* was no death knell for alchemy but a call (and not the first) for it to clean up its act.

Boyle was himself not always skeptical enough. He was seemingly taken in by a French conman named Georges Pierre des Clozets who claimed to know the secret of the philosopher's stone, lavishing him with gifts in the hope that he would reveal all. Boyle even wrote a manuscript called "Dialogue on Transmutation" (which was never published, and of which only fragments survive) in which he considered the possibility of chrysopoeia.

Although in the late seventeenth century there was plenty of skepticism about transmutation, it was by no means embarrassing or disreputable to have an interest in the subject. Yet while Boyle published some of his studies openly in the *Philosophical Transactions* of the Royal Society, both he and his colleague Isaac Newton often observed the old alchemical tradition of secrecy,

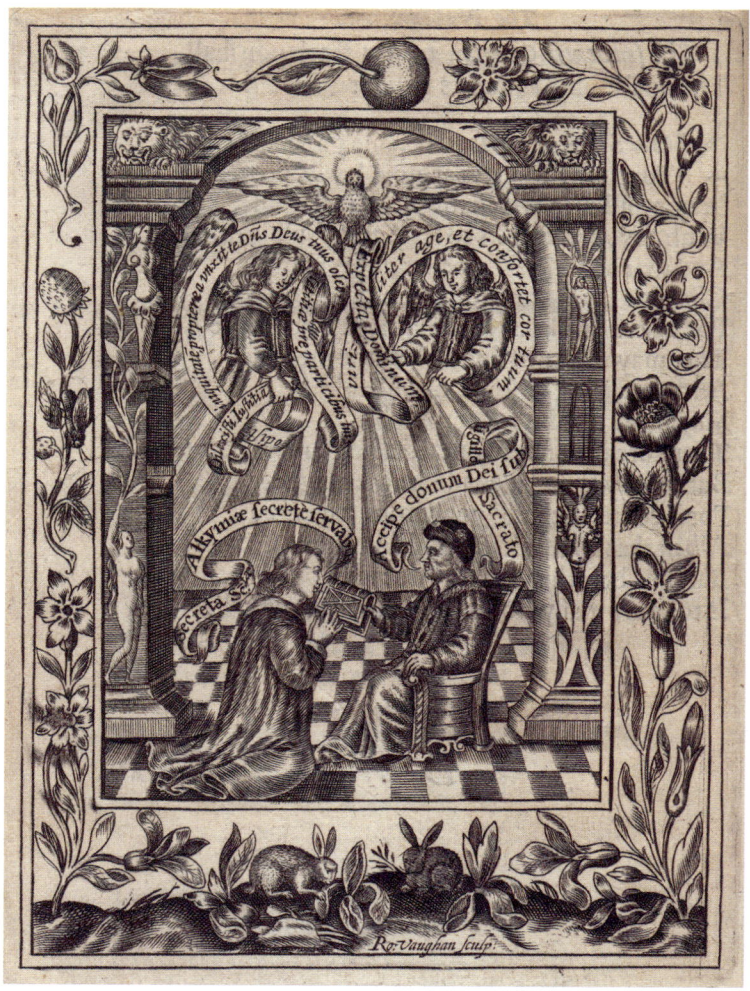

RIGHT Secret lore: an engraving depicting the passing on of alchemical knowledge from adept to apprentice, from Thomas Norton's *The Ordinall of Alchimy* (London, 1652). The master commands: "Accept the gift of God under a sacred seal," and the student replies "I shall preserve the secrets of holy alchemy in secrecy." Norton was a practicing alchemist of the 15th century, and his *Ordinall* was an influential text at that time.

RIGHT Isaac Newton's manuscript copy of Michael Sendivogius's 1614 book *Novum lumen chymicum* (translated as *A New Light of Alchymie*). Judging from the state of its pages, Newton consulted the text extensively in his quest for the philosopher's stone.

seemingly believing that such powerful knowledge should be kept within a circle of elite adepts. They were not "modern," but their alchemical interests did not make them backward-looking either. Like most great thinkers, they were simply of their time.

Part of the appeal of the Boyle myth may be that it promised a clean break: a waymark at which alchemy was cast aside and chemistry took over. But such changes rarely if ever happen in science, and the transition of alchemy to chemistry—often denoted as having happened via the intermediate discipline of *chymistry*—was certainly not like that.

Alchemy was transformed to chemistry in the manner of all alchemical transmutations: through a process of separation, distillation, and purification. This transformation was already underway in the sixteenth century, particularly in the reception of the works of Paracelsus. Some Paracelsians, such as Michael Sendivogius, pursued the alchemical quest at full throttle. Sendivogius worked for a time in the court of Rudolf II, and on a diplomatic mission in Poland he was said to have demonstrated a successful "projection" of base metals into gold, witnessed by the Polish king. A copy of Sendivogius's 1614 book *Novum lumen chymicum* (translated as *A New Light of Alchymie*) was much thumbed in Newton's library.

CRUCIBLES: FROM ALCHEMY TO CHEMISTRY

Others, such as Andreas Libavius, while sharing the Paracelsians' enthusiasm for chemical medicine and their rejection of the old dogmas of Galen and Aristotle, wanted to strip chymistry of mysticism, speculation, and theology and make it a robustly practical art. Such a down-to-earth approach to chemistry was reflected in Hieronymus Brunschwig's books on distillation (see page 122) and in the treatises on metallurgy and mining *De la pirotechnia* (On pyrotechnics; 1540) by Vannoccio Biringuccio and *De re metallica* (On the nature of metals; 1556) by Georgius Agricola (see page 76). There were no appeals to divine inspiration or magical forces here—and as to claims of transmutation, Agricola remarks,

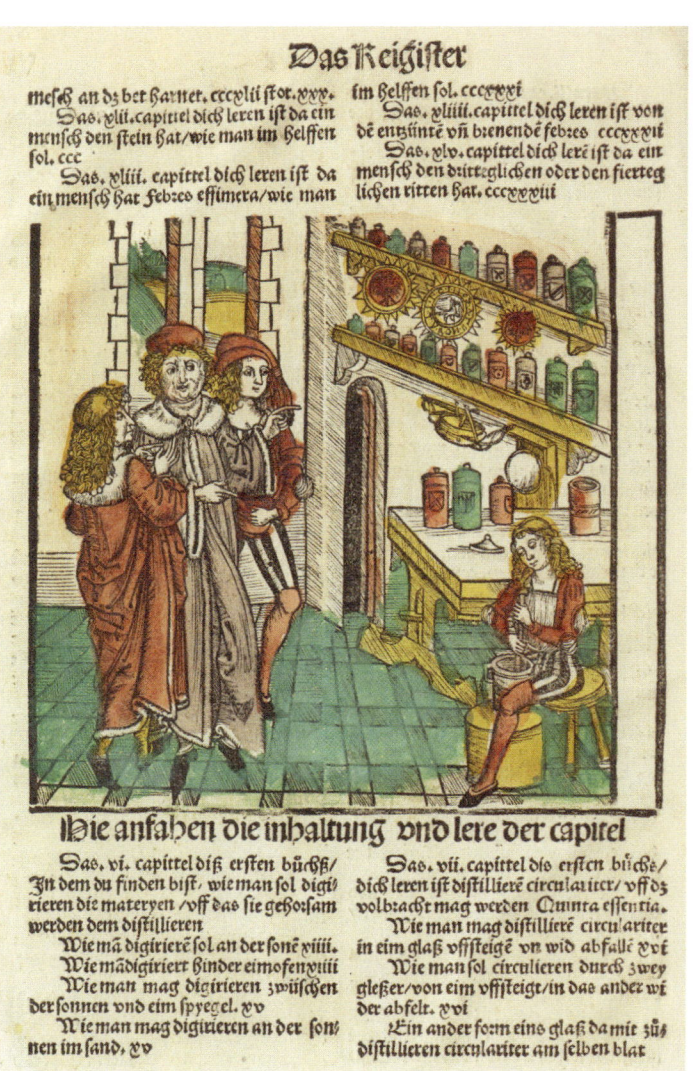

RIGHT A distiller's shop, from a hand-colored edition of Hieronymus Brunschwig's book on distillation *Liber de arte distillandi* (Little book on the art of distillation; Strasbourg, 1512). The book was a practically oriented manual of early chymistry, free from the cryptic symbolism of many alchemical works.

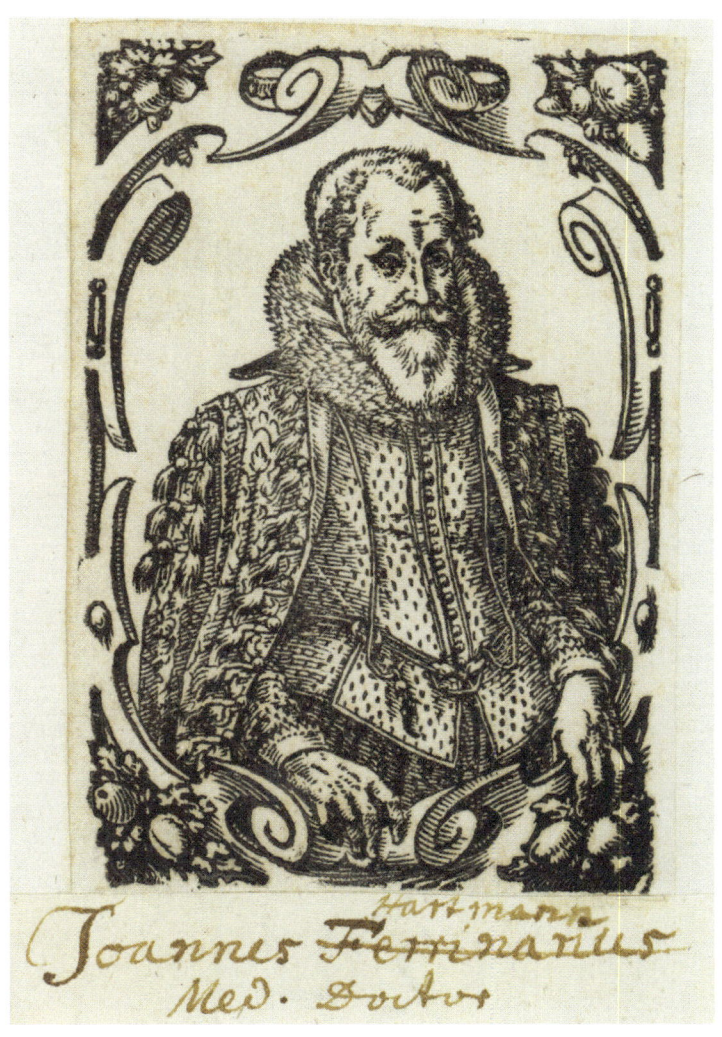

RIGHT The German Paracelsian Johannes Hartmann, shown here in a woodcut from c. 1615, was appointed professor of "chymiatria" at the University of Marburg in 1609. Hartmann's position is sometimes said to be the first academic chair for chemistry, and was a reflection of the subject's newfound professionalization.

"I should say the matter is dubious." Some alchemists, he says, simply deceive others (or themselves) with fraudulent schemes.

By the start of the seventeenth century, chymistry was deemed to warrant a place in the academic curriculum. In 1609 the German Paracelsian Johannes Hartmann was appointed professor of chemistry (actually "chymiatria") at the University of Marburg—some say this was the first such post in the world, although there is evidence that university instruction in aspects of chymistry goes back further. Over the course of that century, the textbooks by Jean Béguin, Nicaise Lefebvre, and Nicolas Lemery (see page 185) marked the emergence of chemistry as a respected and practically oriented discipline.

The chymical chimera

This professionalization of chymistry went hand in hand with efforts to make it a *useful* discipline, to make the materials that society needed, such as medicines, metals, oils, dyes, and pigments. We saw earlier how Libavius and the German chymist Johann Joachim Becher (see page 138) called for the establishment of formal chemical laboratories—something between research institutes and factories—both as places of manufacture and to systematically discover new processes.

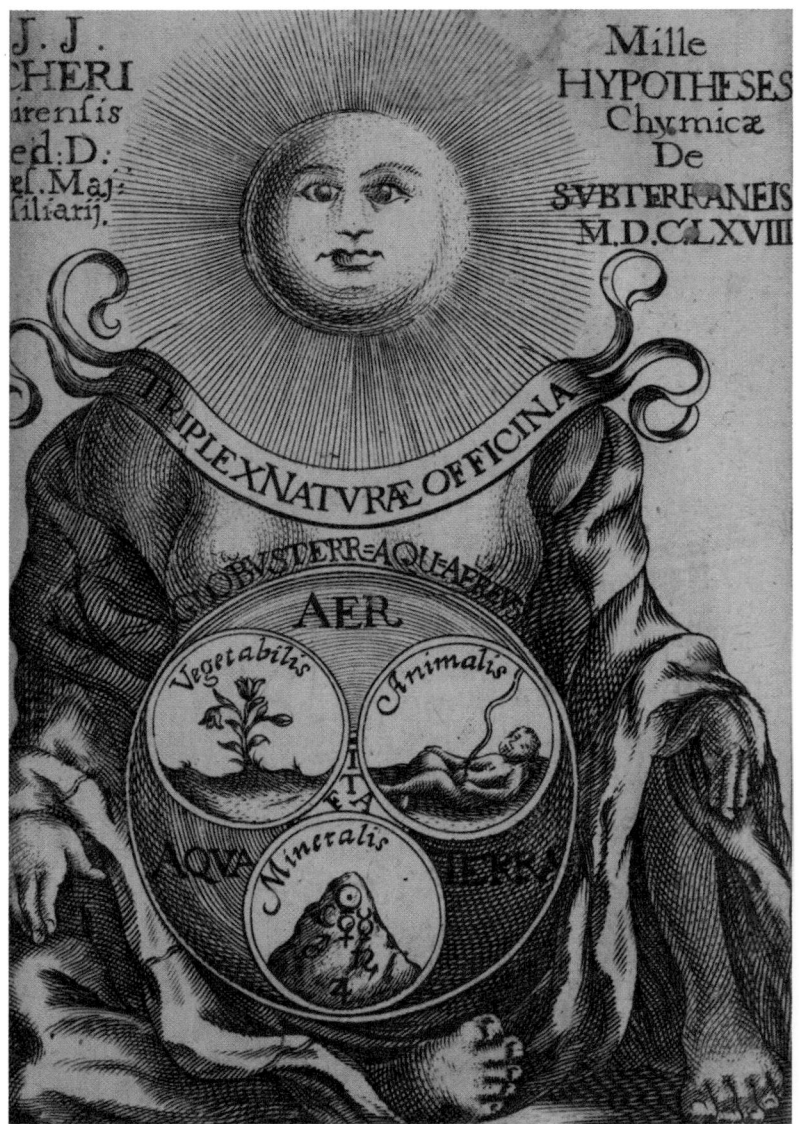

RIGHT Engraved frontispiece of the 1669 edition of German alchemist Johann Becher's book *Physica subterranea* (Underground physics; Frankfurt, 1669), depicting the classical elements of air, water, and earth. Becher considered there to be three varieties of earth, derived from the Paracelsian *tria prima* of mercury, sulfur, and salt.

As well as being a decidedly practical (if ultimately unsuccessful) alchemist, Becher put forward a new theory of the constitution of matter. He acknowledged only two classical elements: water and earth. But he asserted that there are in fact three distinct earths, which we can recognize as the Paracelsian *tria prima* of sulfur, mercury, and salt under new names. *Terra fluida* (or *mercurialis*) was the principle of fluidity, based on mercury; *terra lapida* or "vitreous earth" represented solidity, derived from salt; and *terra pinguis*, "fatty earth," basically fiery sulfur, made matter oily and combustible.

Becher laid out this scheme in his 1669 book *Physica subterranea* (Underground physics). In a new edition of the book published and edited by the chemist and physician Georg Ernst Stahl, chair of medicine at the University of Halle, Stahl gave *terra pinguis* a new name: phlogiston, derived from the Greek *phlogistos*, "burning."

In phlogiston, the rather vague and immaterial notion of a "principle" of fire and combustion was elevated into a physical substance—one that, for the rest of the century, chemists sought to isolate and measure. The phlogiston hypothesis could seemingly make sense of, and indeed unify, a range of chemical processes. Stahl asserted that when a material such as wood burns, it releases phlogiston into the air. It was known that such substances, if ignited and placed in a sealed vessel, would cease burning after a short time: Stahl's theory explained this on the basis that the air in the container becomes "saturated" with phlogiston and can accept no more. Meanwhile, the reason nothing will burn in a vacuum, as Boyle had shown in his experiments with a vacuum pump in the 1650s, was because there was no air to take up the phlogiston. Metals, when heated in air, lose their phlogiston and transform to a dull residue called a calx. Charcoal can release metals such as iron from their ores because it is rich in phlogiston and so will add this substance to the ore (which is like a calx) to restore the metal.

The professionalization of chymistry went hand in hand with efforts to make it a *useful* discipline, to make the materials that society needed, such as medicines, metals, oils, dyes, and pigments.

While phlogiston was not universally accepted by eighteenth-century chymists, few questioned its central premise that fire and combustion relied on some flammable substance. The influential Dutch scientist Hermann Boerhaave, professor of botany, medicine, and chemistry at the University of Leiden, made no mention of phlogiston, but put in its place a combustible ingredient called the *pabulum ignis*, a "matter feeding fire." Boerhaave's 1732 *Elementa chemiae* (Elements of chemistry; he maintained the increasingly outmoded practice of writing and lecturing in Latin) is perhaps the closest thing to a foundational textbook of a modern form of

RIGHT Frontispiece of *Index planarum*, 1710, a list of plants to be found in the academic garden of Lyons, written (in Latin) by Hermann Boerhaave, professor of botany and chemistry at Leiden. Boerhaave was the author of an influential textbook of chemistry, *Elementa chemiae* (Elements of chemistry), in which he identified the combustible principle he called *pabulum ignis*, a "matter feeding fire."

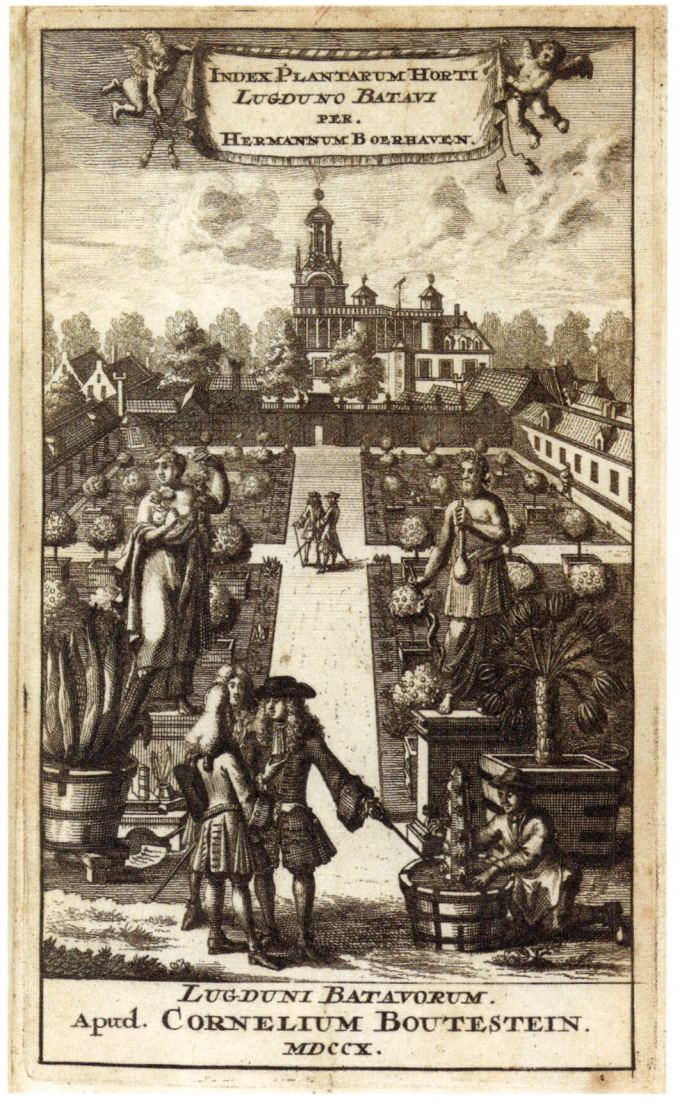

chemistry, establishing the idea that different chemical substances unite according to their affinity—which Boerhaave called a kind of "love"—for one another. He had little patience for the speculative excesses of alchemists, writing, "How I wish... these raving men had restrained themselves and not wished to interpret the Sacred Scriptures in terms of chymical principles and elements."

The phlogiston theory is often derided now as a flawed idea that held chemistry back for the best part of a century. That is a Whiggish view: if only we hadn't persisted with a wrong idea, we could have found the right one! In fact, phlogiston, much like chrysopoeia, was precisely the kind of concept that scientists have always needed to enable them to keep going when their understanding is scanty. Phlogiston allowed chemists (we can call them that by the mid-eighteenth century) to think about processes such as the smelting and corrosion of metals and the burning of fuel using the same framework with which they pondered respiration. Phlogiston theory was wrong—there is no such substance—but it was so nearly right as to be fruitful. Things burn not because they release some combustible substance into the air, but because they take something *from* the air: the gas that Antoine Lavoisier christened oxygen in the 1780s. Metals do not lose anything when they form a calx; rather, they have *combined* with oxygen.

Stimulated by phlogiston theory, eighteenth-century chemistry became the study of "airs"—or gases, a word coined by Jan van Helmont. So long as air itself was regarded as an indivisible element, it remained hard to make sense of all the different kinds of airs that chemists identified: fixed air, mephitic air, inflammable air (the latter, identified by the Swedish chemist Wilhelm Scheele, was suspected by some of being pure phlogiston). We now recognize these as different gases: carbon dioxide, nitrogen, hydrogen. It was Lavoisier's oxygen theory, in which ordinary air was understood as a mixture of oxygen and nitrogen (which Lavoisier called azote, as French chemists still do), that finally made sense of it all—to the chagrin of staunch advocates of phlogiston, as some remained even into the early nineteenth century.

Lavoisier also clarified the idea of a chemical element—not by defining it (he more or less kept Boyle's definition of a substance that couldn't be separated into simpler ones) but by drawing up a list of them in his seminal 1789 textbook *Traité élémentaire de chimie* (Elementary treatise on chemistry), which helped to secure his oxygen theory within France. He listed thirty-three elements—

the roster grew ever longer through the nineteenth century—including all the known metals. If these elements were indeed fundamental and irreducible, most scientists decided that there was no longer any point in attempting transmutation.

The last of the alchemists

But not all scientists agreed. Alchemy's Great Work might have been generally discarded and mocked by the late eighteenth century, but it was not abandoned entirely.

One of the late would-be alchemists was English chemist James Price, who in 1782 claimed to have transmuted mercury into silver and gold. He demonstrated his results at his home near Guildford in Surrey to a distinguished audience that included several lords. Despite skepticism from others, Price was awarded an honorary degree from the University of Oxford for his "chemical labours."

Price was no fringe figure but a Fellow of the Royal Society—which made his claims all the more outrageous to many of his peers. Faced with demands that Price be expelled from the ranks, the president of the Royal Society Joseph Banks demanded that the chemist—whom Banks sardonically dubbed "our Paracelsus of Guildford"—demonstrate his experiments before them. Eventually, Price agreed to perform a transmutation for other Fellows invited to his home—but on the day of that event, he committed suicide by drinking poison. (Some say he dropped dead in front of his audience.)

According to John Timbs, author of *English Eccentrics and Eccentricities* (1866), "the last true believer in alchemy was not Dr Price, but Peter Woulfe, the eminent chemist, and Fellow of the Royal Society, who made experiments to show the nature of mosaic gold [tin sulfide, an old form of 'artificial gold']." Woulfe was an Irishman who lived in London in the late eighteenth century, and his chemical prowess brought him impressive accolades: he was given the Royal Society's Copley Medal in 1768, its oldest and most prestigious award, and delivered the society's Bakerian lectures for three years in a row in the 1770s. Older chemists today might know him as the inventor of the triple-necked Woulfe bottle for collecting gases. Yet, to the dismay of his colleagues, in later life he became fixated on the transmutation of metals. The nineteenth-century English chemist William Brande reported that Woulfe's rooms in Holborn, London, "were so filled with furnaces and apparatus that it was difficult to reach his fireside . . . He had long vainly

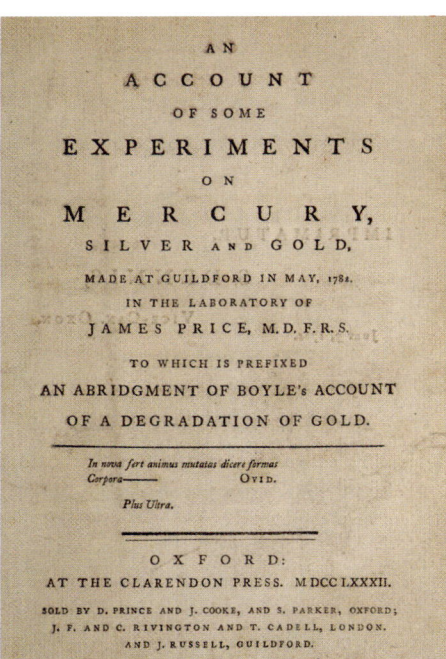

ABOVE James Price's *Account of Some Experiments . . . Made at Guildford in May, 1782* (Oxford, 1782). Price, a respected Fellow of the Royal Society, scandalized his colleagues when he claimed to be able to carry out the alchemical transmutation of metals.

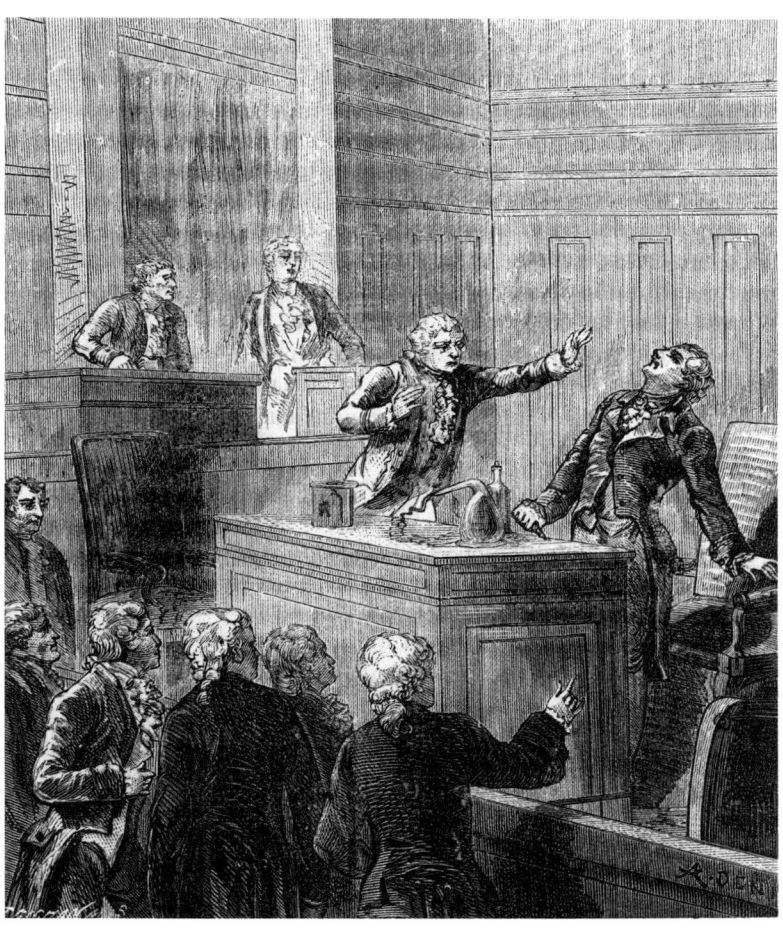

RIGHT An engraving depicting James Price poisoning himself with Prussic acid before a gathering of members of the Royal Society, convened to witness Price's alchemical experiments, in Alexis Clerc's *Physique et chimie populaires* (Popular physics and chemistry; Paris, 1881–83).

searched for the Elixir, and attributed his repeated failures to the want of due preparation."

Woulfe's pursuit of alchemy probably arose from his involvement in esoteric movements such as Rosicrucianism (see page 178), Freemasonry, and the heterodox religious movement begun by the theologian and mystic Emanuel Swedenborg (whom he met in 1769). So thoroughly discredited was alchemy by this stage that Woulfe was thought by his peers to have lost his mind, perhaps because of his isolation or because of chemical poisoning from his earlier career. The eminent French chemist Antoine François de Fourcroy, a colleague of Lavoisier, wrote in 1799 that "The famous Woulfe is in a state of mind which must cause much sorrow to friends of philosophy and science... [He] is no longer interested in chemistry; in his experiments he could no longer use any iron that would not suddenly turn into copper or lead."

Transmutation at last

Despite such ridicule of would-be modern chrysopoeians, the nineteenth-century French chemist Marcellin Berthelot rehabilitated alchemy's image by claiming that a great deal of useful chemistry was discovered in the futile quest to transform metals to gold. The German chemist Justus von Liebig offered a more generous (one might say *too* generous) assessment than did many later historians of chemistry when in the 1850s he wrote that

Alchemy was never at any time different from chemistry. It is utterly unjust to confound it, as is generally done, with the gold-making of the sixteenth and seventeenth centuries ... Alchemy was a science and included all those processes in which chemistry was technically applied.

All the same, it was generally agreed by this time that transmutation of the elements was a futile enterprise. And yet ... were we so sure of that? After all, the idea that there might be some fundamental matter, like the Aristotelian *protē hylē*, from which all substances are made, was still alive and well. After the English chemist John Dalton expounded in the 1800s his theory that all substances are composed of atoms, chemists began to measure the relative weights of the different elements and discovered that these seemed to be whole-number multiples of the weight of the lightest, hydrogen. In 1815 the chemist William Prout proposed that the atoms of all elements are indeed compounded from hydrogen atoms squeezed together.

This is sort of true. The New Zealand–born scientist Ernest Rutherford showed in the early twentieth century that atoms are not after all the unsplittable (Greek *a-tomos*) objects long supposed but are made of yet more fundamental particles: a dense nucleus containing subatomic particles called protons (and, as was later discovered, neutrons too), surrounded by a cloud of much lighter electrons. Hydrogen atoms, which have only a single proton in their nucleus, are thus in a sense the constituents of all others. Indeed, we now know that all other elements are made by the fusion of

By bombarding atoms with high-energy beams of fundamental particles in particle accelerators, other elements have indeed now been transformed into gold.

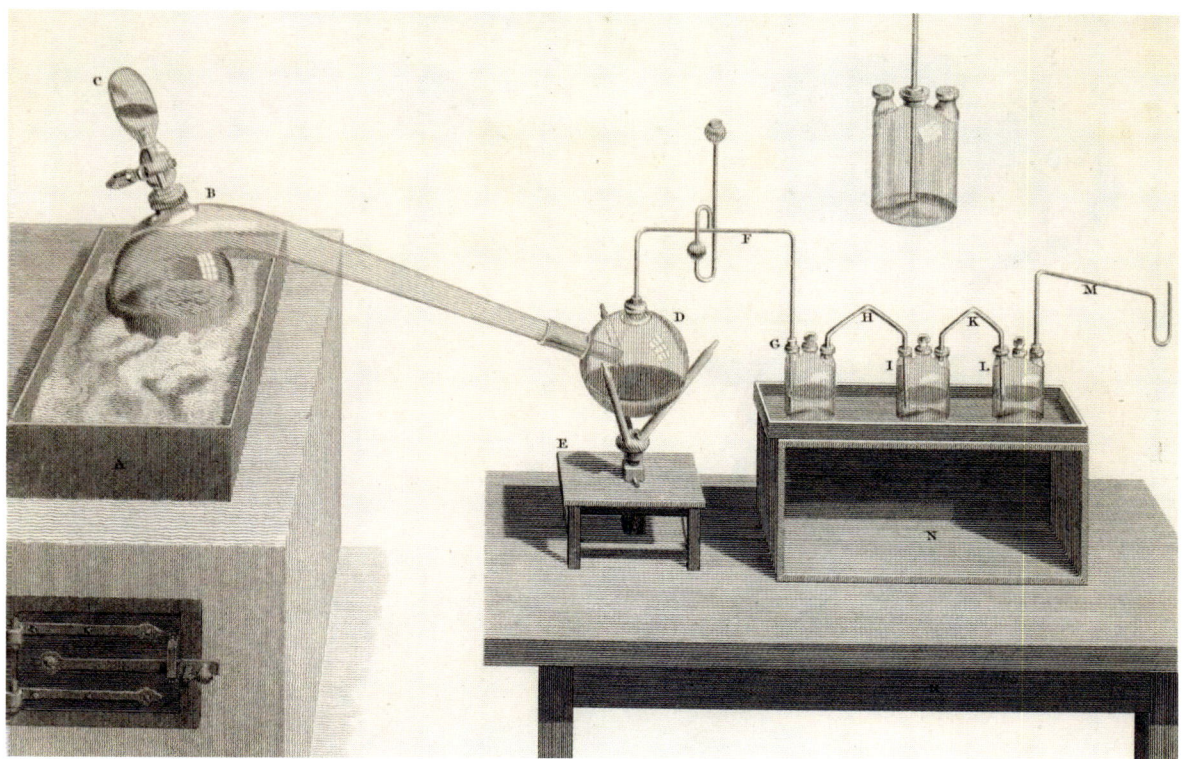

ABOVE The triple-necked apparatus (top right) devised by Peter Woulfe, a 19th-century chemist who claimed in later life to be able to transmute metals, illustrated in Abraham Rees's *Cyclopædia; or, Universal Dictionary of Arts, Sciences, and Literature* (London, 1819). Woulfe's rooms were said to be "so filled with furnaces and apparatus that it was difficult to reach his fireside."

hydrogen nuclei, a process that happens continually in the interiors of stars and generates their tremendous output of energy. Rutherford and others deduced that heavy atoms can be split apart by nuclear fission reactions—the process that releases heat in nuclear reactors and in the first atomic bombs. In both fusion and fission, one element is turned into another. By bombarding atoms with high-energy beams of fundamental particles in particle accelerators, other elements have indeed now been transformed into gold.

When Rutherford and his collaborator, the English chemist Frederick Soddy, first discovered the transmutation of elements by nuclear fission in 1901, he was alarmed at how it might look to their colleagues. Forced to the conclusion that the radioactive decay of the heavy element thorium transformed it into a different element (now recognized as radon), Soddy is said to have exclaimed "Rutherford, this is transmutation"—whereupon Rutherford shot back "For Mike's sake, Soddy, don't call it transmutation. They'll have our heads off as alchemists." For what could be more heretical in modern science than that?

CRUCIBLES: FROM ALCHEMY TO CHEMISTRY

PROFILE

George Starkey

(1628–1665)

Despite living for less than four decades, George Starkey was one of the most influential alchemists in England in the seventeenth century. He benefitted from being an outsider, arriving from the American colonies with a reputation that owed more to his remote and exotic origins than to his achievements. He exploited that renown to considerable effect.

Starkey's birthplace of Bermuda (then called the Somers Islands) was, in the early seventeenth century, a backwater of the nascent British colonies in America. He was of Scottish descent, the son of an evangelizing minister for the Church of England, and his family name was Stirk—Starkey changed it around 1650. He was just nine or so years old when his father died, and in 1639 he was sent for schooling to the recently founded Harvard College in Massachusetts, New England.

Like Paracelsus, Starkey was unimpressed by what the academic curriculum contained—and likewise he turned to "practical experience" as the better teacher in the ways of nature. From 1644 he began exploring the chemical philosophy; four years later he was writing to the alchemist John Winthrop, governor of Connecticut, begging for some spare mercury or antimony. Having set up a medical practice, he married in 1650 and at the end of that year he sailed for England, apparently in the hope of acquiring better laboratory equipment. A good-quality furnace was particularly important to the aspiring adept.

It seems that Starkey arrived in London with a reputation, due in part to Winthrop's acquaintance with the German émigré Samuel Hartlib, who maintained a circle of progressive intellectuals receptive to the utopian tenor of the Rosicrucian movement. "There is one Stirke or Starkie a young Physitian in [New England] ... of a most rare and incomparable universal Witt," Hartlib recorded in his diary early in 1650. "Hee is also very chymical." Just the person, then, to introduce to the chymist Robert Boyle, a prominent member of Hartlib's circle.

Thus, within months of Starkey's arrival Boyle was eagerly seeking his advice on matters alchemical, never mind the American's youth. Starkey also soon established a promising medical practice but gave it up by the spring of 1651 as it was keeping him from his alchemical work, saying to Boyle: "I find that in my searching after secrets, for which I greatly long, I am detained and dissipated." Starkey seems genuinely to believe himself the recipient of divinely gifted knowledge about nature's hidden workings.

In that same missive, Starkey tells Boyle of a substance he has prepared that he believes is

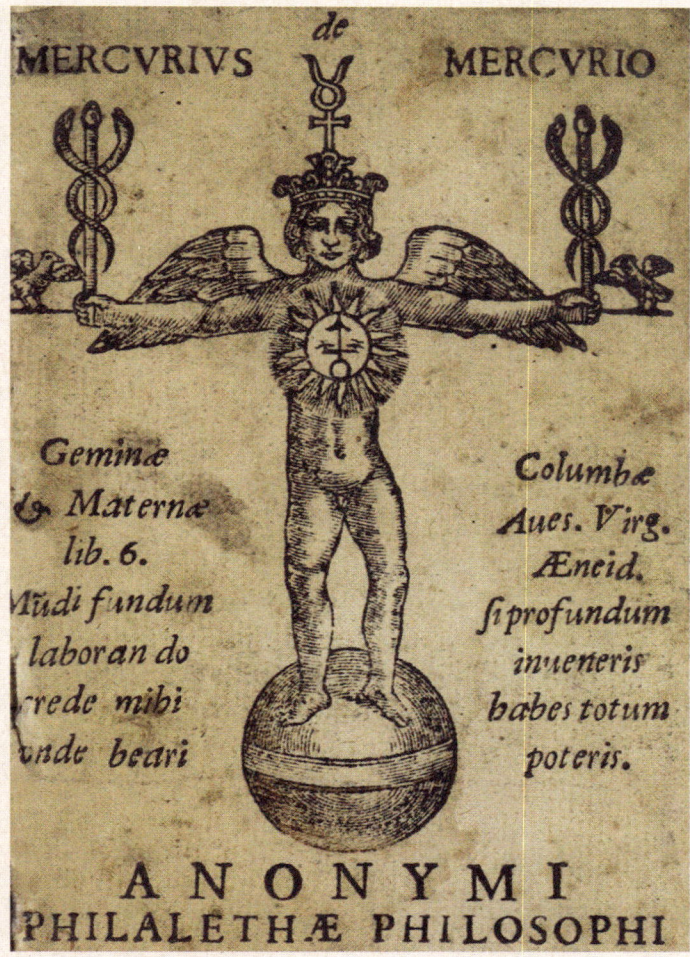

RIGHT Woodcut from Eirenaeus Philalethes's *Anonymi Philalethae philosophi Opera omnia* (The works of the anonymous philosopher Philalethes; Modena, 1695). Eirenaeus was the pseudonym of the alchemist George Starkey. The image here shows Mercury (Hermes), the winged messenger, crowned with John Dee's "Hieroglyphic Monad" symbol and carrying his trademark caduceus staff entwined by twin serpents.

"the mercury [stone] of the philosophers, which was given to me by an adept." This is the first reference to an American alchemist from whom Starkey claimed to have learned all his secrets, and whom he later named as one Eirenaeus Philalethes, meaning "peaceful lover of truth."

That Eirenaeus Philalethes was merely a pseudonym used by Starkey to publish his own alchemical works had long been suspected, but it was proved recently by historian of science William Newman based on a close analysis of Starkey's letters. It was not merely a device for anonymity. Starkey invented an entire history for his alter ego, making him an adept who possessed miraculous secret knowledge. His goal seems to have been to boost the allure of his books—if so, it was a successful ploy. The writings of "Philalethes" became very popular; they were collected into a volume of complete works in 1695 and eagerly consulted by the likes of Boyle and Isaac Newton. But Starkey himself did not benefit from such acclaim. His medical practice and trade in perfumes were not terribly lucrative, and he personifies the caricature of the alchemist in spending time in a debtor's prison before dying, aged thirty-seven, in the Great Plague of London in 1665 after dissecting a plague victim to study the affliction.

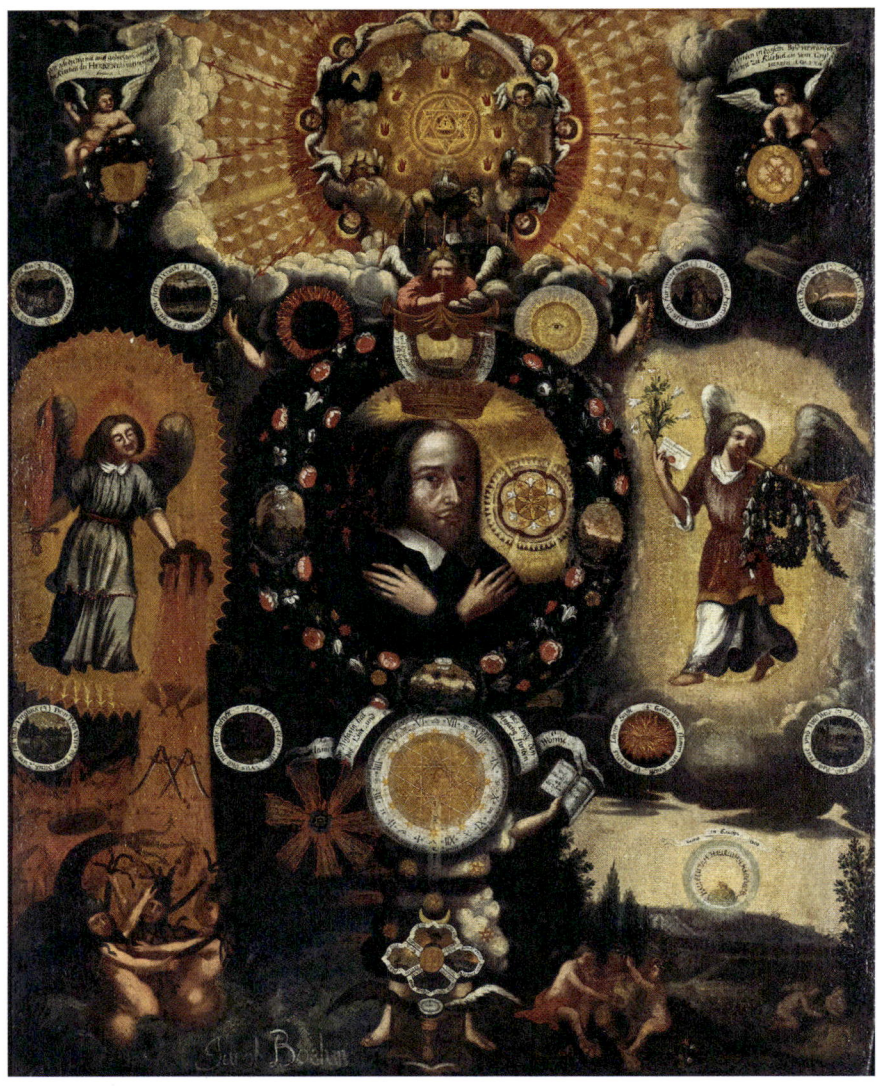

ABOVE Jacob Boehme, shown here in a 1675 portrait by Nicolaus Häublin, was a Protestant theologian and mystic born in Bohemia in what is now part of Poland. He developed a philosophical world-view based on Neoplatonic and alchemical ideas, particularly those of Paracelsus (for example, in his doctrine of signatures). Boehme's work influenced the Rosicrucian movement and German Romantic philosophers such as Schelling and Goethe, as well as William Blake and other literary writers.

OPPOSITE An alchemical cosmology, taken from *The Works of Jacob Behmen, the Teutonic Theosopher* (1764), by the Reverend William Law. Boehme was a key figure in the evolution of a spiritual interpretation of alchemical transformations, comparing the body of Christ to the philosopher's stone and presenting transmutation as a kind of rebirth. This image was originally created by the Christian mystic Dionysius Andreas Freher (1649–1728) and was admired by William Blake, who allegedly compared it to the works of Michelangelo.

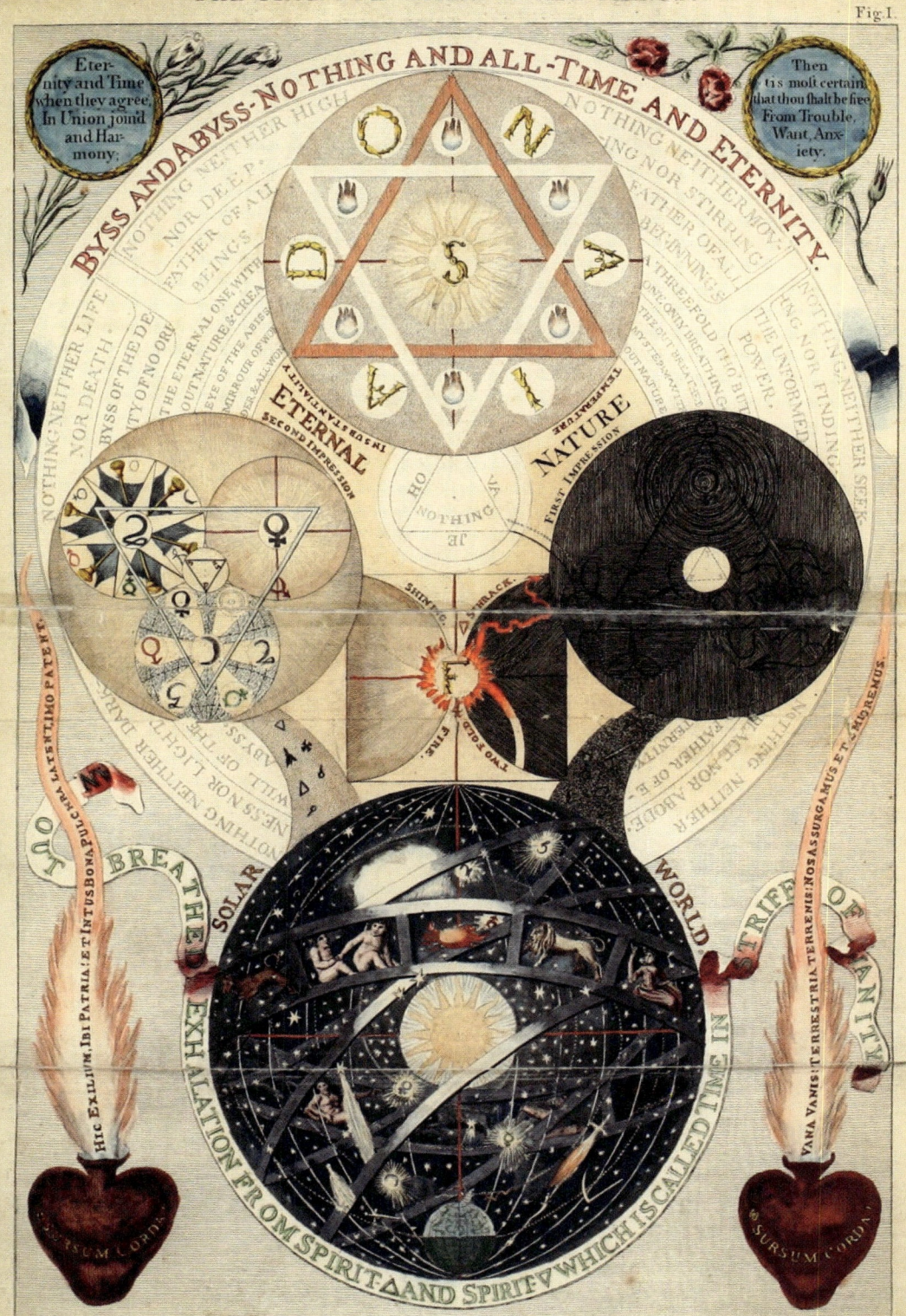

RIGHT Antoine Lavoisier's experiments with respiration in the late 18th century, shown in a drawing by his wife Marie-Anne Paulze Lavoisier, who acted as an assistant and translator to her husband. Here she writes and takes dictation on the right. Lavoisier constructed a mask and suit that allowed him to study how air was transformed during respiration. By such experiments he identified oxygen as an elemental component of air, mixed with nitrogen (which he called azote), and clarified the relationship between respiration and burning.

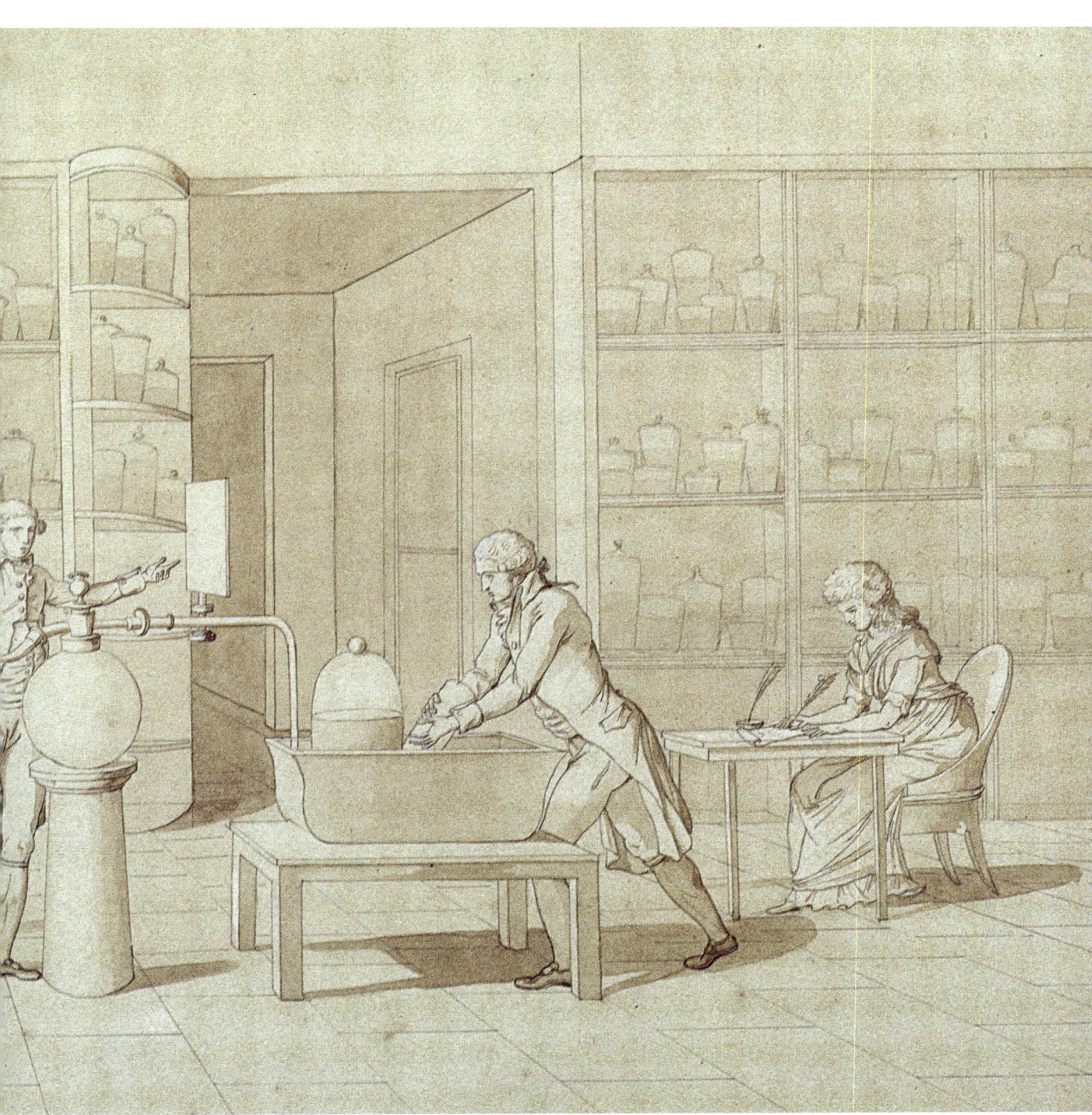

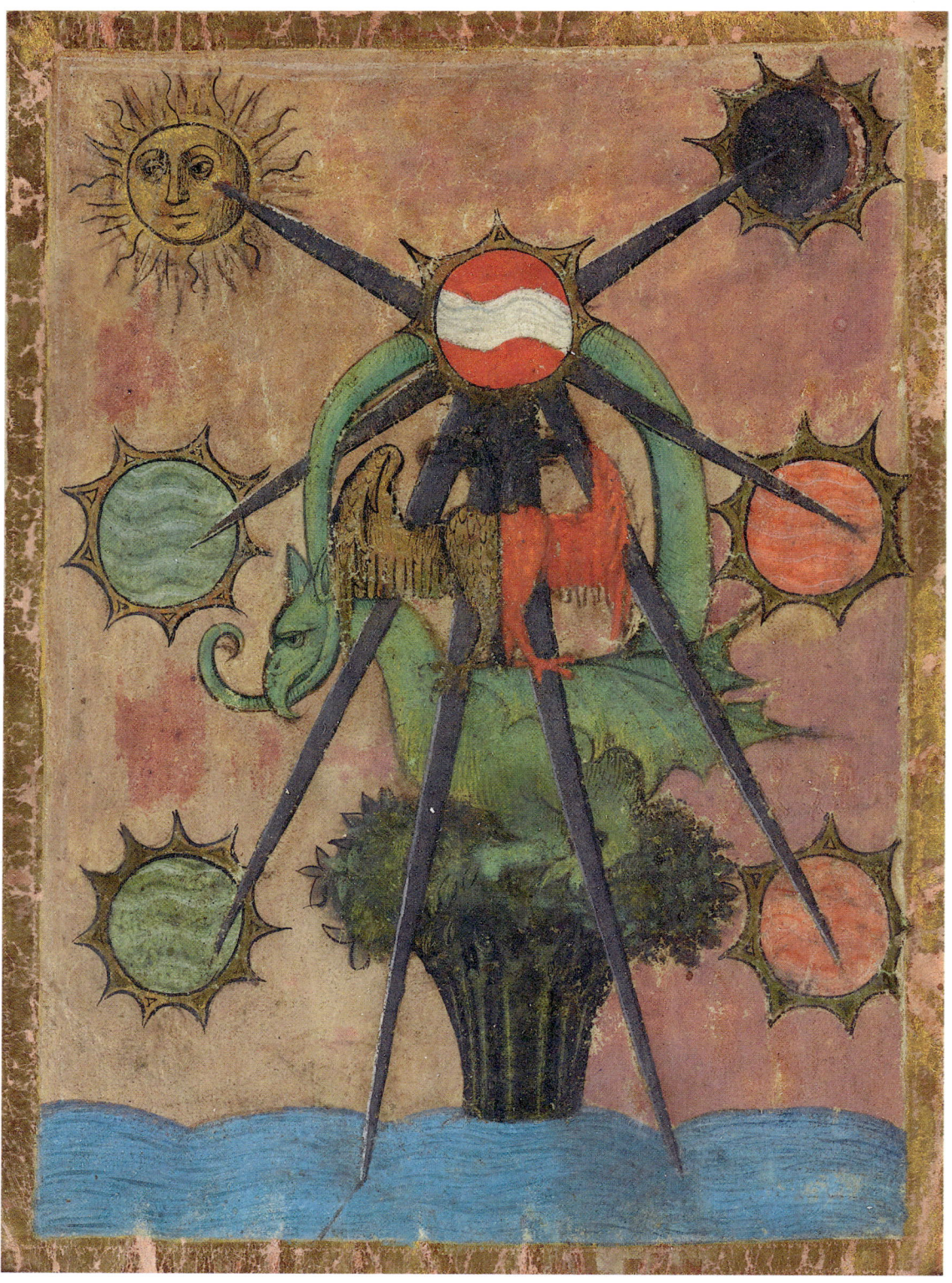

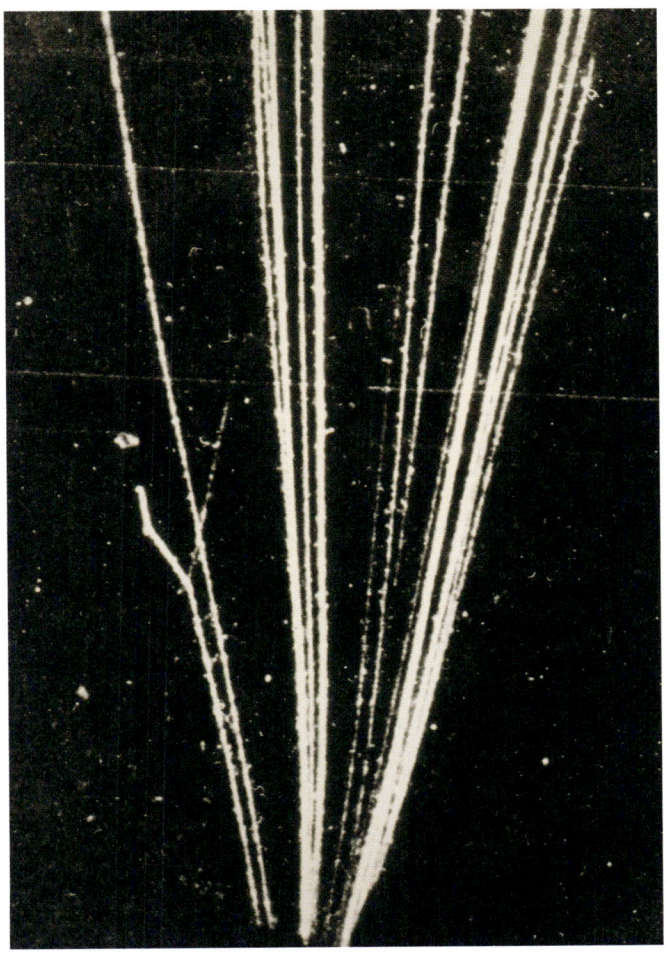

OPPOSITE *The Seven Forms of Transmutation*, based on iconography found in *Das Buch der heiligen Dreifaltigkeit* (The book of the Holy Trinity), the first alchemical text to be published in German, c. 1450–75. The text is attributed to a Franciscan named Brother Ulmann, and it draws analogies between alchemical processes and the death and resurrection of Christ. The book, apparently reconciling alchemy with the Christian faith, became popular among the German nobility. It was one of the first to contain emblematic illustrations, and may have influenced the later, popular illustrated treatise *Rosarium philosophorum* (Rosary of the philosophers; 1550).

ABOVE With the advent of nuclear physics in the early 20th century, it became possible to artificially transmute one element into another by inducing nuclear reactions using high-energy particles. Such transmutations were pioneered by the New Zealand–born physicist Ernest Rutherford and his English collaborator, chemist Frederick Soddy. Such a transformation is shown here in a cloud-chamber experiment by Rutherford's Cambridge colleaague Patrick Blackett in the early 1920s, in which alpha particles from a radioactive source bombard nitrogen molecules in the chamber. The particles leave bright, straight tracks, but the one on the left forks when the particle collides with a nitrogen atom and transmutes it into oxygen, with the ejection of a proton.

CRUCIBLES: FROM ALCHEMY TO CHEMISTRY

PROFILE

Isaac Newton

(1643–1727)

That Sir Isaac Newton—an icon of early modern science, the man who deduced the laws of motion and gravity and devised the calculus—also had a deep interest in alchemy still puzzles and unsettles some people. But this only happens when we try to impose today's ideas and categories onto an age they do not fit.

"I believe that Newton was different from the conventional picture of him," wrote the British economist John Maynard Keynes in 1946. Keynes had seen manuscripts written by Newton that had previously been kept secret until they were auctioned in 1936. They revealed that Newton—"the first and greatest of the modern age of scientists, a rationalist, one who taught us to think on the lines of cold and untinctured reason," as Keynes put it—had devoted an enormous amount of his time to studying alchemy and seeking the philosopher's stone. He "was clearly an unbridled addict," the perplexed Keynes wrote. Newton, he said, "was not the first of the age of reason. He was the last of the magicians."

The remark says more about the time of Keynes than that of Newton. He was in fact not even the last of the "magicians"—in his interest in alchemy he was no anomalous recidivist, even if Newton's view of that art was, like his religious beliefs, idiosyncratic. It did not place him, as Keynes claimed, "with one foot in the Middle Ages and one foot treading a path for modern science." It merely placed him in the seventeenth century, when the goal of natural philosophy was to find a system of the world in its entirety and to uncover hidden connections between its components. If early modern figures like Newton do not fit neatly into our modern conception of a scientist, that is because they did not share the goals of modern science. They had something more in mind.

Isaac Newton was born in Lincolnshire in 1642 (in the old Julian calendar). He studied mathematics at Trinity College, Cambridge, where his precocious intellect was quickly recognized: in 1669, aged just twenty-six, he was awarded the prestigious Lucasian Chair of Mathematics. Legend has it that he devised his theory of gravity, as well as the calculus, while in "lockdown" at home during the Great Plague of 1665, although the historical evidence shows that those epochal breakthroughs were developed over a longer period. Newton's 1687 book on mechanics, *Philosophiae naturalis principia mathematica* (Mathematical principles of natural philosophy), is considered the foundation of modern physical theory, while his 1704 work on the nature of light, *Opticks*, established the basis of that discipline as a modern science. In 1696 he became warden of the Royal Mint, and in 1703 he was made president of the Royal Society; he continued in both positions until his death in 1727.

Newton's alchemical interests ran deep. Around one in ten of the books in the catalog of his library had connections with alchemy. He left behind many pages of notes on the subject, which were deemed "not fit to be printed" and remained hidden until the 1936 sale. These notes describe his efforts to make gold, sometimes using the old, cryptic alchemical terminology for ingredients—"Neptune's Trident," "Mercury's Caducean Rod," the "Green Lyon"—along with allegorical drawings. Alchemy should, Newton insisted, remain a secret business. When in 1676 his colleague Robert Boyle published a paper with the Royal Society reporting the results of some alchemical experiments with a "sophic mercury," Newton wrote to the society's secretary Henry Oldenburg to advise "high silence" on such matters in the future, lest such knowledge should cause "immense damage to the world."

Newton evidently spent hours trying to unravel the codes in the early alchemical texts in his library. According to historian of chemistry William Newman, Newton's alchemical program was largely focused on "the literary decipherment of riddles." But his work was by no means solely theoretical. He conducted many chemical experiments, attested by copious notes like these:

Jan 15 I sublimed 80 grains of this precipitate of ♂ [iron] mixed with thrice as much bole Armonack [red clay] poudered. These I dried well. And afterward in the sublimation when the matter was almost as hot as could be made without bringing it to a dark red there ascended in white fumes an humidity which setled in cleare water.

Some have suggested that the mental breakdown he seems to have experienced around 1693 might have been due to poisoning by heavy metals such as mercury, lead, and antimony incurred from his experiments.

When Boyle died, Newton believed that he had bequeathed to his friend John Locke a red chrysopoeian elixir, and he wrote Locke letters desperately imploring him to share Boyle's secret. Locke sent Newton a sample of the red stuff, but to no avail: despite all his searching, Newton records no successful transmutation, nor any sight of the philosopher's stone.

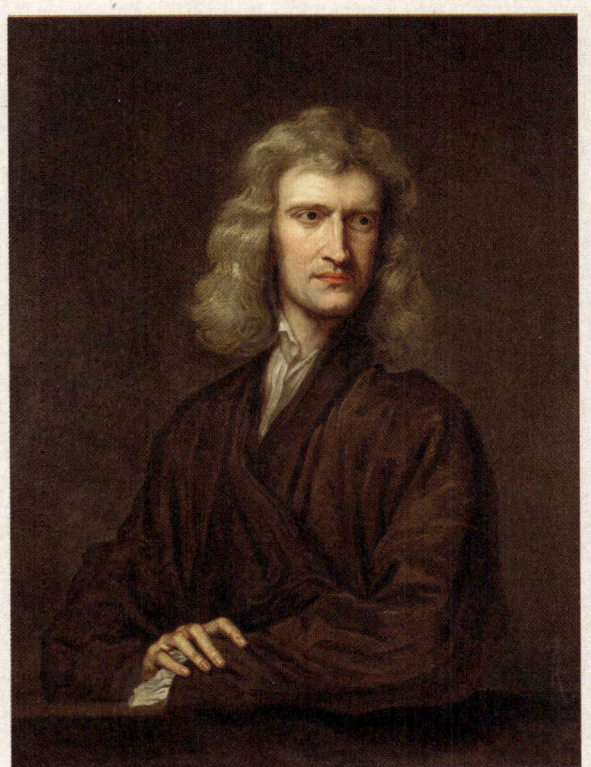

LEFT Sir Isaac Newton, in a portrait by Sir Godfrey Kneller, 1689. Newton is generally considered to be one of the most important scientists (as natural philosophers later came to be called) of all time, and was a key figure in the so-called Scientific Revolution of the 17th century, which laid the foundations of modern science. Newton performed pioneering work on mechanics, gravity, and optics, outlining the laws of planetary attraction that accounted for the motions of the celestial bodies and explaining the colors of the rainbow and of the spectrum created by passing sunlight through a prism. He was also deeply interested in alchemy, believing the transmutation of metals to be possible and seeking all his life for the philosopher's stone.

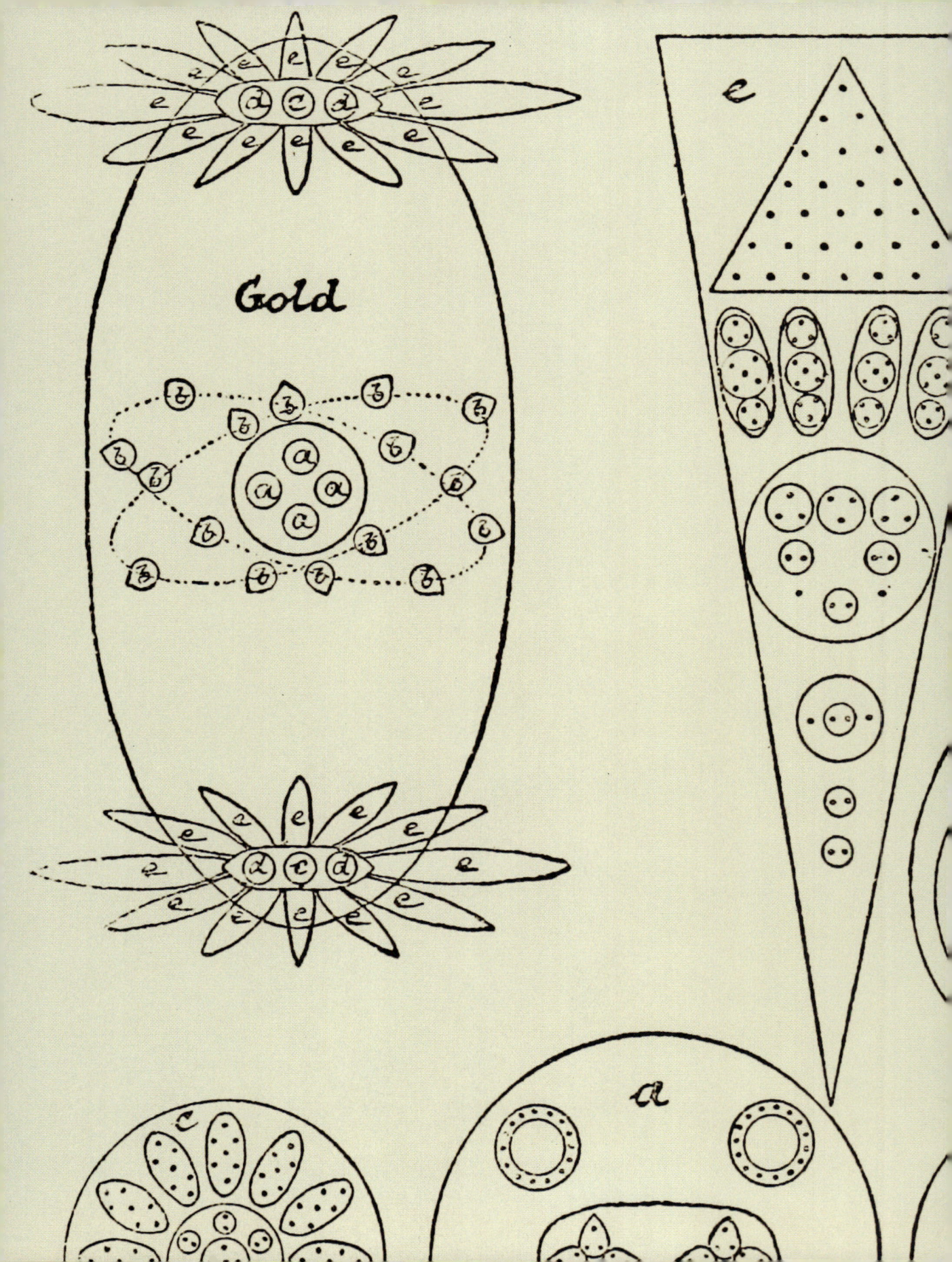

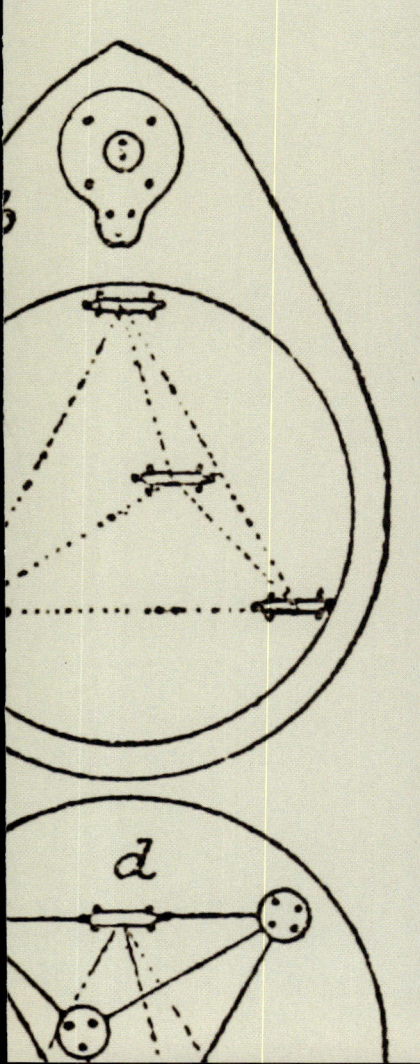

CHAPTER NINE

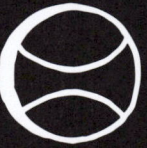

Transmutations

ALCHEMY IN CULTURE

LEFT "Gold," from *Occult Chemistry* (1908) by the theosophists Annie Besant and Charles Leadbeater, who enjoyed a large following in the early 20th century. Besant was a social campaigner who argued for women's rights, birth control, and better conditions for workers, as well as being a leader of the Theosophical Society. Her book with Spiritualist medium Leadbeater depicted the structures of atoms that the pair claimed to be able to discern using occult powers. While these images were totally fantastical, some bear an uncanny resemblance to the complex shapes of electron orbitals in atoms later calculated using quantum mechanics, as well as to the subatomic structures of atomic nuclei.

CHAPTER NINE

By the late eighteenth century, many scholars felt that alchemy was little more than a pantomime display conducted by frauds to dazzle and seduce the credulous. But as alchemy lost its scientific respectability, its magical and esoteric aspects were brought to the fore. During the nineteenth century there was a resurgence in the popularity of occult ideas, creating a new arena within which alchemy could sustain a cultural presence and become emblematic of mystery and wonder in an age increasingly dominated by rationalism.

In 1758 King Louis XV of France was approached by one Claude-Louis-Robert, Comte de Saint-Germain, who claimed to know how to manufacture an elixir of life and the philosopher's stone. The king was impressed enough to employ Saint-Germain for diplomatic missions, during which he apparently met the Italian adventurer Casanova and allegedly turned a coin into gold for him. Saint-

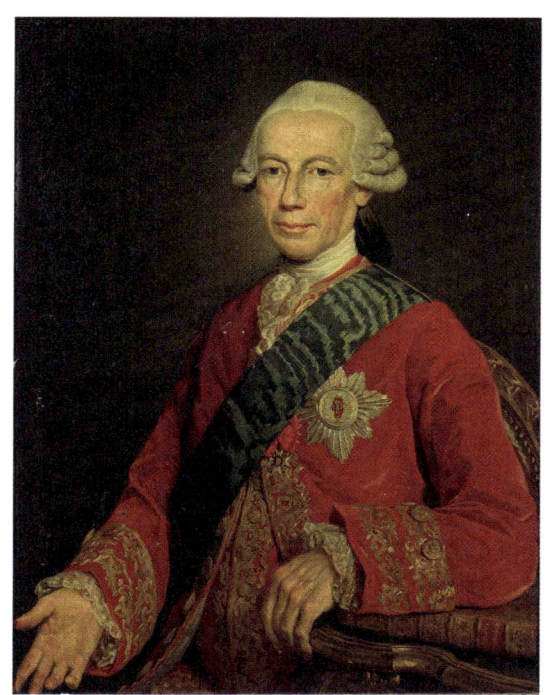

RIGHT Portrait of the "immortal" Claude-Louis-Robert, Comte de Saint-Germain, by Jean-Joseph Taillasson, 1777. In 1758, Saint-Germain tried to sell Louis XV of France his elixir of eternal life. Lionized by some later occultists, he might be generously described as an adventurer, and his name became associated with Casanova, Voltaire, and Mozart.

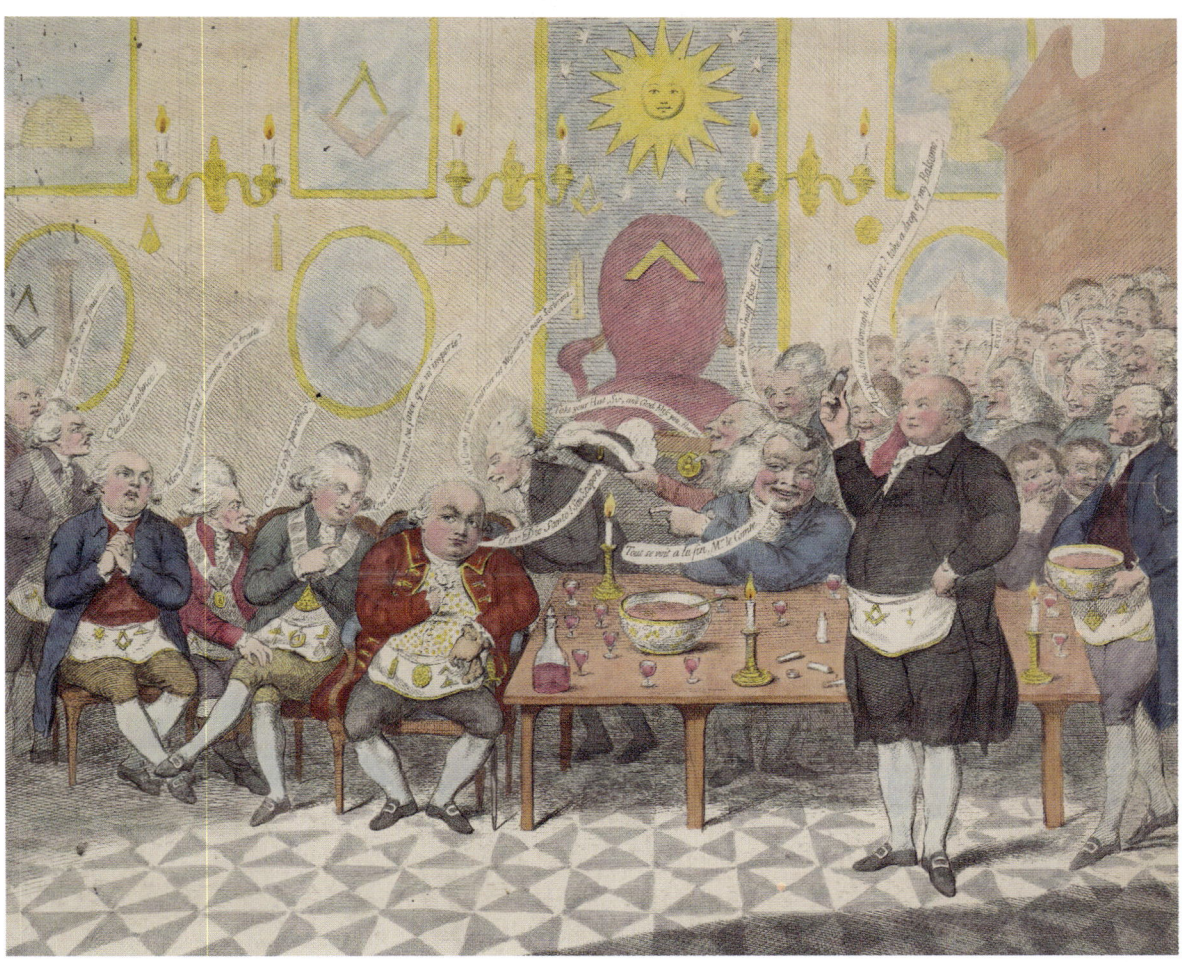

ABOVE In his etching "A Masonic Anecdote," satirist James Gillray pokes fun at the attempt by Italian alchemist Giuseppe Balsamo, the self-styled Count Cagliostro, to found a Masonic "Egyptian Lodge" in London in 1786. Cagliostro was something of an impresario who attempted to exploit the fad for Egyptology at that time.

Germain later emerges practicing alchemy in the court of Frederick the Great, King of Prussia: he was one of the great impresarios of the eighteenth century, conjuring a precarious but occasionally lucrative career by spinning tall tales to nobles.

The most notorious of these individuals was the self-styled Count Alessandro di Cagliostro (1743–95), who has been (inaccurately) described as the "last alchemist." There is no reason to suppose that Cagliostro's interest in alchemy was informed or profound—it was just another dazzling parlor trick, like the beauty creams and youth-giving elixirs he touted to wealthy admirers. What chemistry he knew, he learned as an assistant to a monastic apothecary in Palermo under his real name, Giuseppe Balsamo (he reinvented himself as a "count" in the 1770s). Alert to the enthusiasm in those times for Egyptian esoterica, in 1784 Cagliostro founded the Egyptian Rite, an order that, like the Freemasons and the Rosicrucians, claimed to possess secret magical lore.

Cagliostro was a charlatan and conman who played a dangerous game. He had to flee France in the 1780s after becoming implicated in a notorious scandal involving a diamond necklace purloined by someone posing as Louis XVI's queen Marie Antoinette. He spent the last five years of his life imprisoned by the Inquisition for alleged heresy due to his association with Freemasonry.

Cagliostro played to a counter-Enlightenment passion for the occult. In 1776 the German professor of law Adam Weishaupt founded a society called the Illuminati of Bavaria—a model adopted a decade later by the French writer and mystic Antoine-Joseph Pernety for a Masonic-style order called the Illuminati of Avignon. Both secret societies made use of alchemical imagery and ideas in their pursuit of radical social and political ideals.

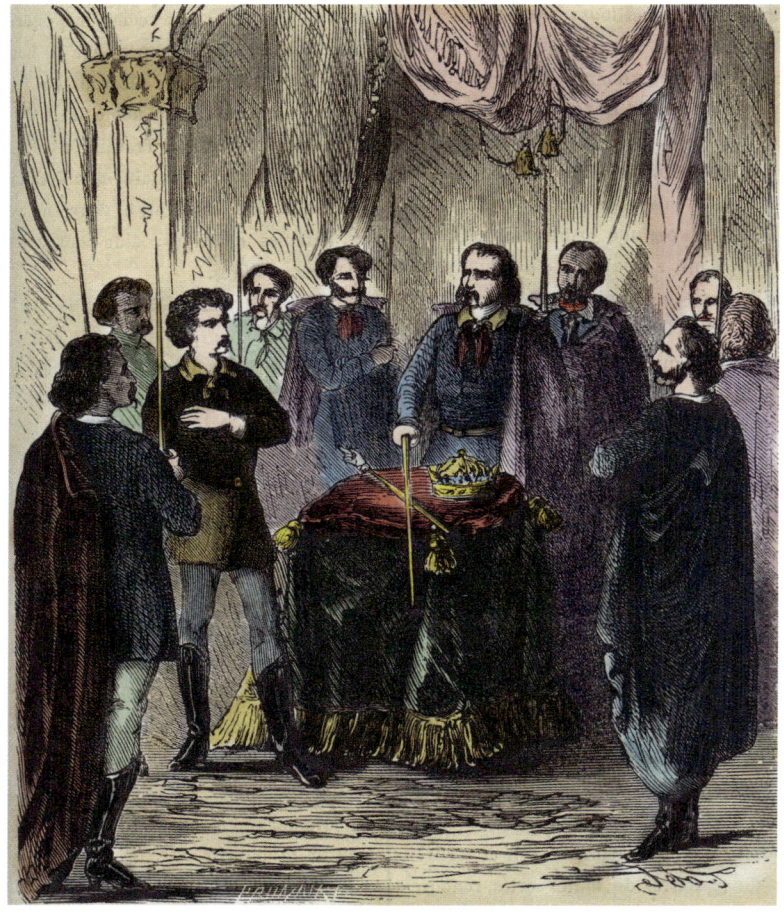

RIGHT The reception of an Illuminatus into a secret society, from a colored copy of Pierre Zaccone's *Histoire des sociétés secrète, politiques et religieuses* (History of secret, political and religious societies; Paris, 1886–79). Masonic-style secret societies flourished in the late 18th and 19th centuries, and they often made use of alchemical imagery in the tradition established by the Rosicrucian movement.

The occult revival

Such movements were the precursors to a full-blown resurgence of the occult in the nineteenth century, heralded in the 1840s by Spiritualism, the belief that the spirits of the dead could be contacted and engaged in dialogue. The works of the Swedish theologian Emanuel Swedenborg, particularly his 1758 book *Heaven and Hell*, provided intellectual grounding for the notion that souls in the afterlife dwell in a hierarchy of spiritual planes where they may be contacted by the living. Trained in science and considered an expert on mining, metallurgy, and engineering, Swedenborg evinced an interest in both the practical and the symbolic aspects of alchemy.

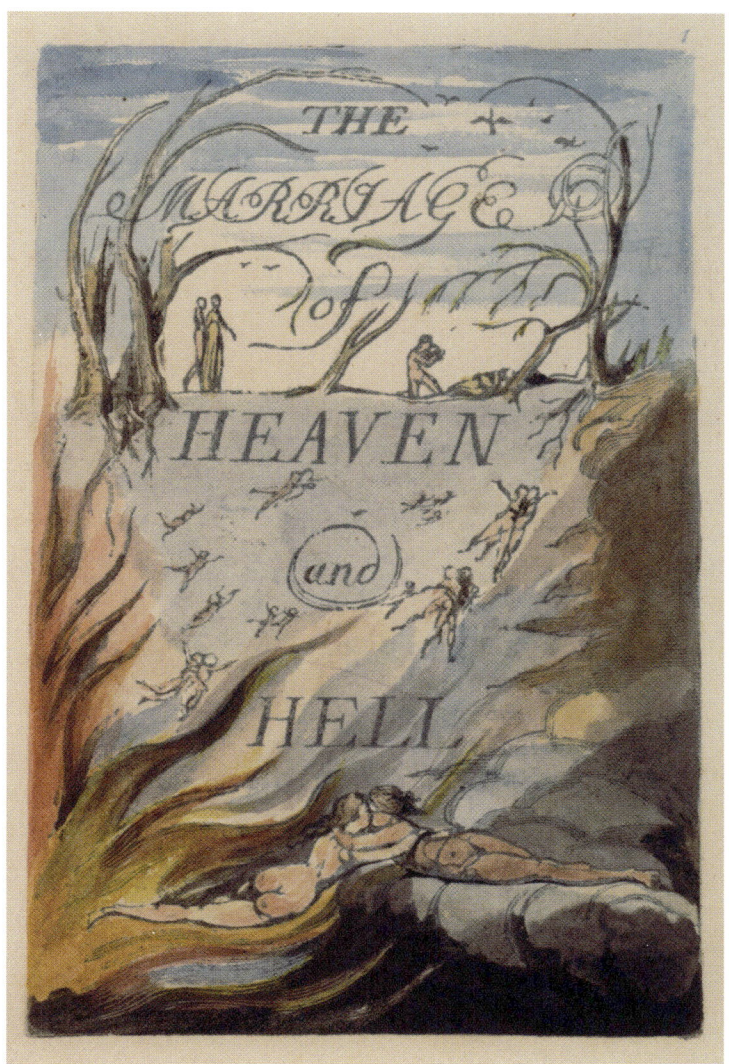

RIGHT The Swedish mystic Emanuel Swedenborg asserted a hierarchy of spiritual planes and a polarized distinction of good and evil. That picture was criticized by William Blake in his *Marriage of Heaven and Hell*, c. 1794, its title an ironic allusion to Swedenborg's 1758 theological work *De coelo et eius mirabilibus, et de inferno* (Of heaven and its wonders, and of hell), a book about the afterlife.

The Golden Dawn studied aspects of esoterica including the Kabbalah, astrology, and geomancy as well as alchemy.

This new, esoteric conception of alchemy was exemplified in *A Suggestive Inquiry into the Hermetic Mystery* (1850) by Mary Anne South (later Atwood), which presented the history of the topic filtered through the lens of early Victorian occultism: clairvoyance, mesmerism, animal magnetism, and spirit worlds. Atwood argued that the goal of the alchemist was to regenerate the human soul in a more perfect state and ultimately to ascend to a higher plane of existence. Atwood prepared her book in collaboration with her father, Thomas South, who amassed an impressive library of alchemical texts. Having paid for the book's publication, South had a change of heart: deciding that it revealed too much about the occult art, father and daughter recovered all the copies they could find (fewer than a hundred were printed) and burned them on the lawn of South's house.

BELOW The English writer Mary Anne Atwood, author of A *Suggestive Inquiry into the Hermetic Mystery* (1850). The book presented a history of alchemy in which the art was portrayed as a primarily spiritual quest. Atwood had second thoughts about the endeavor and tried to destroy all published copies, but the book was posthumously republished in 1918 with this portrait as its frontispiece.

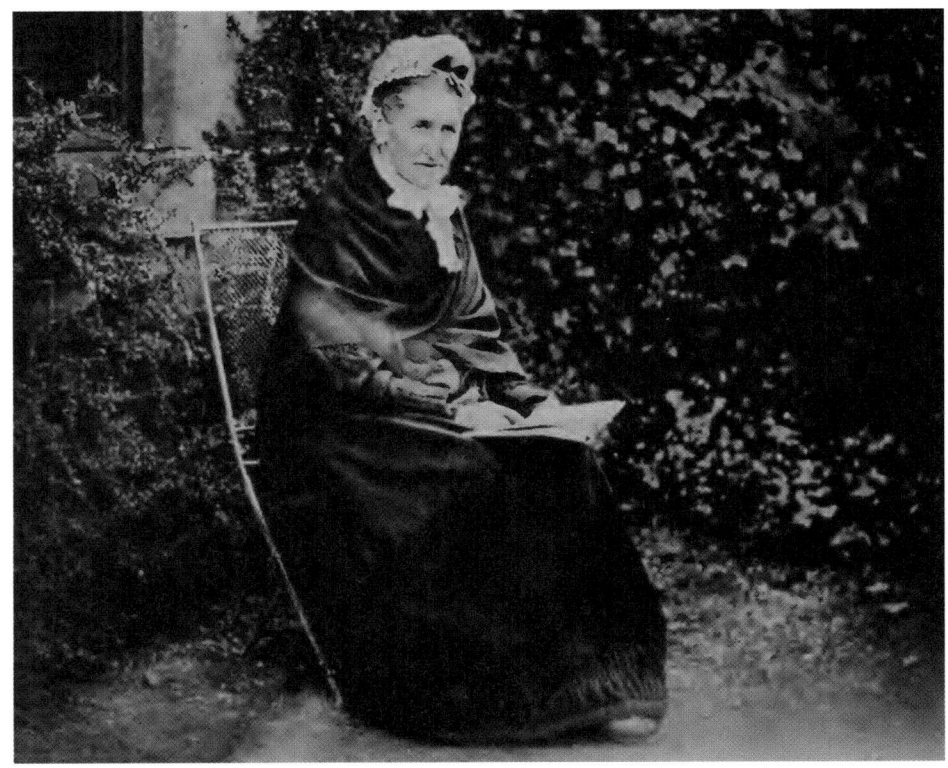

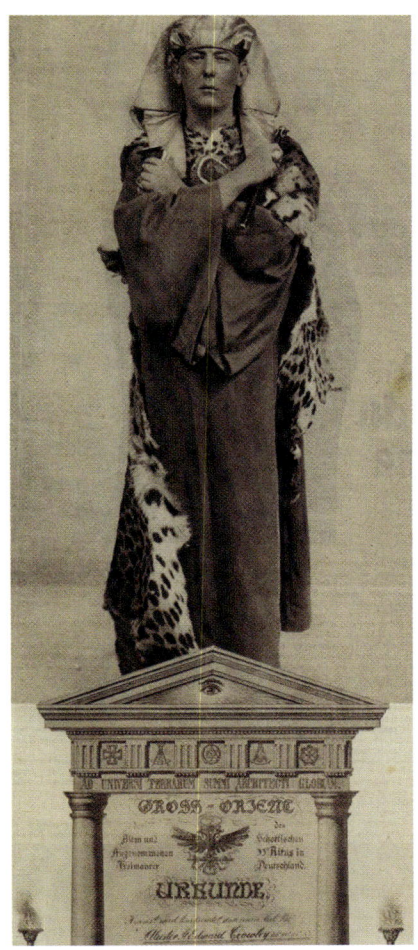
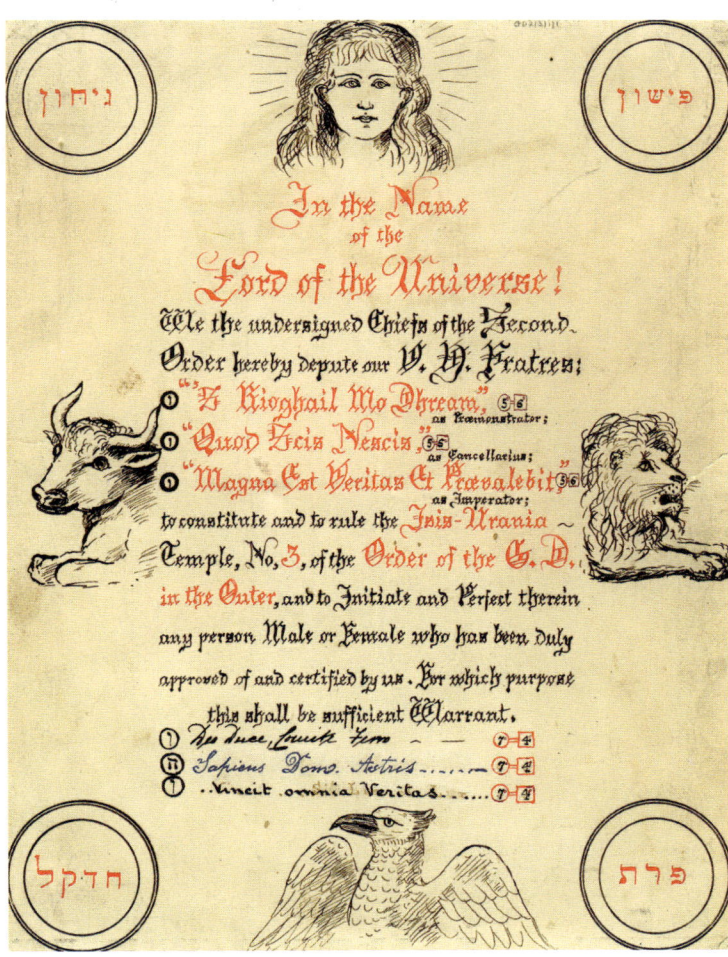

ABOVE The occultist Aleister Crowley dressed as Osiris, god of the dead, in the journal *Détective*, May 2, 1929. Such "Hierophantic Egyptian" costume was popular within the esoteric society known as the Hermetic Order of the Golden Dawn.

ABOVE RIGHT Warrant of the Isis-Urania Temple, London, illustrated by the artist Moina Mathers (born Mina Bergson), a member of the Hermetic Order of the Golden Dawn and wife of its co-founder Samuel Liddell MacGregor Mathers, c. 1888. After her husband's death, she headed the successor group that he established, the Rosicrucian Order of Alpha and Omega. Moina Mathers was the sister of the French philosopher Henri Bergson.

The occult movement reached its apotheosis in the Theosophical Society of the Russian-American mystic Helena Blavatsky (who proclaimed Saint-Germain a reincarnation of legendary alchemists including Roger Bacon and Christian Rosencreutz) and the Hermetic Order of the Golden Dawn, founded in 1888 by the English Freemasons William Wynn Westcott and Samuel Liddell MacGregor Mathers. The Golden Dawn studied aspects of esoterica including the Kabbalah, astrology, and geomancy as well as alchemy, and it attracted Victorian intellectuals such as the poet William Butler Yeats and the writers Arthur Machen and Aleister Crowley. With its ornate ceremonies (complete with Egyptian costume), bogus history, grandly named officials, and petty internal spats, the Golden Dawn now looks an easy target of mockery. But it also produced some worthy scholarship, especially by the poet and mystic Arthur

Edward Waite, an expert on the tarot tradition who translated and published many of Paracelsus's alchemical works. Waite's lack of academic training diminishes the historical value of those works today, but they argued the case for giving the "occult sciences" serious attention at a time when many academics disdained them.

Some of these occult enthusiasts seem to have engaged in practical alchemy—although Westcott warned that "purely chemical experiments" were useless without "simultaneous operation on the astral plane." Crowley claimed to have prepared an elixir after two and a half years of "careful and strenuous research." He says that after consuming it he experienced "an attack of youth in its acutest form," which apparently meant that he became "a mere vehicle of physical energy."

A psychic science?

The occult revival was not entirely orthogonal to science. Some scientists of the fin de siècle, such as Pierre and Marie Curie, the chemist William Crookes, and the psychologist Frederic Myers, took an interest in Spiritualism. Myers was a founder of the Society for Psychical Research, which attempted to investigate paranormal phenomena with scientific rigor. But esoteric ideas found an audience primarily among poets, artists, writers, and public figures, and it was in this sphere that interest in alchemy persisted.

Meanwhile, the idea that alchemy was primarily a spiritual pursuit that sought to transform the psyche resonated with the

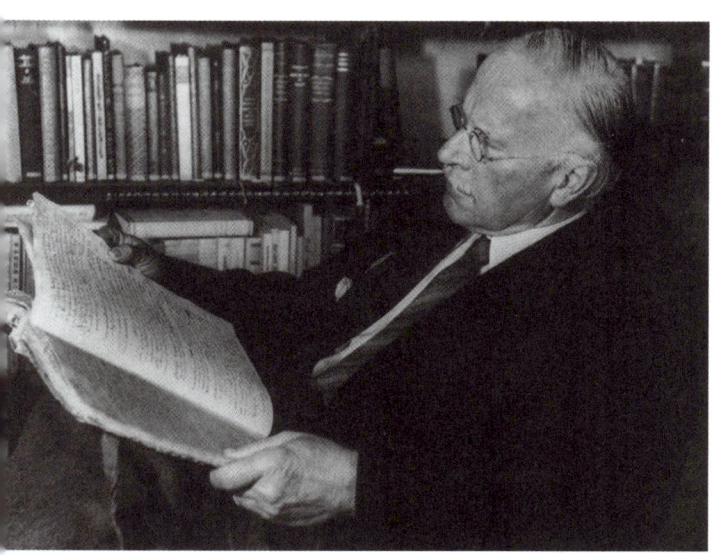

BELOW The Swiss psychoanalyst Carl Gustav Jung in his study, in a photograph dated 1955. Jung collected a vast library of alchemical texts and wrote extensively on the topic, including a character study of Paracelsus. Jung evinced little interest in the practical chemical aspects of the Hermetic art, seeing it mainly as a process of spiritual transformation.

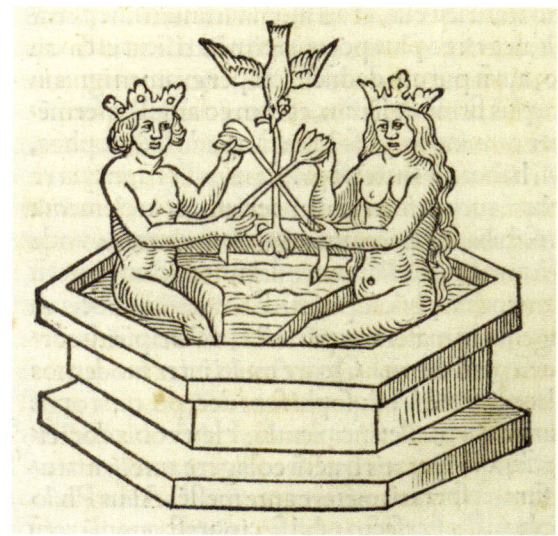

BELOW RIGHT The king and queen (representing the Sun and Moon) overseen by the dove (Spirit), from *De alchimia (On alchemy)*, a collection of texts published in Frankfurt in 1550. The woodcut was one of several alchemical images used by Carl Jung to illustrate his theory of the process of psychological transference between analyst and patient. Jung describes the ritual bath as representing "a descent into the unconscious."

230 CHAPTER NINE

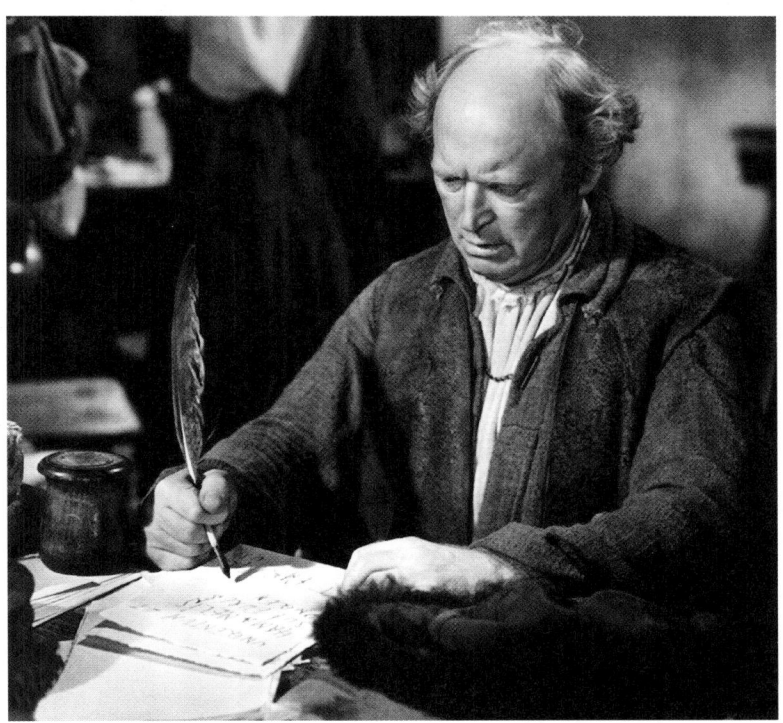

RIGHT Paracelsus was represented as a *Völkische* hero in the 1943 biopic by the German director Georg Wilhelm Pabst. The alchemist was played by Werner Krauss, a stage and film actor who participated in many Nazi propaganda films.

emerging practice of psychoanalysis, particularly for the Swiss psychiatrist Carl Jung. Jung's vision of alchemy as having "a great deal to do with the structure of the unconscious" spoke to the spirit of his times. Here, he wrote in 1942, "we find the true roots, the preparatory processes deep in the psyche, which unleashed the forces at work in the world today." If this is an allusion to Nazism, it is a complex one—for Jung expressed some admiration for the energy of the Nazi movement, although his attitude toward the Third Reich remains debated. The appeal of esotericism to the Nazis themselves is well known; Paracelsus was enlisted as a *Völkische* hero in a 1943 biopic directed by Georg Wilhelm Pabst that was basically a piece of Nazi propaganda.

Jung was right, however, to recognize that alchemical adepts, from Zosimos to John of Rupescissa to Isaac Newton, had considered that chemical changes reflected deeper truths about the world. "The purpose of distillation in alchemy," Jung wrote, "was to extract the volatile substance, or spirit, from the impure body. This process was a psychic as well as a physical experience." It's not too much of a stretch, then, to accept his claim that "Alchemy is not only the mother of chemistry, but is also the forerunner of our modern

OPPOSITE *Attirement of the Bride*, 1940, by the Surrealist Max Ernst. This is one of several paintings by Ernst that seems to allude to alchemical themes; the artist had a strong interest in the subject, seeing it as a quest to liberate the human imagination.

psychology of the unconscious." Yet Jung's perspective on alchemy was very much of its time and largely unmoored from any sense of historical evolution, as if alchemy was in all times and places fundamentally the same thing. And when he suggests of Paracelsus that "like all medieval alchemists, he seems not to have been aware of the true nature of alchemy," using it just to heal the sick and believing it could make gold rather than seeking personal growth, he seems to be condemning Paracelsus for not understanding what Jung had decided alchemy was really about.

Alchemical art

The psychic aspects of alchemical transformation appealed to several artists of the early and mid-twentieth century who approached art as a means of accessing new psychological states. The Surrealists in particular were drawn to Hermeticism and alchemy, which they saw as an alternative mode of thought to the Enlightenment rationalism they sought to challenge. The movement's leader André Breton (1896–1966) included several alchemical allusions in his works. His book *Arcanum 17*, written in 1944, was a symbolic, alchemically inflected exploration of mineralogy. It was illustrated by the Surrealist painter Roberto Matta, who shared Breton's sense of the geological earth as "something terrific, burning, changing, transforming, flowing."

Breton returned to an alchemical view of geology in an article titled "Langues des pierres" (Language of stones; 1957), where he mentions the magical qualities attributed to stones and gems by Paracelsus and by the late eighteenth-century Romantic mystic Novalis. In truth, connections between Breton's text and the historical traditions of alchemy are rather tenuous; as with their interest in modern physics, the Surrealists had a decidedly personal and idiosyncratic perspective on their inspirational subject matter. It was what the tradition represented—a return to the multivalent, imaginative occultism of the Middle Ages as an antidote to bourgeois morality—that mattered most to them.

Some Surrealist references to alchemical imagery are more direct. Max Ernst (1891–1976) gained a reputation as the "magician" of the movement, and declared that alchemy was a metaphor for his own working process. He believed that, as a native of Cologne, he had inherited the legacy of its two famous alchemists: Albertus Magnus and Cornelius Agrippa. Ernst's pictures, often created by collage, abound with astrological eclipses, hermaphrodites, and

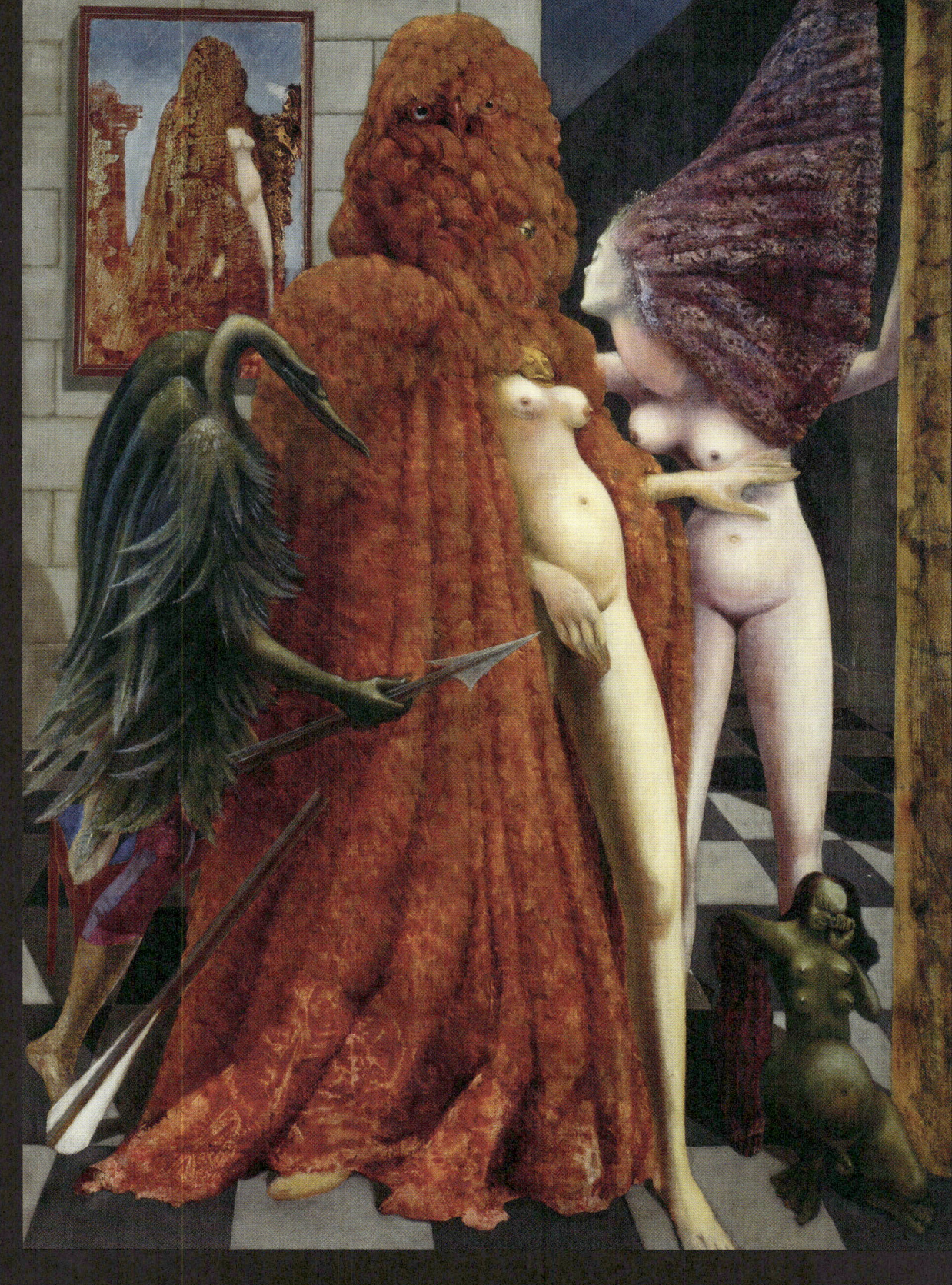

references to alchemical symbolism and apparatus. *Attirement of the Bride* (1940) could almost be one of the allegorical images of Michael Maier's *Atalanta fugiens*, with its color coding, human–animal unions, nuptial associations, and Renaissance-style use of perspective. Perhaps the most explicitly alchemical work of this era is Salvador Dalí's 1976 series *Alchimie des philosophes* (Alchemy of the philosophers), which consists of ten prints depicting alchemical themes (the Emerald Tablet, the King and Queen, the Ouroboros, and so on) inspired by texts from the third to the seventeenth centuries, housed within an elaborate portfolio box decorated with alchemical symbols.

These uses of alchemical imagery by twentieth-century artists had ample precedent. For the Surrealists alchemy embodied the sense of irrationality with which they hoped to activate the imagination, but earlier artists found in alchemy a rich web of themes that would have been common knowledge to an informed audience, providing a shared cultural language of allusion and metaphor. This approach has been claimed to be evident in the paintings of the fifteenth-century Dutch artist Hieronymus Bosch, whose exuberant and fantastical works have been interpreted as displaying a profound interest in and knowledge of alchemical theory and practice. (At least one member of Bosch's family is thought to have been an apothecarist.)

Bosch's *Adoration of the Magi* (c. 1500) presents a rather unusual and puzzling interpretation of a popular subject for artists of that time, perhaps alluding to a common analogy between the incarnation or resurrection of Christ and the transmutation effected by the philosopher's stone. Balthazar's silver orb topped by a golden bird has alchemical associations, while Melchior presents a golden sculpture of the biblical sacrifice of Isaac that rests on the backs of three black toads, symbols of the alchemical process of *nigredo* (blackening). The odd-shaped buildings in the distance, meanwhile, are like none in any Dutch city but look very much like contemporary illustrations of alchemical furnaces.

Bosch's most famous work, *The Garden of Earthly Delights* (c. 1495–1505), is filled with weirdly shaped structures that resemble alchemical equipment—flasks, pelicans, alembics, eggs, glass tubes. This vast, phantasmagorical work literally has an alchemical framing: the rear of the triptych panels shows the world enclosed in a transparent orb like a glass flask, within which the water encircling the land is evaporating and condensing into clouds,

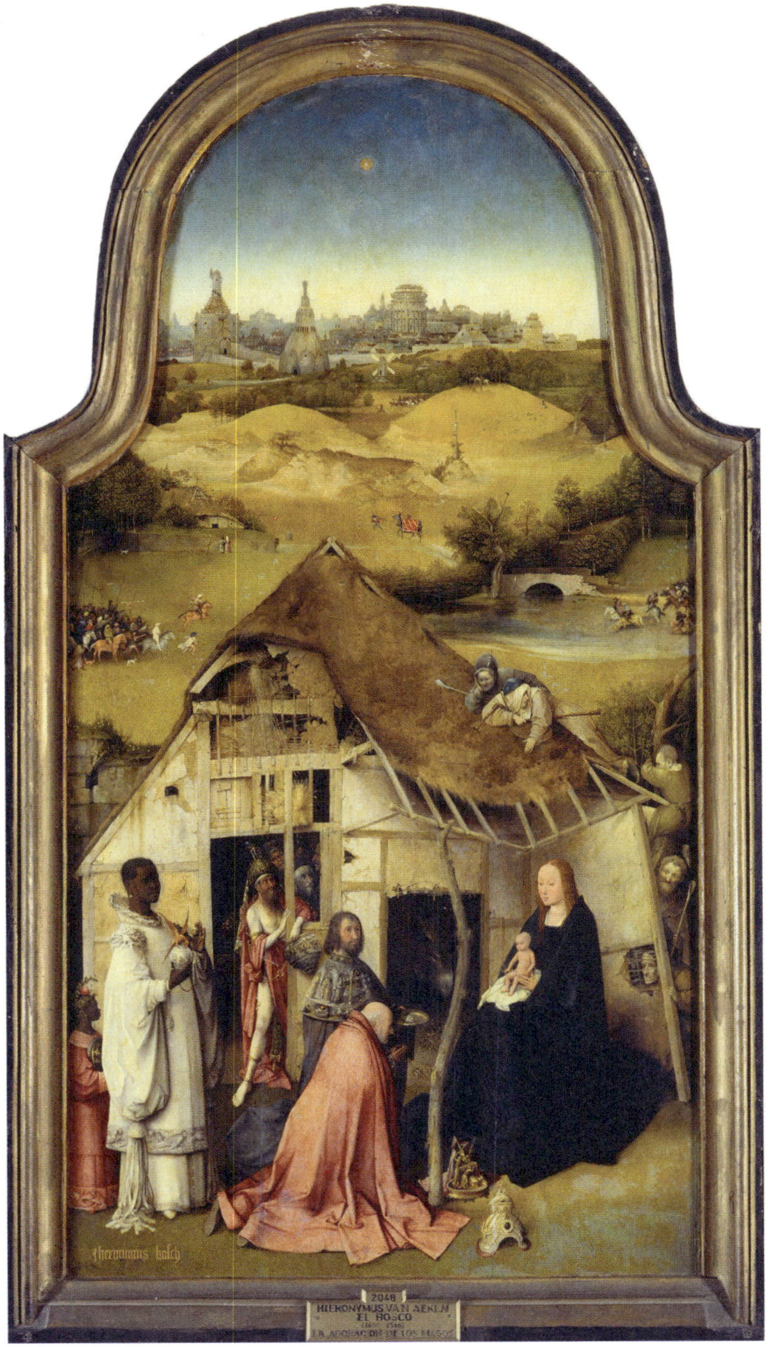

LEFT The *Adoration of the Magi*, 1510, by Hieronymus Bosch, and two details (below). The golden bird resting on Balthazar's silver orb seems to refer to alchemical themes, and the golden sculpture depicting the Sacrifice of Isaac, one of the gifts presented to the infant Christ, rests on three black toads, symbols of the alchemical process of *nigredo*.

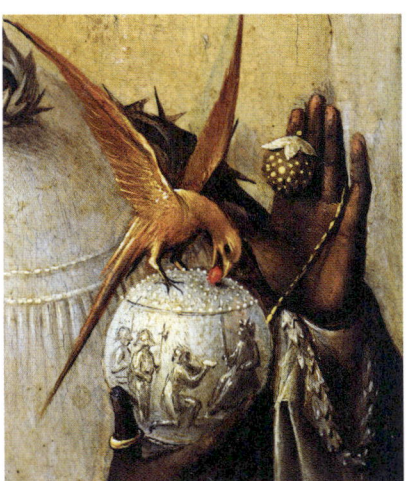

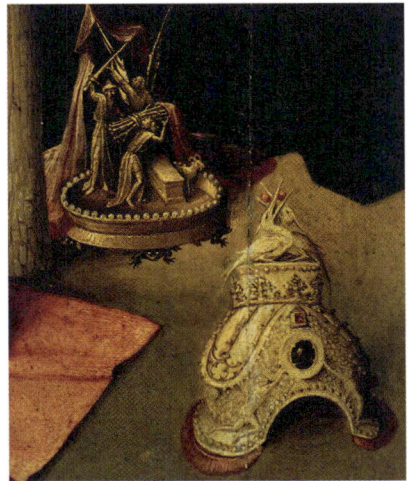

TRANSMUTATIONS: ALCHEMY IN CULTURE

just as alchemists at that time spoke of the processes within their apparatus as a microcosm of the macrocosmic process of the divine Creation. By such means, Bosch might have used alchemical references to weave religious, philosophical, cosmological, and sexual themes into a coherent whole.

Alchemy in literature

Alchemy was expressed in the literary arts as well. The fifteenth-century alchemist George Ripley (see page 242) wrote his *Compound of Alchemy* in verse. Ripley's protégé Thomas Norton followed the same model for his *Ordinall of Alchimy*, and in 1652 the English

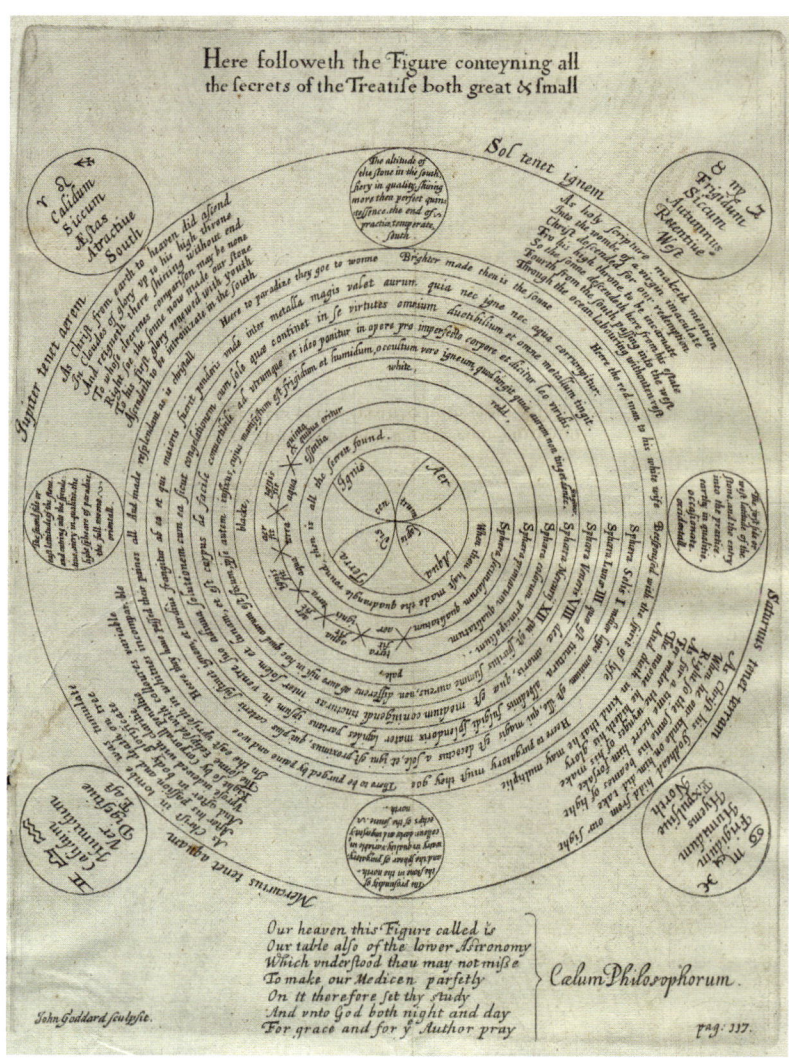

RIGHT George Ripley's Wheel, depicting concentric planetary spheres along with his alchemical recipes, from Elias Ashmole's compilation of poetic alchemical texts *Theatrum chymicum britannicum* (London, 1652). Ripley was a highly influential English alchemist of the 15th century.

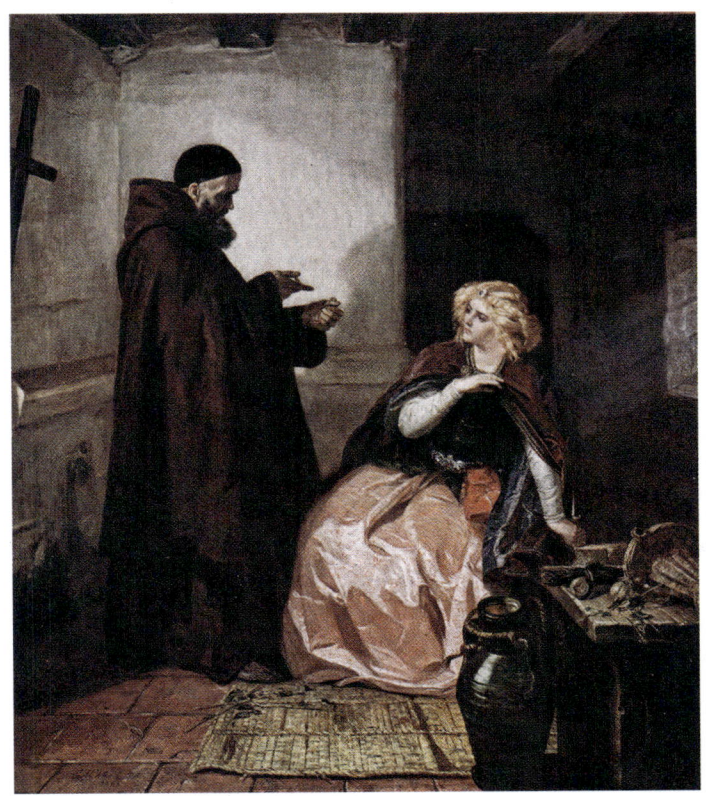

RIGHT "Take thou the phial... and this distilled liquor drink thou off": in his cell littered with herbs and a distillation vessel, Friar Laurence offers Juliet a potion in a lithograph from J. Coker & Co.'s *Complete Works of William Shakespeare*, after a painting by Edward M. Ward, 1867. The friar shows hallmarks of possessing alchemical knowledge, and is among the many allusions to alchemy in Shakespeare's works.

scholar Elias Ashmole included Ripley and Norton in a collection of British poetic alchemical texts called *Theatrum chemicum britannicum*. The use of allusion and metaphor in alchemy made it inherently akin to poetry. When the author of *The Hunting of the Green Lyon*, one of the works in Ashmole's anthology, explains that practitioners of the Hermetic art "vaile their secrets with mistie speech," he could almost be talking instead about the craft of the poet.

According to literary scholar Charles Nicholl, the popularity of iatrochemistry in England in the late sixteenth century "was quickly picked up by the literary antennae." Poets found in alchemy a fertile metaphor for romantic love: John Donne's love poetry draws on the image of the chymical wedding, while Shakespeare's love sonnets make several references to alchemical principles. In his sonnet 33, the morning sun is described as "Kissing with golden face the meadows green, gilding pale streams with heavenly alchemy." *All's Well That Ends Well* casually name-checks Paracelsus, a reference Shakespeare could assume his audience would understand. Friar

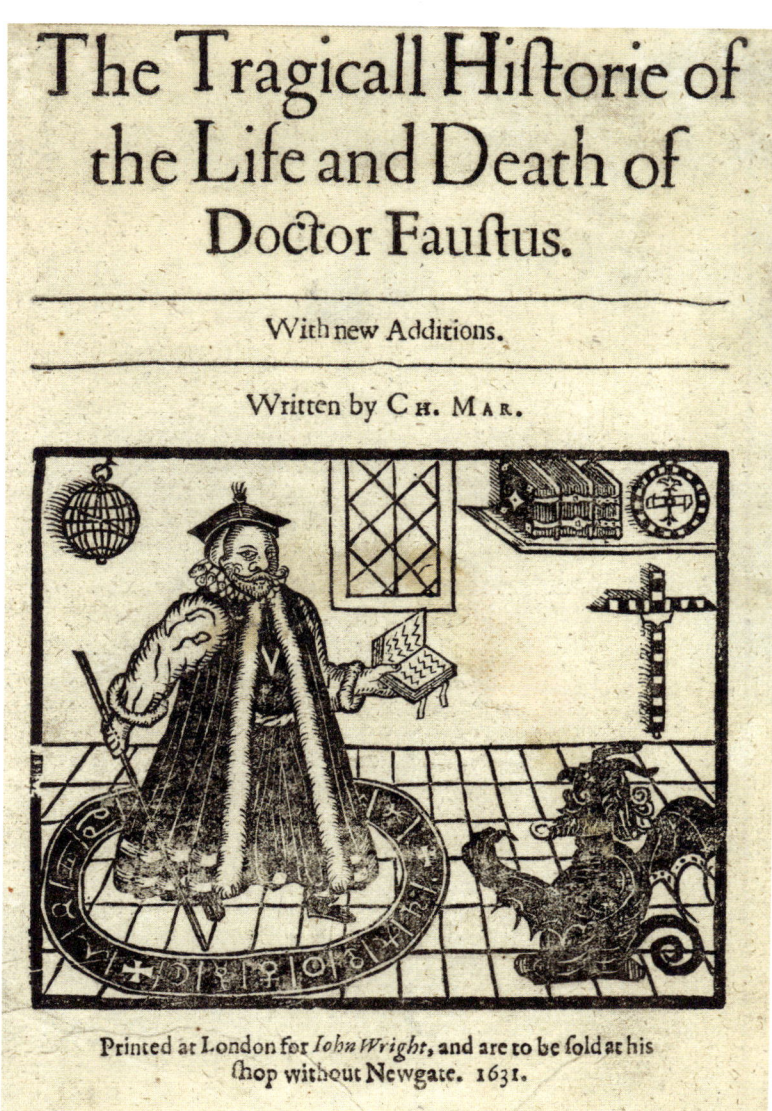

RIGHT Christopher Marlowe's *Tragicall Historie of the Life and Death of Doctor Faustus* (London, 1631). Marlowe's Faustus, here standing in his magic circle of arcane symbols, with a yoked dragon at his feet, provided a template for later representations of alchemists as dealing in forbidden knowledge.

Laurence in *Romeo and Juliet* is a Paracelsian herbalist, extracting medicines from nature by distillation, while Prospero in *The Tempest* is the archetypal Renaissance magus (John Dee has been proposed as the model). Nicholl argues that *King Lear* is structured on alchemical themes: the various ordeals endured by Lear can be seen as a representation of the processes in the Great Work by means of which the raw materials are transmuted gradually into the "Red King" and the philosopher's stone.

Christopher Marlowe's *Doctor Faustus* (c. 1592), based on an old legend, provides a literary template for the hubristic scientist. Alchemy has served as an off-the-shelf signifier of (sometimes disreputable) magic arts for at least two hundred years or so. Mary Shelley's Victor Frankenstein becomes obsessed, as a young student, with legendary alchemists and natural magicians such as Agrippa, Paracelsus, and Albertus Magnus, only to be told by his crabby professor Krempe that these are "exploded systems, and useless names." Shelley's tale is clearly influenced by the Faust legend, and informed too by her father William Godwin's knowledge of the old natural magicians, the subject of his 1834 book *Lives of the Necromancers*. In Johann Wolfgang von Goethe's retelling in the late eighteenth and early nineteenth centuries, Faust becomes a tragic hero who yearns for knowledge. Like Victor Frankenstein, Goethe himself studied alchemy in his youth, including the works of Paracelsus and Agrippa.

We have lost access now to this allusive, multifaceted fluidity of meaning in alchemy. In recent times it tends to feature in literature only as a rather crude signifier of magic or inner knowledge. The bestselling novel *The Alchemist* (1988) by Brazilian author Paulo Coelho doesn't really contain any alchemy at all: it is a tale of spiritual self-realization, a riff on old stories about a person who undertakes a quest for treasure only to discover that it is to be found back at home. As such, the book is a logical endpoint of the efforts in the nineteenth and twentieth centuries to strip alchemy of all chemical content and to make it a process enacted solely in the psyche.

In J. K. Rowling's *Harry Potter* novels, meanwhile, alchemy provides a shorthand for fairy-tale conjuring. Paracelsus appears on the walls of Hogwarts school amidst portraits of great wizards, credited as having discovered the language of snakes. The "philosopher's stone" ("sorcerer's stone" in the US edition) of the first book in the series is an agent of immortality, and the only person who knows how to make it is the French wizard and alchemist Nicolas Flamel. The historical Flamel was a fourteenth-century scribe who probably did little more than dabble in alchemy; a legend later became attached in which he and his wife Perenelle prepared the elixir and become immortal. Yet if the alchemy in the *Harry Potter* series is rather tokenistic, Rowling's interest in it was genuine. "I've never wanted to be a witch," she said in an interview, "but an alchemist, now that's a different matter."

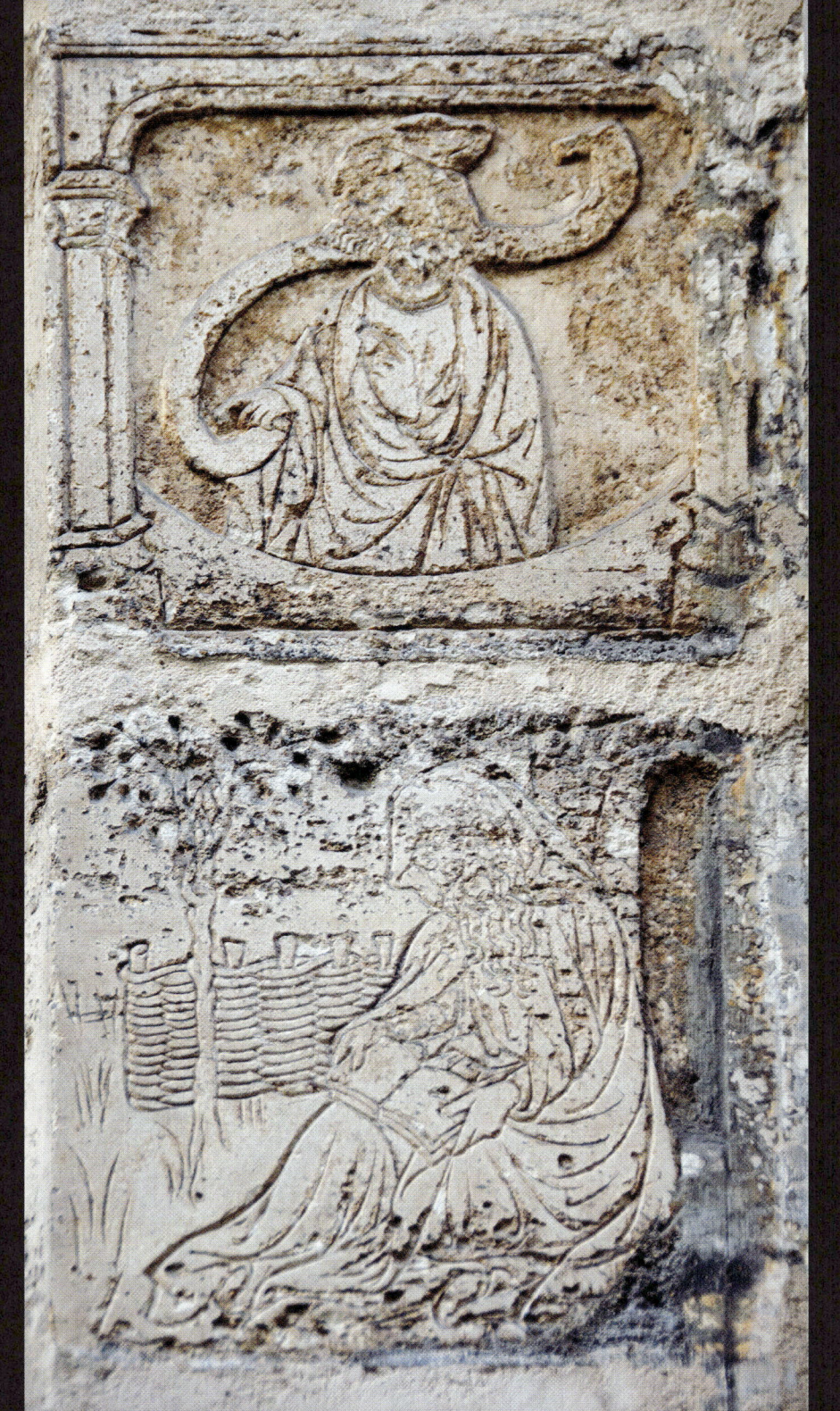

OPPOSITE A carved stone face from the house of Nicolas Flamel, said to be the oldest stone house in Paris, built in 1407. The historical Flamel was a 14th-century scrivener who probably had no connection to alchemy, but he accrued a posthumous reputation as an adept who found the secret of immortality. He features as such in the *Harry Potter* stories.

It is rare to find deeper engagement with alchemy in contemporary literature. One exception is English writer Lindsay Clarke's 1989 novel *The Chymical Wedding*, loosely based on the life of Mary Ann Atwood. And there is plenty of the older, creative, and allusive spirit in the Japanese manga series *Fullmetal Alchemist*. The creator, Hiromu Arakawa, borrows freely from alchemy for his fictional steampunk world Amestris, but with flair and imagination. Here alchemy is mainstream science, practiced by government-sponsored State Alchemists who are adepts at transmutation but forbidden to apply their arts to transform humans or to make gold. In general, to transmute an object alchemists must offer up something of equal value—a natural law that can only be bypassed using a philosopher's stone, which some have used to create artificial humans called Homunculi. By weaving together the philosophical and practical facets of historical alchemy, *Fullmetal Alchemist* shows how the imagery and concepts of the discipline can still be mined to imaginative effect.

Transforming the world

Perhaps the clearest legacy of alchemy in contemporary culture is simply the widespread use of the word itself to refer to any act of transformation. Scientific reporting is liberally scattered with references to "alchemy" in relation to the development of new drugs, materials, and inventions of all sorts: to speak of alchemy is to imply an almost magical power to create the unexpected from seemingly mundane ingredients.

During the Covid-19 pandemic, the *Times of India* claimed that "A vaccine that can prevent or cure Covid-19 has become the new 'philosopher's stone', the latter-day 'elixir of life'." The vaccines were, of course, nothing of the sort, but rather, regular pharmaceuticals developed using well-researched scientific principles. But the allusion speaks to something more—to the almost magical life-saving properties that such agents seem to have even while so few who benefit from them grasp the principles on which they work. And more than this, the comparison highlights the sense of the alchemist as a savior who can satisfy our deepest yearnings and desires. As long as there is still a gold, still an elixir, that we crave, alchemy's currency will persist.

PROFILE

George Ripley

(c. 1415–1490)

Despite constant concerns that alchemy was not only disreputable but allied to impious magical arts, the most influential scholars of alchemy in the Middle Ages tended to be clerics, such as the English Franciscan Roger Bacon and (at least in popular tradition) the German Dominican friar Albertus Magnus. This is true too of probably the most renowned English alchemist of the fifteenth century: George Ripley, a canon of the town of Bridlington on the Yorkshire coast.

Rather little is known for certain about Ripley's life, although one account describes him as "a man of quick and curious wit who spent almost his whole life in searching on the occult and abstruse causes and effects of natural things." It is said that Ripley took leave from his monastic life to travel throughout Europe, allegedly from France to Italy and as far as Rhodes, where the Knights of the Order of Saint John were struggling to hold back the Turks.

Ripley's quest for secret knowledge culminated in his 1475 treatise *The Compound of Alchemy*, dedicated to the English king Edward IV. Although not a particularly original work, it offered an appealing twelve-step "how to" systematization of the preparation of the philosopher's stone, which Ripley compared to passing through the twelve gates of a castle. Each of the stages has a name that refers to a chemical process: calcination, solution, separation, putrefaction, and so forth.

He is said to have instructed Thomas Norton of Bristol, another influential alchemist of the fifteenth century, who wrote the poem *The Ordinall of Alchimy* (1477) and avers that from Ripley "I learned all the secrets of alchemy."

The Compound of Alchemy is unusual in that it is written entirely in rhyming verse. Ripley's meaning is often hard to follow, couched in the cryptic symbolism of the alchemist—a purposeful strategy, he explained, "to discourage the fools, for although we write primarily for the edification of the disciples of the art, we also write for the mystification of those owls and bats which can neither bear the splendour of the sun nor the light of the moon." Those who knew the code, however, were offered the tantalizing promise of the precious red material that could "multiply" metals into gold:

Pale and black with false citrine, imperfect white and red,
The Peacock's feathers in gay colours, the rainbow which shall go over,
The spotted panther, the lion green, the Crow's bill blue as lead.
These shall appear before you perfect white, and many more others.

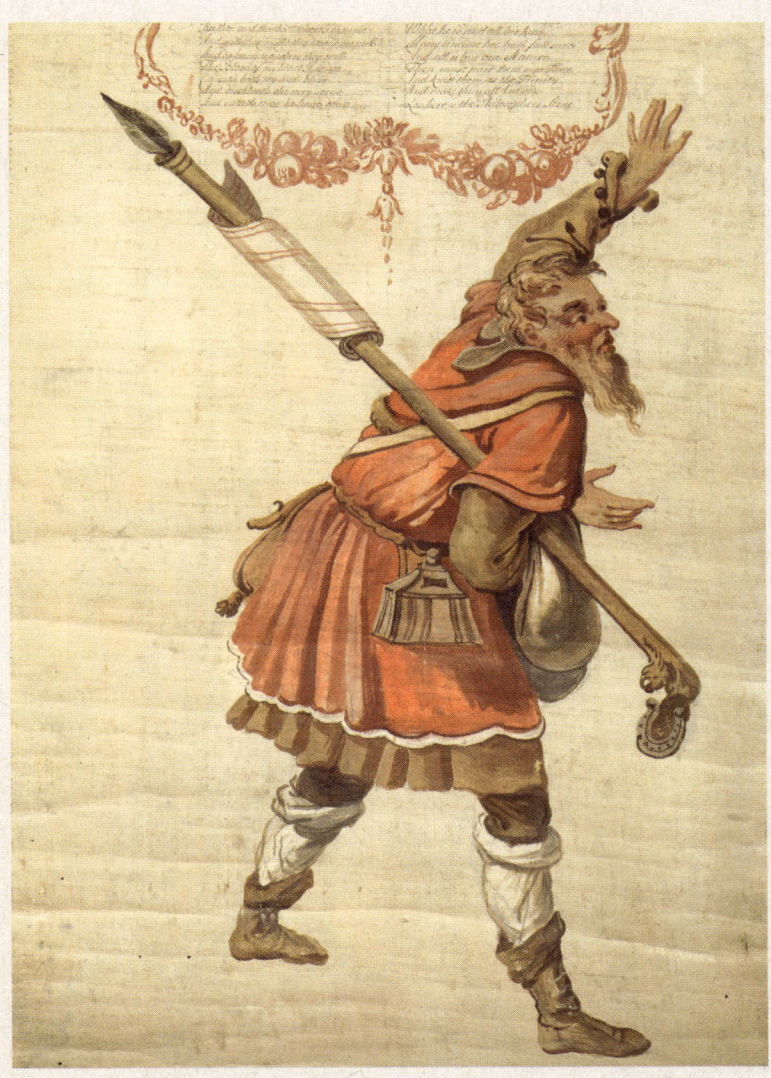

RIGHT A detail from the foot of the Ripley Scroll, c. 1600, apparently depicting the author of the work, George Ripley, himself. This extraordinary manuscript, nearly 20 feet (6 m) long and full of mystical symbolism, describes how to make the philosopher's stone. Several copies of these richly illustrated, poetical scrolls still exist, but it is not known who first produced them.

And after the perfect white, grey, false citrine also,
And after these, there shall appear the red body invariable,
Then you have a medicine of the third order of his own kind multipliable.

Some of Ripley's verses were reproduced on elaborate illustrated vellum scrolls made in the sixteenth and seventeenth centuries, of which twenty-three copies still survive. The imagery of the scrolls is still not fully understood, and the creators are not known; although there was probably a fifteenth-century original, there is no reason to suppose that Ripley himself was its author.

Ripley's works are the epitome of medieval alchemy's cryptic nature, promising powerful secrets if only one can crack the code. As such, they proved irresistible to later enthusiasts of alchemy: the American George Starkey (see page 212) consulted them closely, as did Isaac Newton (see page 220).

TRANSMUTATIONS: ALCHEMY IN CULTURE

RIGHT An early satire on alchemy by Petrus van der Borcht, c. 1580: here an alchemist's laboratory is inhabited by apes who call at the poorhouse, left destitute after their fruitless quest to make gold. This was a popular theme in the late 16th and 17th centuries—but such imagery does not necessarily indicate skepticism about the possibility of gold-making itself. Rather, it served as a moral comment on the foolishness of those who waste their lives obsessively seeking the philosopher's stone while neglecting their more prosaic obligations. These images often displayed considerable familiarity with the instruments and practices of the alchemical workplace.

Docta etiam sanos Plantis haurite liquores,
Queis ægro medicam sedula præstet opem.

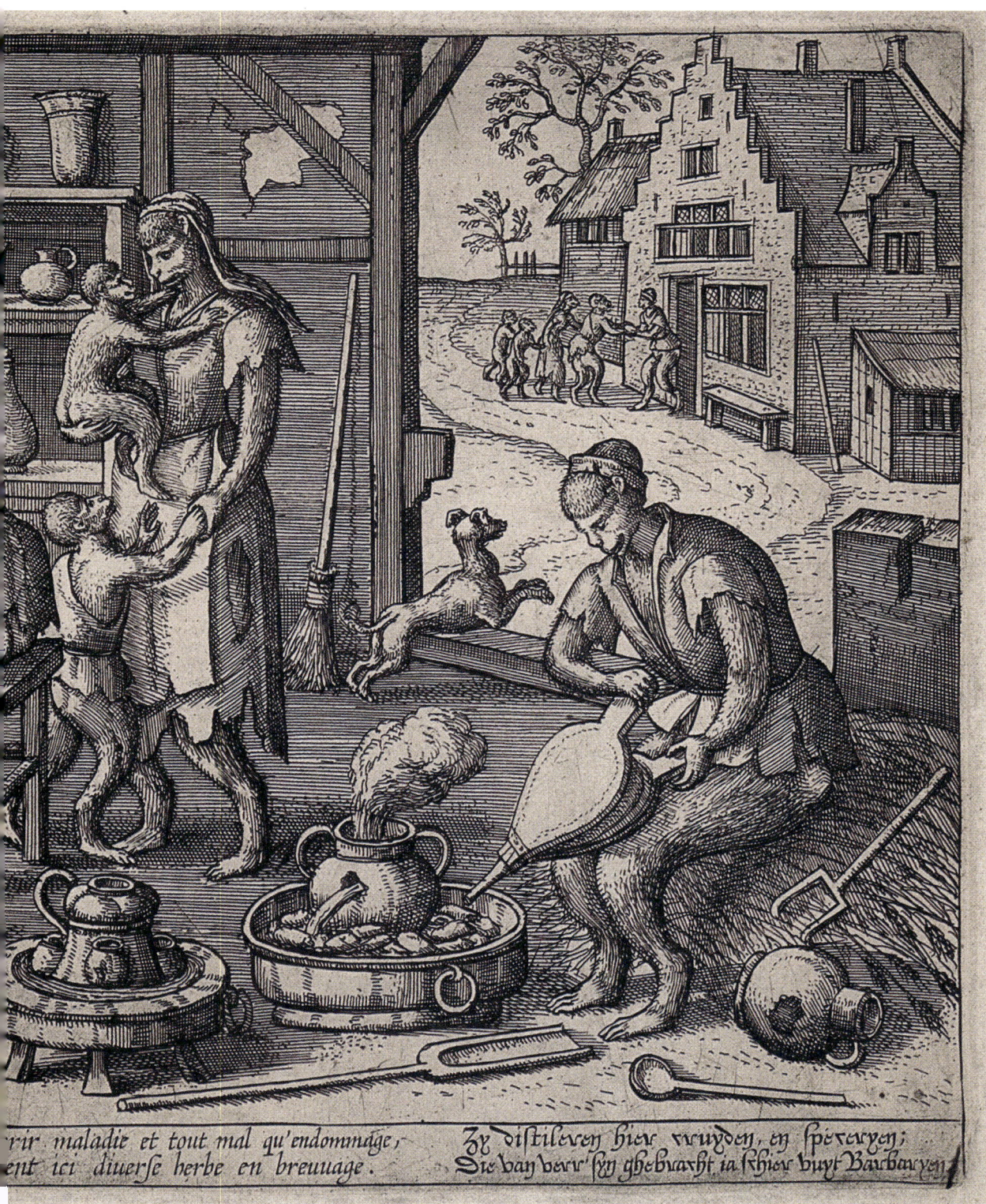

rir maladie et tout mal qu'endommage, Zy distileren hier cruyden, en specerijen;
ent ici diuerse herbe en breuuage. Die van verr syn ghebracht is schier buyt Barbaryen

TRANSMUTATIONS: ALCHEMY IN CULTURE

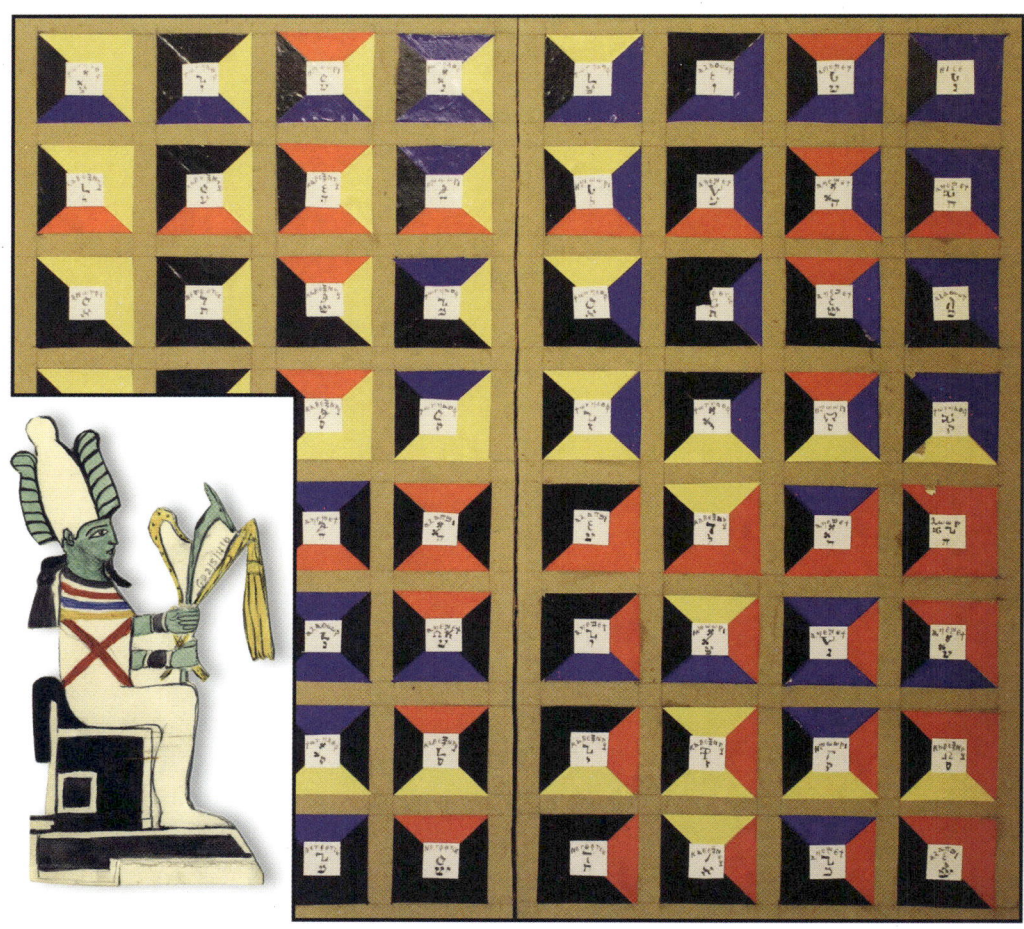

OPPOSITE In this early 19th-century satire on alchemy by Edmund Bristow, a *Monkey Alchemist* attends to his furnace, one foot propping open his workbook, his experiment closely watched by his assistant. By this time, the theme was largely a generic trope. Bristow generally painted rustic and domestic scenes, often including animals—the monkey was one of his favorite subjects.

ABOVE An "Enochian chess board" created by the 19th-century occultists William Wynn Westcott and Samuel MacGregor Mathers. The game, for four players, was inspired by the system of Enochian magic allegedly revealed by angels. It was taught to members of the Inner Order of the Golden Dawn for competition, instruction, and divination. The king piece is represented by the Egyptian god Osiris, and the four colors represent the four Classical elements.

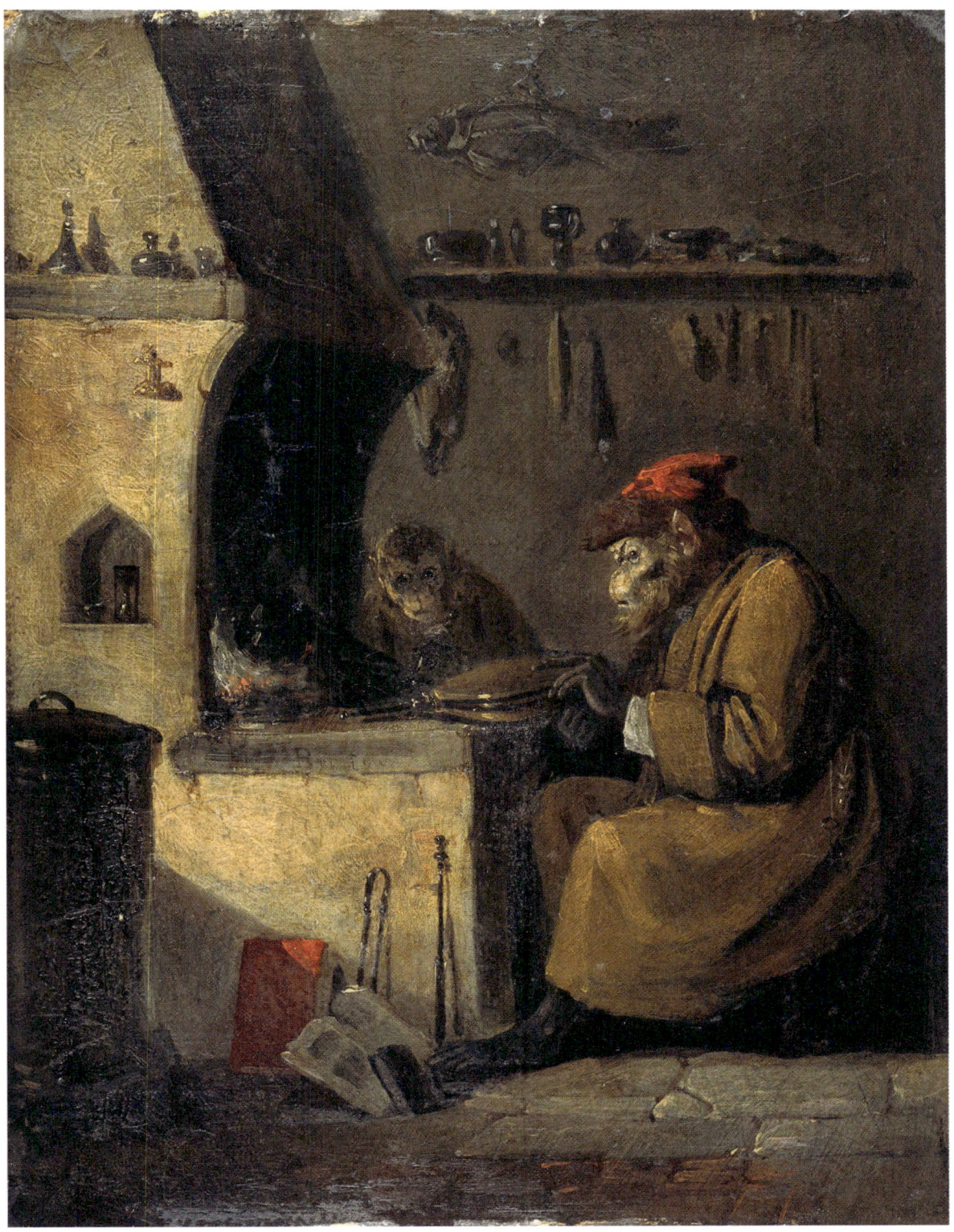

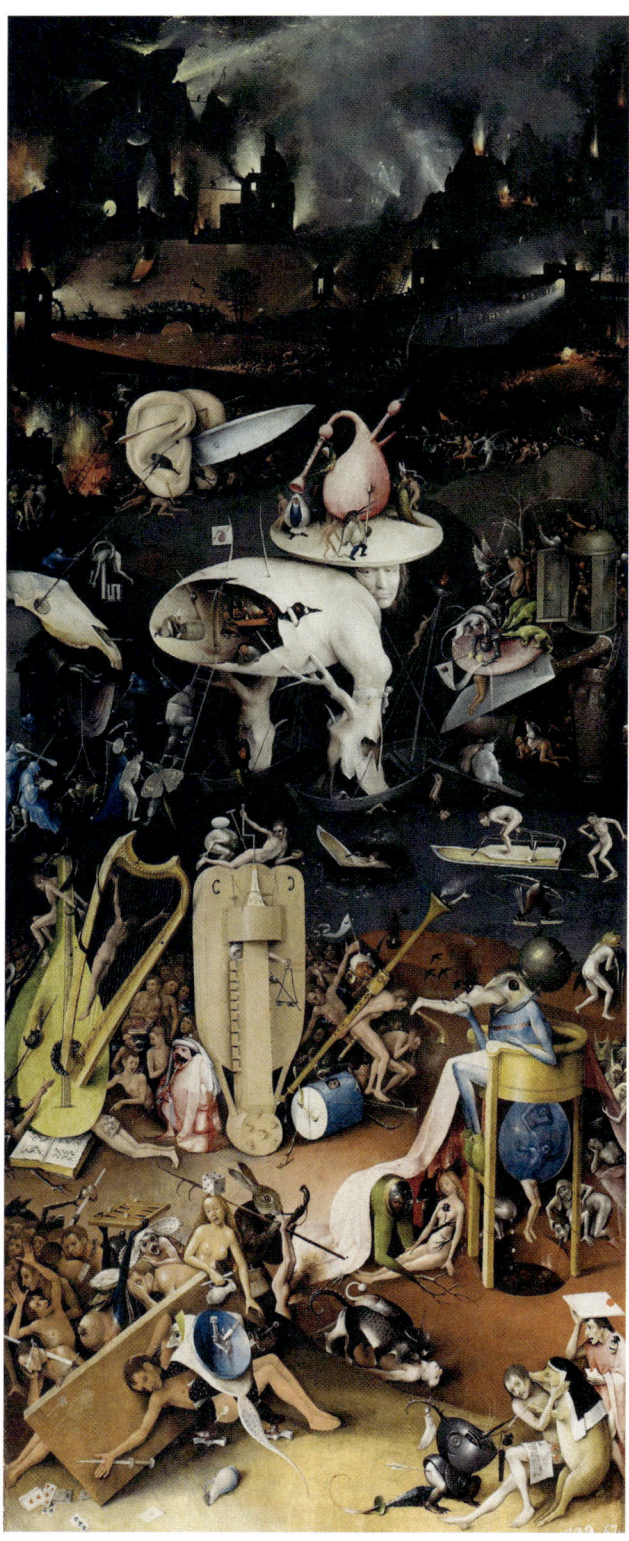

LEFT Hell as depicted in Hieronymus Bosch's *Garden of Earthly Delights*, 1490–1500. The triptych has been said to be replete with alchemical imagery: eggs, spheres, flasks, and furnaces. However, it remains unclear how direct the connection is, and there is still some scholarly disagreement about the extent of the Netherlandish painter's alchemical interests. Did he wish to allude to alchemical allegories, or was he simply using a visual language familiar to his audience?

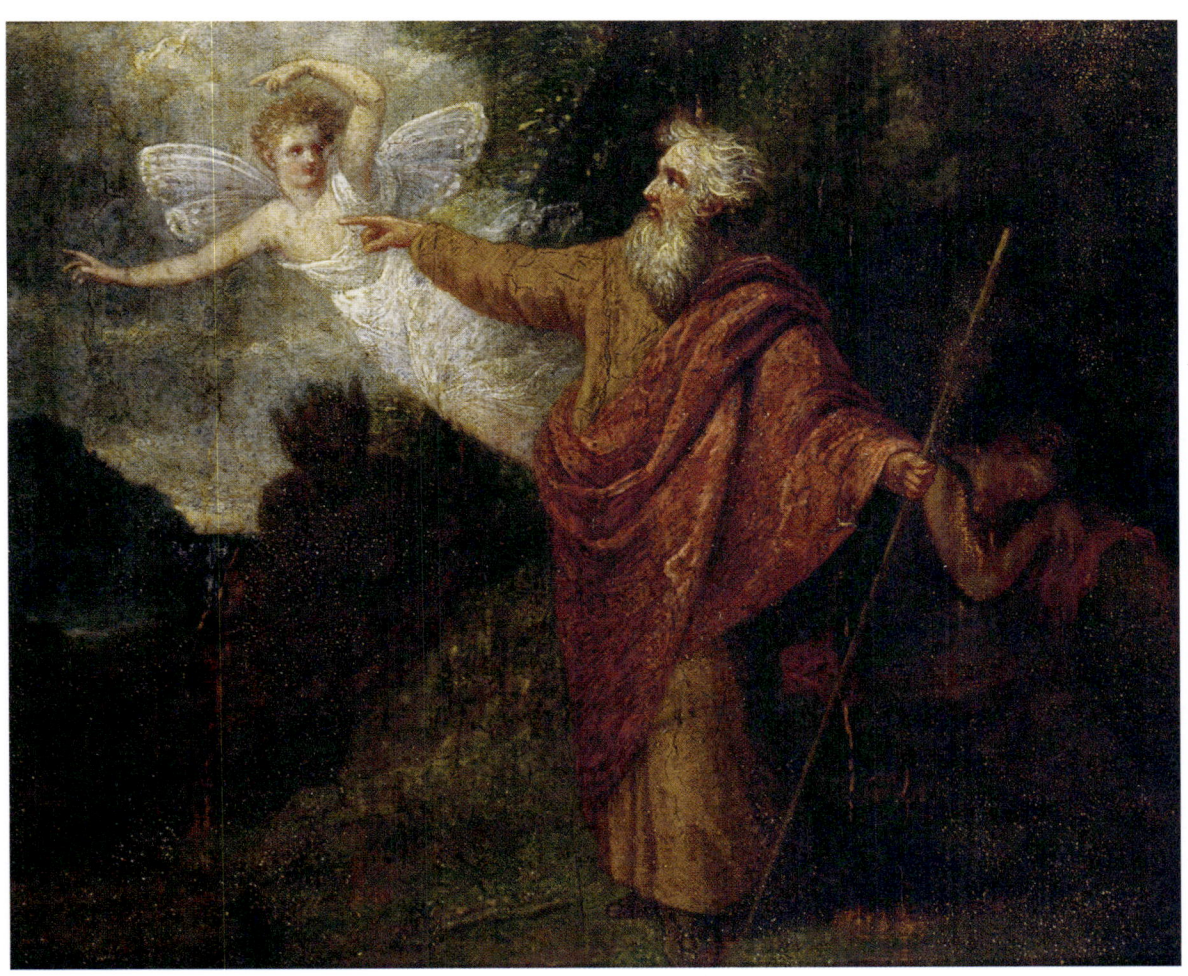

ABOVE Alchemy was a fertile source of imagery and inspiration for playwrights, poets, and artists. Shakespeare's Prospero in *The Tempest*, shown here commanding Ariel in a painting by John White Abbott in 1829, is considered an archetypal magus who some believe was modeled on the Elizabethan mathematician and astronomer John Dee.

Index

Page numbers in *italics* refer to illustration captions.

Abbott, John White *Prospero* 249
Abū al-Qāsim al-'Irāqi
  Kitāb al-aqālim al-sa'bah 32, *34*, *132*
Accademia Nazionale dei Lincei, Rome 98, 194
Act Against Multipliers (England) 1404 170
Agricola, Georgius 85, 125–126
  De re metallica 76, 202–203
al-Haytham, Ibn 25
al-Khwārizmī 25
al-Manṣūr, Abū Ja'far Abd Allāh ibn Muḥammad 24
al-Rāzī 25, 70, 116, *117*
  Sirr al-Asrār 27, 89, *89*
Albertus Magnus 7, 70, 78–79, 232, 239, 242
alchemy 6–13
  alchemists' laboratories *113*, 114–139
  alchemy in China 39, 40–61
  alchemy in culture 223, 224–249
  alchemy in India 53
  chemical philosophy *141*, 142–165
  controversies *167*, 168–195
  from alchemy to chemistry 197–221
  gold-making 63–85
  origins of alchemy 16–37
  uses of alchemy 87–111
alembics *117*, 123
Alexander the Great 59, 89, *110*
Alhazen *see* al-Haytham, Ibn 25
aludels 118–119
Amenenham II of Egypt 32
Amenhotep II of Egypt 19
Amman, Jost *11*
amulets 32
Andreae, Johann Valentin *179*, 179–181
  Reipublicae Christianopolitanae 181, 181–182
antimony 185
  Basil Valentine *Triumph-wagen antimonii* 185, *186*
Apollonius of Tyana 28, *35*
  Secret of Creation 27–29
aqua vitae 121, 130–131
Arakawa, Hiromu *Fullmetal Alchemist* 241
Arcimboldo, Giuseppe 7, 178
  Vertumnus 192
Aristotle 24, 31, 54, 74, 78, 90, 102, 122, 130, 150, 182, 202, 210
  Secret book of Secrets 89–90
  theory of four elements 65–67
Arnald of Villanova 121, 130, 194
art 232–236

Ashmole, Elias 173
  Theatrum chemicum britannicum 169, *236*, 237
astrology 146–147, 168, 229
Atharvaveda 48
atomic theory 210–211, 213
Atwood, Mary Anne 228, 241
  A Suggestive Inquiry into the Hermetic Mystery 228, *228*
Augustus II of Poland, Elector of Saxony 100–101, *101*
Aurora consurgens 107
Averroes (Ibn Rushd) 78
Avicenna *see* Ibn Sīnā
azote 207, *216*
Azoth 107

Bacon, Francis *New Atlantis* 181–182
Bacon, Roger 74, 229, 242
bain-marie 114, *115*, 123
Balinus 28
Balsamo, Giuseppe *see* Cagliostro, Alessandro di
Bandiera, Pietro Antonio 195
Banks, Joseph 208
Bao Gu 54
Bao Liang 54
Basil Valentine 13, 108, 181, 186–187
  Last Will and Testament 73, 186
  Révélation des mystères des teintures essentieles 187
  Triumph-wagen antimonii 185, *186*
  The Twelve Keys of Basil Valentine 167, *187*
basilisks 110
Bayt al Ḥikmah (House of Wisdom), Baghdad 26, 30
Becher, Johann Joachim 138–139, 204–205
  Chymisches laboratorium 126–127
  Physica subterranea 138, 204, *205*
Beck, David *Queen Christina of Sweden* 195
Bega, Cornelis *The Alchemist* 72, *118*
Béguin, Jean 203
  Tyrocinium chymicium 183, *184*
Bergson, Henri 229
Bertholet, Marcelin 127–128, 210
Bertuch, Friedrich *Justin Bilderbuch für Kinder* 110
Besant, Annie *Occult Chemistry* 223
Bible 146, 149, 150, 234
Biringuccio, Vannocio *De la pirotechnia* 202
Blackett, Patrick 219
Blake, William 214
  Marriage of Heaven and Hell 227
Blavatsky, Helena 229
Boas Hall, Marie 198
Boehme, Jacob *141*, 214
Boerhaave, Hermann
  Elementa chemiae 206, 206–207
  Index plantarum 206
Bohemia *141*, 178, 214

Bolos of Mendes 23
Boniface VIII, Pope 71
Borri, Giuseppe Francesco 194–195
Bosch, Hieronymus 234
  Adoration of the Magi 234, *235*
  Garden of Earthly Delights 234–236, *248*
Böttger, Johann Friedrich 101
Boyle, Robert 128–129, 157, 170, 187, *198*, 199, 200–201, 212–213, 221
  The Sceptical Chymist 98, *98*, 198–199
Brahe, Tycho 7, 176–178
Brande, William 208–209
Brandt, Hennig 127–129, *129*, 195
Breton, André *Arcanum 17* 232
Bristow, Edmund *Monkey Alchemist* 246
Bronze Age chemistry 16–20
Bruegel, Pieter the Elder *The Alchemist* 123, 123–124
Brunschwig, Hieronymus
  Liber de arte distillandi 122, 122–123, 202, 2c2
Bry, Theodor de
  Doctor of Fools 188
  Paracelsus 102

Cagliostro, Alessandro di 225, 225–226
Caligula 95
Cantara, Simeon ben *Cabala mineralis* 83, *143*
carbon dioxide 207
Casanova 224
Cassini, Giovanni 194
Cavendish, Margaret 195
Cennini, Cennino *Libro dell'arte* 88, 92, 95
charlatans 170–173
Charlemagne 30
Chaucer, Geoffrey
  "The Canon's Yeoman's Tale" 170–171, *171*
chemical philosophy *141*, 142–144, 160–165
  alchemy and religion 150–157
  alchemy as a world model 147–150
  Khunrath, Heinrich 158–159
  Plato's return 144–147
chemistry from alchemy 197, 198–203, 214–217
  chymical chimera 204–208
  last of the alchemists 208–209
  Newton, Isaac 226–227
  Starkey, George 212–213
  transmutation at last 210–211
chemistry in the ancient world 16–20
  mysteries 23–24, 27–29
  recipes 20–22
China 39, 40–42
  artefacts *41*, 56
  eternal life 46–52
  varieties of gold 43–46
Christ 214, 219
Christian Rosencreutz *180*, 181, 229
Christianity 152–153
Christina of Sweden 192, 194–195, *195*
chrysopoeia *see* gold-making

Chymical Wedding of Christian Rosencreutz, The 180, 181
chymistry 134, 139
Clarke, Lindsay *The Chymical Wedding* 241
Claudio de Domenico *Book of Alchemical Formulas* 164
Clement V, Pope 71
Clement VIII, Pope 178
Coelho, Paulo *The Alchemist* 239
commercial secrets 98
Compositiones ad tingendae/variae 91, 92
Confessio fraternitatis 180–181
controversies 167, 168–170, 188–193
    alchemical charlatans 170–173
    alchemical courts 173–178
    antimony wars 182–185
    Basil Valentine 186–187
    Christina of Sweden 194–195
    Rosicrucian Enlightenment 178–182
Cornelius Agrippa 7, 143–144, 232, 239
    De occulta philosophia libri tres 142, 144
Corpus hermeticum 144–145, 145
correspondences 145–146
Covid-19 pandemic 241
Croll, Oswald 178, 194
Crookes, William 230
Crowley, Aleister 229, 229, 230
crucibles 119, 119–120, 121
culture 223, 224–226, 244–239
    alchemy in art 232–236
    alchemy in literature 236–241
    alchemy in modern culture 241
    occult revival 227–230
    psychic science? 230–232
    Ripley, George 236, 242–243
cupellation 22, 76–77, 123
Curie, Pierre and Marie 230
currencies 169–170

Dali, Salvador *Alchimie des Philosophes* 234
Dalton, John 210
Dante *Inferno* 171
Daoism 39, 41, 45–47, 47, 52, 59, 60
Davidson, William 185
    Philosophis pyrotechnia 185
De alchimia 230
De La Brosse, Guy 185
Dee, John 8, 124–125, 158, 173–176, 175, 189, 238, 249
    Monas Hieroglyphica 180, 213
Democritus 23
Des Clozets, Georges Pierre 200
Descartes, René 155, 194
Détective 229
Diesbach, Johann Jacob 95
Diocletian 169–170
Dippel, Johann Konrad 94, 95
distillation 117–120
    spirit and essence 120–123
doctrine of signatures 146, 214

Donne, John 237
Douglas, William Fettes 191
Dubs, Homer 40
Duchesne, Joseph 184
dyes 18, 19

Earth 147–148
Edward IV 242
Egypt, ancient 15, 16, 17, 17–20
    recipes 20–22
Egyptian Rite 225–226
elemental theory 65–67
elixirs 46–47, 69
    elixirs of immortality 46–53, 239
Elizabeth I 173–176
Empedocles of Akragas 65, 65–66
Enlightenment 31, 178–179, 232
Enochian chess board 246
Ercker, Lazarus 125, 125–126
Ernst, Max *Attirement of the Bride* 232, 232–234
essences 120–121
Euclid 24, 175
Evans, Robert 176

Fama fraternitatis 180–181
Ferdinand II of Styria 178
Ferdinand III, Holy Roman Emperor 138
Ferdinand Maria of Bavaria 138
Ficino, Marsilio 144–146
    Pimander 145
Figuier, Louis *Vies des savants illustres du moyen age* 117
Filelfo, Francesco 172
Flamel, Nicolas and Perenelle 239, 241
Fludd, Robert 149–150, 155–156, 182
    Tomus secundus...de superrnaturali, naturali, praeternaturali, contranaturali 150
Forman, Simon 173, 173
Forshaw, Peter 159
Fourcroy, Antoine François de 209
Franck, Johannes 194
fraudsters 168–170
Frederick the Great of Prussia 225
Frederick V, Elector Palatine of the Rhine 178
Freemasonry 182, 209, 226, 229
Freher, Dionysius Andreas 214
furnaces 70, 114, 114–116, 169

Galen 24, 96, 105, 150, 182, 184, 202
Galileo 98, 194
Gan Bozong *Ge Hong* 55
gases 207, 208
Gassendi, Pierre 155
Ge Hong 43–46, 46, 54–55, 56
    Baopuzi 54
Geber *see* Jābir ibn Ḥayyān; Pseudo-Geber
Gilbert, William 147
    De magnete 148

Gilray, James "A Masonic Anecdote" 225
glass-making 17–18
Glauber, Johannes 116, 195
Godwin, William *Lives of the Necromancers* 239
Goethe, Johann Wolfgang von 214
    Faust 7, 10, 142, 239
gold-making 63–85, 126–127
    disreputable occupation 168–170
    gold-making in the Middle Ages 72–74
    it works in theory 64–71
    what really happened? 76–77
Golden Lane, Prague Castle 6, 6–7
Great Work 32, 94, 208, 238
Greek fire 67
Gustav Adolphus of Sweden 194

Harry Potter 239, 241
Hartlib, Samuel 100, 179, 199, 212
Hartmann, Johannes 203, 203
Hārūn al-Rashīd 30
Häublin, Nicolaus 141, 214
He Wei 45
Hellmerich, Georg
    Musaeum hermeticum 13, 68, 107, 108
Henri IV of France 183–184
Henry IV 170
Henry VI 170
Henry VIII 173
Heraclius *Mappae clavicula* 91–92
Hermes Trismegistus 29, 144, 145
    Emerald Tablet 28–29, 35, 108, 145
Hermetic art 64, 88–89, 114, 126, 158, 237
Hermetic philosophy 149–150
Hinduism 48
Hippocrates 12, 24, 96, 105, 184
Hohenhein, Philippus Aureolus Theophrastus von *see* Paracelsus
Hokusai, Katsushika 43
homunculi (homunculus) 10, 102, 188, 241
House of Wisdom (Bayt al Ḥikmah), Baghdad 26, 30
humoral theory (four humors) 95, 96, 105
Hunting of the Green Lyon, The 237
Hutchinson's History of the Nations 168
hydrogen 207, 210–211

iatrochemistry 97–98, 237
    antimony wars 182–185
    Basil Valentine 186–187
Ibn Rushd (Averroes) 78
Ibn Sīnā 25–26, 37, 70, 78
Illuminati 226
India 53, 59, 241
iron acetate 186
Isidore of Seville *Etymologiae* 65
Islamic alchemists 24–27, 32
    elixirs 46–47
    Jābir ibn Ḥāyyān 30–31
Ismailis 89

Jābir ibn Ḥayyān 26–27, 30–31, 67, 69, 78, 89, 92–93, 96, *107*
   Geber 31, 69–70
   Kitāb al-Raḥma al-kabīr 30
Jacob, Willem *157*
James I and VI *178*
James le Palmer *Omne bonum* 93
Janneck, Franz Christoph *The Medical Alchemist 137*
Japan *4*, *43*, 45, *45*
Jardin du Roi, Paris *184*, *185*
Jingdi *170*
John of Rupescissa 70–71, 130–131, 231
   De consideratione quintae essentiae 121–122, *130*
John XXII, Pope *170*
Johnson, Samuel 31
Jonson, Ben *The Alchemist 172*, 172–173
Jung, Carl *108*, 214, 230, 231–232

Kabbalah *143*, 146, 175, 229
Kelley, Edward *8*, *175*, 175–176, *176*, 189
Kepler, Johannes *7*, 150, 176–179
Kerseboom, Johann *Robert Boyle 198*
Keynes, John Maynard *220*
Khunrath, Heinrich *152*, 158–159, *159*, *178*
   Amphitheatrum sapientiae aeternae 152–153, *153*, *163*, *158*
Killian, W. P. *Johann Becher 138*
Kircher, Athanasius 150, 194
   Mundus subterraneus 148–149, *149*
Kneller, Sir Godfrey *Sir Isaac Newton 221*
Krafft, Daniel 128
Kraus, Paul 30
Krauss, Werner *231*
Kunckel, Johann 128
Kuniyoshi *51*

laboratories *113*, 114–116, 130–139
   bearer of light 127–129
   devices 117–21
   equipment *132*, *134*
   invention of the laboratory 123–127
   spirit and essence 120–123
Lancilloti, Carlo *Der brennende Salamander 132*
Lange, Johann *Der brennende Salamander 132*
lapis philosophorum see philosopher's stone
Lavoisier, Antoine *198*, 207, 209, 216
   Traité élémentaire de chimie 207–208
Lavoisier, Marie-Anne Paulze 216
Law, William *The Works of Jacob Behmen 4*, *214*
Le Bailiff, Roch 183
Leadbeater, Charles *Occult Chemistry 223*
Lefebvre, Nicaise 203
   Cours de chimie 185
Leibniz, Gottfried 128, 139
Lemery, Nicolas 203
   Cours de chimie 185, 194
Leopold I, Holy Roman Emperor *139*

Leyden Papyrus 20–22
Libavius, Andreas *159*, 194, 202
   Alchymia 97–98, *98*, *126*, *126*
Liebig, Justus von 210
Lippmann, Edmund Oscar von 40
literature 236–241
Liu An *Huainanzi* 45
Liu Bang 45
Locke, John 221
Lorenzetti, Pietro *Crucifixion 91*
Louis XV of France 224
Louis XVI of France 226
Lull, Ramon 70, 194
   Liber de secretis naturae seu de quinta essentia 70–71, *80*
Luyken, Jan *Figures in a Laboratory 197*

Machen, Arthur 229
Maier, Michael *7*, *108*, *178*, 182, 187
   Atalanta fugiens 153–155, *155*, *163*, *234*
   Symbola aureae mensae 116
   Themis Aurea 178, *178*
Manget, Jean Jacques
   Bibliotheca chemica curioso 137
Maria Judaea (Maria the Jewess) *114*, *115*, *116*, *117*
Marie Antoinette of France 226
Marlowe, Christopher *Doctor Faustus 238*, *239*
Masonic Rosicrucian Society 182
Mathers, Moina 229
Mathers, Samuel Liddell MacGregor 229, *246*
Matta, Roberto *232*
Mayerne, Théodore Turquet de 184
Medici, Cosimo de' 145
medicines 95–98, 131
   Chinese medicines 45–52
   Paracelsus 102–105
Meissen ware *101*
Mercurius 74
Mersenne, Marin 155–156, *159*
Mesopotamia 17, 17–18, 120
metallurgy 19–20, 125
   China 40–42, *42*
Michelangelo 214
Mirandola, Pico della 146
Miscellanea Alchemica 134
Modern Chemical Apparatus, The 4
Mögling, Daniel *199*
monkey scenes *244*, *246*
Moses 146
Mozart, Amadeus *The Magic Flute* 182
multipliers 169
   Act Against Multipliers (England) 1404 170
mummy casings 36
Myers, Frederick 230

natural magic 142–144
   alchemy as a world model 147–150
   Neoplatonism 144–147, 175
Needham, Joseph 41, 44, 45, 46–47, 54, 77

Neoplatonism 29, 144–147, 175, 214
Newman, William 213, 221
Newton, Isaac 77, 147, 148, 157, 187, 200–201, *201*, 213, 226–227, 231
   Opticks 220
   Philosophiae naturalis principia mathematica 220
   Physices elementa mathematica 157
Nicholl, Charles 237–238
nitrogen 207, 216, 219
Nizami *Khamsa* 59
Norton, Thomas 114
   Ordinall of Alchymy 169, *200*, *236*, *242*
Novalis 232
nuclear fission 211

Oberstosckstall, Austria 119
occult 90–91, 142–143, 173–176, 182, 223, 224–236
   occult revival 227–230
   psychic science? 230–232
oils 120–123
Oldenburg, Henry 221
Order of the Golden Dawn 229, 229–230, *246*
Order of the Inseparables 179
Order of the Rosy Cross *see* Rosicrucianism
Osiris *246*
Ouroboros *32*, *71*
oxygen 207, 216, 219

Pabst, Georg Wilhelm *231*, 231
Paracelsus *7*, 97, 102–105, *137*, 138–139, 142, 146, 147, 158–159, 201, 212, 230, 231, 232, 237–239
   Der grossen Wundartzney 104
   doctrine of signatures 146, 214
   "don't make gold, make medicines" 96–98
   Paracelsians 119, *141*, 179, 182–185, 194, 201–203
   Rosicrucianism 179
   tria prima 96, 150, *152*, 156, *163*, *204*, 205
Paul of Taranto 69
Pérez, Gonzalo *164*
Pernety, Antoine-Joseph 226
Pettus, Sir John *Fieta minor 125*
Philalethes, Eireanaeus *see* Starkey, George
Philip III, Landgrave of Hesse-Butzbach *199*
Philip of Tripoli *90*, *110*
philosopher's stone 69, 79, 130, 187, 194
   gold-making in the Middle Ages 72–74
   Newton, Isaac 221
Philosophical Mercury 66, 68–69
Philosophical Sulfur 68–69, 130
phlogiston 205–207
Phocylides *85*
phosphorus 128, 195
Physica et Mystica 23
Piedmontese, Alessio *Secreti* 98
pigments 11, *15*, 19, 36, 52, 54, 71, 88
   recipe books 91–95

Plato 65. 74
  Timaeus 144
Platonic Academy, Florence 144
Pleydenwerff, Wilhelm *131*
Pliny the Elder 16, 95
Plot, Robert 120
Plotinus 144–145
porcelain 100–101, *101*
Price, James 208, *209*
  Account of Some Experiments ... Made at Guildford 208
Principe, Lawrence 40, 186
Prout, William 210
Prussian blue 94, 95
Pseudo-Albert
  Libellus de alchimia 78–79
Pseudo-Aristotle
  Secretum secretorum 90, 110
Pseudo-Arnald
  Rosarium philosophorum 71, 219
Pseudo-Geber 69–70, 114, 194
Pseudo-Lull *see* Lull, Ramon

Qian Xuan, Wang Xizhi watching geese 56
Qin Shi Huang 50, *50*–52
Quintessence (aether) 122, 130

Rasarnava 53
Read, John 159
Rees, Abraham *Cyclopaedia* 211
Reisch, Gregor *Margarita philosophica* 95
religion 150–157
Rhazes *see* al-Rāzī
Ribit, Jean 183, *184*
Riolan, Jean 184, *185*
Ripley George 236
  Compound of Alchemy 236, 242–243
  Ripley Scroll *4*, *87*, *108*, *243*, 243
Robert of Chester *De compositione alchemiae* 70
Roestraten, Pieter Gerritsz van *The Alchemist 8*
Romanticism 214
Roquetaillade, Jean de *see* John of Rupescissa
Rosicrucianism 178–182, 199, 209, 212, 214, 226
  Geheime Figuren der Rosenkreuzer 192
  Rosicrucian Order of Alpha and Omega 229
Rowling, J. K. *Harry Potter* 239
Royal College of Physicians 150
Royal Mint 220
Royal Society 128–129, 179, 199, 208, 220–221
  Philosophical Transactions 200
Rudolf II, Holy Roman Emperor 6–7, *7*, 71, 152, 155, 158, 175–178, 179, 189, 192, 194, 201
Rupert, Prince 139
Ruscelli, Girolamo 98
Rutherford, Ernest 210–211, 219
Ryckaert, David III *Painter's Workshop 88*

Saint-Germain, Claude-Louis-Robert 224, *224*–225, 229
Sarton, George 198

Schedel, Hartmann *Nuremberg Chronicle* 64, *131*
Scheele, Wilhelm 207
Schelling, Friedrich 214
Schinkel, Karl Friedrich *182*
Schrotter, Georgius 97
Schweighardt, Theophilus
  Speculum sophicum Rhodostauroticum 199
Scientific Revolution 179, 221
Sendivogius, Michael *7*, 178
  Novem lumen chymicum 176, 201, *201*
Shadwell, Thomas *The Virtuoso* 173
Shakespeare, William 237, *237*–238, 249
Shelley, Mary *Frankenstein* 239
signatures *see* doctrine of signatures
Sima Qian 51–52
Sivin, Nathan 40
Skylitzes, John 67
Smith, Pamela 139
Society for Psychical Research 230
Soddy, Frederick 211, 219
South, Thomas 228
spagyric chemistry 185, 186
spirits (alcohols) 120–121
Spiritualism 213, 230
St Francis 130
Stahl, Georg Ernst 205
Starkey, George 98, 77, 100, 212–213, 243
  Anonymi Philalethae philosophi Opera omnia 213
Stockholm Papyrus 20, *20*–22
Stuart, Elizabeth 178
Summa perfectionis 69
Sun Simiao *Danjing yaojue* 48, *48*
Surrealism 232–234
Swedenborg, Emanuel 209
  Heaven and Hell 227
Swift, Jonathan *Gulliver's Travels* 173
Syria 17, *41*, 59
Tantrism 43, 53
Tao Hongjing 44, 48
Teniers, David the Younger 123–124
  Ein Chemiker in seinem Laboratorium 124
terracotta army, Shaanxi, China 52
Thales of Miletus 64, 65
Themistius of Byzantium 76
Theophilus *De diversis artibus* 92, *92*–94, 110
Theophrastus 102
Theosebeia 34
Theosophical Society 213
Thevet, André
  Les vrais pourtraits et vies des hommes illustres grecz, latins et payens 31
Thirty Years' War 178, 179, 194
Thölde, Johann 181, 186
Thomas Aquinas 78, 79, 107
Thorndyke, Lynn 198
Thurn, Leonhard Thurneysser zum *Quinta Essentia 11*
Thymme, Thomas 152–153

Timbs, John *English Eccentrics and Eccentricities* 208
Times of India, The 241
Traité de chymie 134
transmutation 68, 69–71, 139, *189*, 214, 219
  Price, James 208–209
  transmutation at last 210–211, 219
Trithemius, Johannes 176
Tschirnhaus, Ehrenfried 100–101
Tu Youyou 55

Ulmann, Brother *Das Buch der heiligen Dreifaltigkeit* 219
Urban of Trennbach 119
uses of alchemy 88, 106–111
  making medicines 95–98
  Paracelsus 102–105
  recipe books 91–95
  secrets and commerce 98
  tradition of secrets 88–91
  white gold 100–101

Van der Borcht, Petrus 244
Van der Werff, Pieter *Entombment of Christ* 94
Van Helmont, Jan Baptista 207
  Ortus medicinae 156, *156*–157
Van Helmont, Mattheus *A Savant in His Cabinet 160*
Van Muyden, Jacques Alfed *10*
Verrochio, Andrea del *Virgin and Child with Two Angels 44*
Vos, Marten de 63

Waite, Arthur Edward 229–230
Walther, Albertus *Sermon of Saint Albertus Magnus* 79
Wang Xizhi 56
Wei Boyang 50
Weiditz, Hans *Francisci Petrache 121*
Weishaupt, Adam 226
Weltchronik 85
Westcott, William Wynn 229, 230, 246
Winthrop, John 212
Wolgemut, Michel *131*
Woulfe bottles 208, *211*
Woulfe, Peter 208–209, *211*
Wright, Joseph *The Alchemist* 129, *129*

Yates, Frances 178–179
Yeats, William Butler 229
Yi Zhenren Xingming guizhi 47

Zaccone, Pierre
  Histoire des sociétés secrètes, politiques et religieuses 226
Zoffany, Johann 172
Zoroaster (Zarathustra) *Clavis artis* 152
Zosimos of Panopolis 24, *25*, 34, 67, 69, 114, 115, 116, 117, 231
  Cheiromekta 23–25

Further Reading

Alchemy has been rehabilitated as a serious and significant aspect of the history of science—and of thought and culture—through the diligent work of a contingent of modern historians of science. These include Lawrence Principe, William Newman, Jennifer Rampling, Peter Forshaw, Pamela Smith, Bruce Moran, Didier Kahn, Stephen Clucas, Allen Debus, and Tara Nummendal. Much of the content of this book is indebted to their research and that of their colleagues. In general, that work has been presented in academic papers, but I recommend in particular the following books:

A. G. Debus, *Man and Nature in the Renaissance* (Cambridge University Press, 1978).

A. G. Debus, *The Chemical Philosophy* (Dover, 2003).

W. Eamon, *Science and the Secrets of Nature* (Princeton University Press, 1994).

B. T. Moran, *Andreas Libavius and the Transformation of Alchemy* (Watson Publishing International, 2007).

B. T. Moran, *Paracelsus: An Alchemical Life* (Reaktion Books, 2019).

W. R. Newman, *Gehennical Fire: The Lives of George Starkey* (Harvard University Press, 1994).

W. R. Newman, *Newton the Alchemist* (Princeton University Press, 2018).

L. M. Principe, *The Aspiring Adept: Robert Boyle and His Alchemical Quest* (Princeton University Press, 1998).

L. M. Principe, *Chymists and Chymistry* (Science History Publications, 2007).

P. H. Smith, *The Business of Alchemy* (Princeton University Press, 1994).

C. Webster, *Paracelsus: Medicine, Magic and Mission at the End of Time* (Yale University Press, 2008).

C. Webster, *From Paracelsus to Newton: Magic and the Making of Modern Science* (Barnes & Noble, 1996).

Far and away the best and most authoritative general introduction to alchemy as it is currently conceived by historians is Lawrence Principe's *The Secrets of Alchemy* (University of Chicago Press, 2013). *The Chemical Choir* by P. G. Maxwell-Stuart (Continuum, 2008) is also a nice overview.

Some older texts retain their charm and have plenty to commend them so long as one recognizes that scholarship has often moved on since then. These include:

E. J. Holmyard, *Alchemy* (Dover, 1990; first published 1957).

H. M. Leicester, *The Historical Background of Chemistry* (Dover, 1971; first published 1956).

J. Read, *From Alchemy to Chemistry* (Dover, 1995; first published 1957).

C. J. S. Thompson, *Alchemy and Alchemists* (Dover, 2002; first published 1932).

For those interested in exploring further the glorious visual aspects of alchemy, I suggest *The Mirror of Alchemy* (British Library, 1994) by Gareth Roberts, and Alexander Roob's hefty *Alchemy and Mysticism* (Taschen, 1997)—with the proviso that such imagery cannot be properly understood when detached from its intellectual and temporal contexts.

Picture Credits

Images on the pages listed below are reproduced by kind permission of the owners. The publishers have made every effort to contact the owners and apologize for any unwitting infringement. Specific acknowledgments are as follows: l=left, r=right.

1 The Metropolitan Museum of Art, New York, 10.211.1302 (CC0); **2** Getty Research Institute, Manly Palmer Hall collection of alchemical manuscripts, GRI Special Collections, MS 205 (CC0); **3** Getty Research Institute, Los Angeles; **4** Courtesy of Science History Institute (CC0); **6** Benedikt Sarda/Scheufler Collection/Corbis/VCG via Getty Images; **7** Brandstaetter images/ Imagno/Getty Images; **8** Lobkowicz Palace, Prague Castle, Czech Republic/Bridgeman Images; **9** Edward Kelley and John Dee Wellcome Collection, London (CC0); **10** Public domain, via Wikimedia Commons; **11** Getty Research Institute, Los Angeles (CC0); **12** Wellcome Collection, London (CC0); **13** General Collection, Beinecke Rare Book and Manuscript Library, Yale University (CC0); **14–15** Neues Museum, Berlin. Photo Peter Horree / Alamy Stock Photo; **16** The Metropolitan Museum of Art, New York, 07.228.122 (CC0); **17l** Photo © Christie's Images / Bridgeman Images; **17r** DeAgostini/Getty Images (detail); **18** The Metropolitan Museum of Art, New York, 09.184.217 (CC0); **19** DeAgostini/S. Vannini/Getty Images; **21** Stockholm Papyrus, Royal Library, Stockholm/World Digital Library via Library of Congress; **24** Rivalette, via Wikipedia (CC BY-SA 4.0); **25** Bibliothèque nationale de France, Paris, Ms 2327, fo. 81v. Photo © NPL - DeA Picture Library / Bridgeman Images; **26** Bibliothèque nationale de France, Paris, MS Arabe 5847, fo. 5v. RMN-Grand Palais /Dist. Photo SCALA, Florence; **27** Wellcome Library, London, via Library of Congress, Washington D.C.; **28** Library of Congress, Washington D.C., 11014016; **29** J. Paul Getty Museum, Los Angeles, 83.AM.342 (CC0 1.0); **31** Wellcome Collection, London (CC0); **32** Babylonian Collection, Yale Peabody Museum, YPM BC 038041 (CC BY 1.0); **33** The British Library, Add. 25724, fo.50v. From the British Library archive / Bridgeman Images; **34** The British Library, Add. 25724, fo.18v. From the British Library archive / Bridgeman Images; **35** Topkapi Palace, Istanbul, Ahmet III 2075, via Wikipedia (CC BY 1.0); **36** The J. Paul Getty Museum, Villa Collection, Malibu, California, 81.AP.42; **37** National Museum, Damascus, Syria/Luisa Ricciarini / Bridgeman Images; **38–39** Christie's Images / Bridgeman Images; **40** Musée du Louvre, Paris. © RMN-Grand Palais / Raphael Chipault / RMN-GP / Dist. Foto SCALA, Florence (detail); **41** Lucy Maud Buckingham Collection, Art Institute of Chicago, 1929.644 (CC0); **42** Bibliothèque nationale de France, Département des Manuscrits, MS Chinois 5563, pp.320-232; **43** Freer Gallery of Art Collection, Smithsonian Institution; Gift of Charles Lang Freer, F1905.282 (CC0 1.0); **44 (both)** The National Gallery, London, NG 296 / Fine Art Images/Heritage Images via Getty Images; **45** National Museum - Náprstkovo Museum, Prague, inv. 30252 (CC BY 4.0); **46** Freer Gallery of Art Collection, F1916.140 (CC0 1.0); **47** Wellcome Collection, London (CC0); **48** Wellcome Collection, London (CC0); **49** Wellcome Collection, London (CC0); **50** The British Library, London, Or. 11515, fo.11v. From the British Library archive / Bridgeman Images; **51** Museum of Fine Arts, Boston, Massachusetts, 11.16544-6. Photo © 2025 Museum of Fine Arts, Boston / Bridgeman Images; **52** Palindrome6996, via Wikimedia Commons (CC BY 2.0); **55** Wellcome Collection, London (CC0); **56** Arthur M. Sackler Gallery, National Museum of Asian Art, Smithsonian Institution, Washington D.C., S1987.305; **57** The Metropolitan Museum of Art, New York, 1973.120.6 (detail) (CC0); **58** Ethnologisches Museum - Staatliche Museen zu Berlin, Berlin. Photo Scala, Florence/bpk, Bildagentur fuer Kunst, Kultur und Geschichte, Berlin; **59** The British Library, London, MS Or. 12208, fo. 294. From the British Library archive / Bridgeman Images; **60–61** The British Library, London, Add. 22689, fo.6. From the British Library archive / Bridgeman Images; **62–63** Wellcome Collection, London, 45101i (detail); **64** Morse Library, Beloit College, Wisconsin, via Wikipedia (CC0); **65** Image from the collections of Yale University Library; **67** National Library of Spain, World Digital Library via Library of Congress, Washington D.C.; **66** Free Library of Philadelphia, Lewis E M 28:26-27 (CC BY 1.0); **67** National Library of Spain, World Digital Library via Library of Congress, Washington D.C.; **68** General Collection, Beinecke Rare Book and Manuscript Library, Yale University, Mellon MS 70, fig. 55r. (CC0); **70** Courtesy of Science History Institute, Philadelphia, MS 7, p.36 (PDM 1.0); **71** Courtesy of the Foundation of the Works of C.G. Jung: https://doi.org/10.3931/e-rara-4955; **72** The J. Paul Getty Museum, Los Angeles, 84.PB.56 (CC0); **73** Courtesy of Science History Institute, Philadelphia QD25.B37513 1671 (CC0); **75** General Collection, Beinecke Rare Book and Manuscript Library, Yale University, Mellon MS 110, fo.359r (PDM); **76** Wellcome Collection, London (CC BY 4.0); **77** Courtesy of Science History Institute, Philadelphia, Othmer MS 13, p.6 (CC0); **79** The Metropolitan Museum of Art, New York, 64.215 (CC0) ; **80–81** Courtesy of Science History Institute, Philadelphia, MS 7, pp. 74v.–75r. (CC0), **82** University of Glasgow, Sp Coll MS Ferguson 210, illustration 17; **83** Warburg Library, FGH 6480, fig. 7; **84–85** The J. Paul Getty Museum, Ms. 33 (86.MP.70), fo. 212 (CC0); **86–87** Getty Research Institute, Manly Palmer Hall collection of alchemical manuscripts, GRI Special Collections, MS 205 (detail) (CC0) ; **88** Musée du Louvre, Paris. Photo © Photo Josse/ Bridgeman Images (detail); **89** Wellcome Collection, London, MS Arabic 161, fos 2v-3r. (CC BY 4.0); **90** The British Library, London, Add. MS 47680, fo.48r. From the British Library archive / Bridgeman Images; **91** The Metropolitan Museum of Art, New York, 2002.436 (CC0); **92** The British Library, London, Harley MS 3915, fos 18v–19r. From the British Library archive,/Bridgeman Images; **93** The British Library, London, Royal MS 6 E VI/2, fo.329r. From the British Library archive/ Bridgeman Images; **94l** Courtesy Dr. Senckenbergische Stiftung, Frankfurt am Main, inv. No. 14; **94r** Stiftung Preussische Schloesser und Gärten Berlin-Brandenburg. Photo Scala, Florence/bpk, Bildagentur für Kunst, Kultur und Geschichte, Berlin; **95** Wellcome Collection, London (CC0); **97** General Collection, Beinecke Rare Book and Manuscript Library, Yale University. Mellon MS 68, fo.1r; **98** Wellcome Collection, London, 15007/A; **99** Digitized by the Johann Christian Senckenberg University Library, Frankfurt am Main, 39;125360T, titlepage (CC0 1.0); **100** Digitzed by Universitätsbibliothek Johann Christian Senckenberg, Frankfurt am Main, 11385049, titlepage (CC0 1.0); **101** The Metropolitan Museum of Art, New York, 42.205.17 (CC0); **103** Wellcome Collection, London (CC0); **104** National Library of Finland, Helsinki (CC0); **106** Zentralbibliothek, Zürich, MS Rh.172, fo. 21v; **107** 'General Collection, Beinecke Rare Book and Manuscript Library, Yale University (CC0); **108** General Collection, Beinecke Rare Book and Manuscript Library, Yale University (CC0); **109** Getty Research Institute, Manly Palmer Hall collection of alchemical manuscripts, GRI Special Collections, MS 205 (http://hdl.handle.net/10020/cat357367), (CC0); **110** The British Library, London, Add, MS 47680, fo.8r. From the British Library archive/Bridgeman Images; **111** Heidelberg University Library (CC0); **112–113** Wellcome Collection, London, 45104i (CC0); **114** Courtesy of the Science History Institute, Philadelphia, QD25.G434 1531 (CC0); **115** Foundation of the Works of C.G.Jung, Zurich, via e-rara (https://www.e-rara.ch/cgj/content/zoom/1919246); **116** Foundation of the Works of C.G.Jung, Zurich, via e-rara (https://www.e-rara.ch/cgj/content/zoom/1916363); **117** Leemage/Corbis via Getty Images; **118** The J. Paul Getty Museum, Los Angeles, 84.PB.56 (detail) (CC0); **119** Photo by Andreas Rausch, courtesy of Marcos Martinon-Torres; **121** Medical Historical Library, Cushing/Whitney Medical Library, Printo0922 (CC0); **122** The National Library of Medicine, Bethseda, Maryland; **123** Kupferstichkabinett, Staatliche Museen zu Berlin. Photo Scala, Florence/bpk, Bildagentur für Kunst, Kultur und Geschichte, Berlin; **124** Courtesy of Science History Institute, Philadelphia; **125** Wellcome Collection, London (CC0); **126** Wellcome Collection, London (CC0); **129** Derby Museum & Art Gallery. © Derby Museums / Bridgeman Images; **131** The Parker Library, Corpus Christi College, Cambridge, MS 395, fo. 50r. **132** The british Library, London, Add. 25724, fo. 36v. The British Library archive / Bridgeman Images; **133** Sächsische Landesbibliothek -Staats- und Universitätsbibliothek (SLUB) Dresden, Deutsche Fotothek (4.A.4483,angeb); **134** The J. Paul Getty Museum, Los Angeles, MS: 11, 57, 143, pp.10-11 (CC0); **135** Wellcome Collection, London, MS 524, fo.2v. (CC BY 4.0); **136** Courtesy of Science History Institute, Philadelphia (CC BY 4.0); **137** Courtesy of the Foundation of the Works of C.G. Jung: (via e-rara https://doi.org/10.3931/e-rara-7378); **138** Wellcome Collection, London (CC0); **139** Wellcome Collection, London (CC0); **140–141** Everett V. Meeks, B.A. 1901, Fund, Yale University Art Gallery, New Haven, 2000.7.1.; **142** Library of Congress, Washington D.C., 20007812 (CC0); **143** Warburg Library, FGH 6480, fig. 1; **144** Library of Congress, Washington D.C., 20007812 (CC0); **145** Photo © Christie's Images / Bridgeman Images; **148** Wellcome Collection, London, EPB/D/2830/1 (CC0); **149** Photo © Christie's Images / Bridgeman Images; **151** Wellcome Collection, London, 2326/D (CC0); **152** Biblioteca Civica Attilio Hortis, Trieste, MS-2-27, vol.3, fo.34; **153** Wellcome Collection, London (CC0); **154** Courtesy of Science History Institute, Philadelphia **155** Courtesy of Science History Institute, Philadelphia **156** Wellcome Collection, London (CC0); **157** Wellcome Library, London (CC0); **159** Wellcome Collection, London (CC0); **160–161** Wellcome Collection, London (CC0); **162** Courtesy of Science History Institute, Philadelphia (CC0); **163** Wellcome Collection, London (CC0); **164–165** Manly Palmer Hall collection of alchemical manuscripts, Getty Research Institute, Los Angeles (CC0); **166–167** Charles Walker Collection / Alamy Stock Photo; **168** Bridgeman Images (detail); **169** Courtesy of Science History Institute, Philadelphia (CC0); **171** The Huntington Library, San Marino, California, MSS EL 26 C 9, fo. 194r (detail). (CC0); **172** GL Archive / Alamy Stock Photo ; **173** The British

Library, London, Sloane MS 3846 fo.102v. From the British Library archive / Bridgeman Images; **174** Ashmolean Museum of Art and Archaeology/Heritage Image Partnership Ltd / Alamy Stock Photo; **175** © Trustees of the British Museum, OA.106; **176** Getty Research Institute, Los Angeles (CC0) ; **177** Germanisches Nationalmuseum, Nuremberg, MP 22234, Kapsel-Nr. 380; **178** Wellcome Collection, London (CC0); **179** Württembergisches Landesmuseum, Stuttgart. The Picture Art Collection / Alamy Stock Photo; **180** Getty Research Institute, Los Angeles (CC0); **181**, Getty Research Institute, Los Angeles (CC0) **182** The Metropolitan Museum of Art, New York, 54.602.1(13) (CC0); **183l** Special Collections and Archives, University of Kent, Maddison Collection 1A21, F10448000; **183r** Wellcome Collection, London, 753/A (CC0); **184** © Archives Charmet / Bridgeman Images; **185** Wellcome Collection, London (CC0); **187** Wellcome Collection, London (CC0); **188** Getty Research Institute, Los Angeles, GRI 87-B9592 (CC0); **189** Getty Research Institute, Los Angeles (CC0); **190–191** © CSG CIC Glasgow Museums and Libraries Collections, acc.3188; **192** Skokloster Castle, Sweden, SKO 11615; **193** Opensource, via Archive.org; **195** Nationalmuseum, Stockholm, NM 308. Photo: Magnus Mårtensson/RAÄ (CC0); **196–197** Rijksmuseum, Amsterdam, RP-P-1896-A-19368-984 (detail); **198** Wellcome Collection, London, 47317i (CC0); **199** Central Library, Zürich, SCH R 201, plate 1a; **200** Courtesy of Science History Institute, Philadelphia (CC0); **201** Herman B Wells Library,, Indiana University Bloomington, Indiana; **202** US National Library of Medicine, Bethesda, Maryland (CC0); **203** Germanisches Nationalmuseum, Nuremberg, Mp 10015, Kapsel-Nr. 165; **204**, Wellcome Library, London (CC0); **206** Wellcome Collection, London (CC0); **208** Wellcome Library, London, EPB/MST/288; **209** Stefano Bianchetti / Bridgeman Images; **211** University of California Libraries, via BHL; **213** Wellcome Collection, London (CC0) ; **214** Dr. Senckenbergische Stiftung, Frankfurt am Main, inv. 11; **215** Getty Research Institute, Los Angeles, vol. 3, fig. 1 (17641781); **216–217** Private collection. © Archives Charmet/ Bridgeman Images; **218** Courtesy of Science History Institute, Philadelphia (CC0) **219** Following on from Rutherford and Soddy's work; Patrick Blackett / Science Photo Library; **221** © Christie's Images / Bridgeman Images; **222–223** Gold: plate VII from Occult Chemistry Wellcome Collection, London (CC0); **224** Château de Versailles, France/Bridgeman Images; **225** © The Trustees of the British Museum, BM 148900001; **226** Stefano Bianchetti/Corbis via Getty Images; **227** Library of Congress, Rare Book and Special Collections Division, Washington D.C.; **228** Mary Anne Atwood. Harold B. Lee Library, Brigham Young University, Utah; **229l** Private Collection; **229r** Warrant of the Isis-Urania Temple, London, c.1888, GD 2/3/1/1. © Museum of Freemasonry, London, 2025; **230l** Keystone-France/ Gamma-Keystone via Getty Images; **230r** Wellcome Library, London, 163/B/1; **231** United Archives GmbH / Alamy Stock Photo ; **233** Peggy Guggenheim Collection, Venice/ Cameraphoto/Scala, Florence. © ADAGP, Paris and DACS, London 2025; **235 (all)** Museo Del Prado, Madrid. G. Dagli Orti /© NPL - DeA Picture Library / Bridgeman Images; **236** Courtesy of Science History Institute, Philadelphia (CC0); **237** Private Collection /Bridgeman Images; **238** Folger Imaging Department, Folger Shakespeare Library, STC 17436 (CC0); **240** PjrTravel / Alamy Stock Photo ; **243** Getty Research Institute, Manly Palmer Hall collection of alchemical manuscripts, GRI Special Collections, MS 205 (http://hdl.handle.net/10020/cat357367) (CC0); **244–245** Wellcome Collection, London, 17512i; **246** © Museum of Freemasonry, London, 2025, GD 8/4/7a-d & GD 2/5/1/1b; **247** Wellcome Collection, London, 45113i; **248** Museo del Prado, Madrid, P002823, via Wikipedia; **249** Folger Imaging Department, Folger Shakespeare Library, FPa2 (CC0)

Author Credits

It is always a pleasure to work with the team at Quarto, who do such a splendid job of finding images and assembling them into something beautiful. Sara Ayad Cave has again unearthed some glorious illustrations, and Ruth Patrick has been consistently patient and eagle-eyed in guiding the project along. And I am immensely grateful to Eszter Karpati of Quintessence for having the idea in the first place and persuading me into it. I am grateful to be working with a team who believe that books can and should continue to be objects of desire.

© 2025 Quarto Publishing Plc.

All rights reserved.
No part of this publication may be reproduced, in whole or in part, including illustrations in any form (beyond that copying permitted by Sections 107 and 108 of the U.S. Copyright Law and except by reviewers for the public press), without permission in writing from the publishers.

This book was conceived, designed, and produced by
Quintessence, an imprint of The Quarto Group
1 Triptych Place, Second Floor
London SE1 9SH
www.quarto.com

Senior commissioning editor: Eszter Karpati
Project editor: Ruth Patrick
Design: Blok Graphic
Senior designer: Rachel Cross
Picture research: Sara Ayad
Production manager: David Hearn
Managing editor: Emma Harverson
Art director: Gemma Wilson
Associate publisher: Eszter Karpati
Publisher: Lorraine Dickey

Published in association with
Yale University Press
302 Temple Street
P.O. Box 209040
New Haven, CT 06520-9040
yalebooks.com

Library of Congress Control Number: 2025933923
ISBN 978-0-300-28087-6

Authorized Representative in the EU:
Easy Access System Europe, Mustamäe tee 50,
10621 Tallinn, Estonia, gpsr.requests@easproject.com

10 9 8 7 6 5 4 3 2 1

Printed in China

Front cover (T): An alchemical cosmology from *The Works of Jacob Behmen, the Teutonic Theosopher* (1764), by the Reverend William Law. See page 215 (detail). Front cover (C): "Vitriol," from Georg Hellmerich's *Musaeum hermeticum* (Hermetic museum; 1692). See page 107 (detail). Front and back cover (L and R): A chart of "the whole corporeal and incorporeal nature," from William Davidson's *Philosophia pyrotechnia* (Paris, 1640). See page 185 (detail).

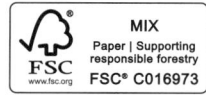